Vegetable Seeds: Production and Technology

Vegetable Seeds: Production and Technology

Gregory E. Welbaum

CABI

CABI is a trading name of CAB International

CABI
Nosworthy Way
Wallingford
Oxfordshire OX10 8DE
UK

Tel: +44 (0)1491 832111
E-mail: info@cabi.org
Website: www.cabi.org

CABI
200 Portland Street
Boston
MA 02114
USA

Tel: +1 (617)682-9015
E-mail: cabi-nao@cabi.org

A catalogue record for this book is available from the British Library, London, UK.

Library of Congress Cataloging-in-Publication Data

Names: Welbaum, G.E. (Gregory E.), author.
Title: Vegetable seeds : production and technology / Gregory E Welbaum.
Description: Boston, MA : CAB International, [2024] | Includes bibliographical references and index. | Summary: "Farmers all over the world share one concern: how to get the best seed and planting material for next season's crops. This book covers it, and is a replacement for CABI's Vegetable Seed Production by R.A.T. George, now out of print"-- Provided by publisher.
Identifiers: LCCN 2023032707 (print) | LCCN 2023032708 (ebook) | ISBN 9781789243246 (paperback) | ISBN 9781789243253 (ebk) | ISBN 9781789243260 (epub)
Subjects: LCSH: Vegetables--Seeds. | Vegetables--Reproduction. | Vegetables--Propagation. | Seed technology.
Classification: LCC SB324.75 .W45 2024 (print) | LCC SB324.75 (ebook) | DDC 635/.04--dc23/eng/20231116
LC record available at https://lccn.loc.gov/2023032707
LC ebook record available at https://lccn.loc.gov/2023032708

ISBN-13: 9781789243246 (hardback)
9781789243253 (ePDF)
9781789243260 (ePub)

DOI: 10.1079/9781789243260.0000

Commissioning Editor: Rebecca Stubbs
Editorial Assistant: Lauren Davies
Production Editor: Rosie Hayden

Typeset by Straive, Pondicherry, India

Contents

Preface

Writing this book completes a lifelong interest in seeds. As a child growing up on a farm, I was amazed how a few bags of seed could transform vast bare fields in spring into a sea of green crops in just a few weeks. This amazing process of crop propagation by seed is taken for granted by most people. Our existence depends on the annual production of seed crops for propagation. We do not have enough stored seed in reserve to grow food without annual production of new seeds. The goal of this book is to educate how seeds for vegetable propagation are produced.

The book begins with six chapters providing information on a broad range of seed-related topics: the importance of seeds, reproductive biology of plants, genetic improvement strategies, quality assurance of seed production, post-harvest seed enhancement, and organic production. The remaining 11 chapters cover seed production in 11 important plant families. These 11 seed crop production chapters include specifics on botany, important types and cultivars, genetic improvement, pollination, soil fertility management, pest management, irrigation, crop production, harvesting, post-harvest handling, and seed yields. It is hoped that this book will be useful as a college textbook as well as a reference for the seed industry.

I developed undergraduate and graduate classes on vegetable seeds, which I have taught for the past 25 years. This book was inspired by my research program on seed biology and technology and content from my vegetable seed classes. Some of my inspiration also comes from the Seed Production short course I taught portions of for several years through the Seed Biotechnology Center at University of California, Davis. These classes made me aware of a need for a plant-focused book covering the biology of seed development as well as the applied aspects of managing commercial seed production and application of enhancement treatments. Working in academia makes it challenging to keep up to date with practices of the vegetable seed industry, especially living in a region with little commercial production. I enlisted the expertise of long-time seed industry specialist, Tim Loeffler, of Seed Vigor Consulting LLC. Tim has over 40 years of international seed experience in production, processing, technology, and quality assurance for both vegetable and field seed crops. Tim filled in missing information, suggested revisions, and provided most of the beautiful pictures of seed production in this book. I also appreciate the helpful comments from Kent Bradford and Peter Marks. Kent also contributed some of his teaching diagrams, which I have included in Chapters 2, 3, and 5 on reproductive biology and seed enhancement technology. Mike Pereira's lectures and organization from the Seed Production short course at UC Davis also influenced the organization and content of this book.

Global climate change poses a new threat to the production of seeds in the areas where they have traditionally been grown. Despite challenges, seed science and technology education and research has been in decline at US universities for decades. This redirection of research priorities away from crop seeds seems misplaced given their importance to our food supply, need for improved sustainability, pending global climate change, opportunities for genetic improvement, and protecting against seed-transmitted diseases. Fortunately, many US universities are hiring seed ecologists that will help fill the information and research needs of the future. It is also reassuring to see outstanding seed research being conducted in Asia and Europe. Hopefully this book will stimulate interest in seeds among a new generation of scientists.

Gregory E. Welbaum, PhD, Visiting Professor, Seed Biotechnology Center, University of California, Davis and Professor, School of Plant and Environmental Sciences, Virginia Tech. Photograph by Whitney Welbaum.

Dedicated to the memory of Chan-Lu Fu

Online Supplementary Content

Supplementary videos and images are available for *Vegetable Seeds: Production and Technology* and can be accessed at:

https://www.cabidigitallibrary.org/doi/book/10.1079/9781789243260.0000

1

Vegetable Seeds and Their Global Importance

What are Vegetables?

Most vegetables grown in temperate climates are annuals that must be replanted each year. A diet rich in vegetables is essential for human health because they contain vitamins, minerals, dietary fiber, folate, and many other nutrients critical to human health. Because human health is dependent on vegetables, production and trade of vegetable seed is a never-ending essential endeavor. "Vegetable" is a somewhat nebulous term based on usage rather than a specific botanical classification. One accepted definition is: "A vegetable is an herbaceous plant, or portion of a plant, that is eaten whole or in part, raw or cooked, generally with an entrée or in a salad but not as a dessert." There are notable exceptions to this definition. Watermelon and rhubarb are often eaten as desserts but considered to be vegetables based on how they are grown. This definition is troubling to some because vegetables can also be fruit. Fruit is a botanical definition for a mature, ripened ovary and its contents that also describes vegetables like tomato and melon. Vegetables are also intensively managed, high value, and perishable crops as opposed to agronomic crops, sometimes called field crops, that are of lower value, extensively grown, and harvested at full maturity as a non-perishable grain or for fiber. Corn, peas, and soybeans can be either agronomic or vegetable crops depending on when they are harvested. They are considered vegetable crops when harvested at an immature stage of development and considered agronomic crops when harvested as mature dry grain (Welbaum, 2015).

Vegetable seed production is much different and more challenging than production of grain crops. Agronomic crops such as wheat, rice, maize, and dry beans are largely self-pollinating and extensively grown annuals harvested by combining when the mother plant dies at the end of its life cycle. Some vegetables need to be grown for a much longer time before seeds are produced. Because many vegetables are biennial, they have a juvenile phase and then vegetative and reproductive cycles before flowering. In carrot, onion, and cabbage for example, plants must be exposed to vernalizing temperatures for a prolonged period to initiate bolting and flower production. Most vegetables grown today are F1 hybrids. Hybrid production involves developing two or more inbred lines, maintaining them, and then carefully controlling the cross-pollination on a large scale to produce seed. Brassica crops like cabbage, broccoli, cauliflower, Brussels sprouts, kale, and others can intercross, so special attention must be made to isolation distances between seed fields. Vegetable seeds tend to be small and irregularly shaped, presenting challenges for harvesting and conditioning. Cucurbit and Solanaceous seeds develop in fleshy fruit and must be harvested and processed at a high moisture content, which presents special problems. Some

DOI: 10.1079/9781789243260.0001

vegetable seeds are prone to shattering near maturity requiring precise timing of harvests and in some cases hand harvesting. Since many vegetables are eaten when immature, special considerations must be made to their vegetative appearance during seed production. Off-types caused by disease or genetics must be removed from fields to make sure the seeds produce the desired phenotype in the next generation. Generic or cytoplasmic male sterility, sex expression manipulation using chemicals, and self-incompatibility systems can prevent selfing for commercial-scale crossing to make F1 hybrid seeds, adding to the complexity of production. For some vegetables, no male sterility genes or other techniques are available to encourage natural crossing, so labor-intensive emasculation and hand cross-pollination must be used. There are many different vegetables and many types and cultivars of each vegetable (e.g. leaf, crisphead, butterhead, romaine, lettuces of various colors and leaf shapes), making tracking production, preventing cross-contamination during cleaning and storage, and keeping inventories a challenge. These are some of the issues that made vegetable seed production unique.

Consumption of fresh vegetables has been increasing in developed countries because of greater year-round availability due to increased trade and realization of their positive health benefits. There is a great diversity of vegetables worldwide. For example, in the USA alone, there are approximately 100 “major” vegetable crops and roughly 350 vegetable crops traded commercially. In Asia, even more plants are consumed as vegetables.

Botanical Classification of Vegetables

Vegetables have many different common names that vary with language and local customs around the world. The botanical classification system was created as a common reference for international communication about plants. Vegetable crops are classified in many ways, but the botanical classification system of Latin names is widely used because it is based largely on flower morphology and Latin language is relatively unchanged over time. This book will discuss seed production by plant family because this is an efficient way to group vegetables with similar reproductive characteristics (Welbaum, 2015).

New genetic sequencing information has drawn questions about the accuracy of traditional botanical classification, which is based on plant morphology (phenotypes) rather than genotypes. Recently alternative names have been proposed for some important vegetables based on their genetic sequence information. This has created confusion among plant biologists, and it will take years before there is a consensus on the correct scientific names of some vegetables, particularly at the family classification level and below. A new or modified classification of some plants will occur over time as differences between morphological characteristics and genetic sequence information will likely be resolved. To date, the old system of classification based on plant morphology is holding up better and requiring less revision than some experts anticipated. We will use current consensus scientific names for vegetables in this book.

The term “variety” in the botanical classification system is often used in the seed trade as well. Variety can be confusing because there are both botanical varieties and cultivated varieties. Botanical variety is a taxonomic rank below species. Botanical varieties often exist in nature as interbreeding, free-living populations that differ by multiple traits. Botanical variety is sometimes referred to as subspecies and may include a collection of cultivated varieties or cultivars.

Cultivar is a taxonomic rank below botanical variety and is derived from a contraction of the words “cultivated variety”. Cultivars differ from one another by at least one important trait such as color or disease resistance, and are maintained through human propagation rather than natural interbreeding. In the seed industry, “variety” is often used as an informal ambiguous substitute for cultivar.

Scope

This book is about production of vegetable seeds as propagules and related topics. Some vegetable seeds, such as dry beans, corn (maize), vegetable soybean (edamame) and peas, are consumed directly as food. Production and human consumption of seed as food are not covered in detail. The later chapters are organized by major vegetable families. Each chapter provides a brief history, description of botanical characteristics, lists of important types and cultivars, discussion of growth and development, soil fertility and nutrient management requirements, pollination biology, irrigation needs, rogueing requirements, genetics, seed production techniques, seed harvesting practices, post-harvest handing, expected seed yields, physiological disorders, weed control practices, common diseases, and insect pests. The goal is to educate the reader about the various steps and fundamental management practices needed to successfully produce vegetable seeds. The broadest possible set of vegetable crops are described in the space allotted. A chapter is devoted to post-harvest seed treatments. However, room doesn't permit addressing other essential seed science topics. Some useful printed references related to vegetable seed production are listed in Table 1.1.

Seeds

A seed is a reproductive unit of a flowering plant. Another simple definition is "a miniature plant in an arrested state." Seeds are very efficient propagules due to their small size, desiccation tolerance, storability, and low cost. They are

Table 1.1. Useful printed references related to vegetable seed production.

Authors	Year	Title and publisher
Agarwal, V.K. and Sinclair, J.B.	1997	*Principles of Seed Pathology*, 2nd edn. CRC Press, Boca Raton Florida.
Baalbaki, R.Z., McDonald, M.B., Copeland, L.O. and Elias, S.G. (eds)	2012	*Seed Testing: Principles and Practices.* Michigan State University Press, East Lansing, Michigan.
Basra, A.S. (ed.)	1999	*Heterosis and Hybrid Seed Production in Agronomic Crops*. Food Products Press, Binghamton, New York.
Basra, A.S. (ed.)	2000	*Hybrid Seed Production in Vegetables: Rationale and Methods in Selected Crops*. Food Products Press, Binghamton, New York.
Basra, A.S. (ed.)	2006	*Handbook of Seed Science and Technology*. Food Products Press, Binghamton, New York.
Bewley, J.D., Bradford, K., Hilhorst, H. and Nonogaki, H.	2013	*Seeds: Physiology of Development, Germination and Dormancy*. Springer, New York.
Black, J., Bewley, J.D. and Halmer, P. (eds)	2006	*The Encyclopedia of Seeds: Science, Technology and Uses*. CAB International, Wallingford, UK.
Copeland, L.O. and McDonald, M.B.	2001	*Principles of Seed Science and Technology*, 4th edn. Kluwer Academic Publishers, Norwell, Massachusetts.
Desai, B.B.	2004	*Seeds Handbook: Biology, Production, Processing and Storage*, 2nd edn. Marcel Dekker, New York.
Egli, D.B.	2017	*Seed Biology and Yield of Grain Crops*, 2nd edn. CAB International, Wallingford, UK.
George, R.A.T.	2009	*Vegetable Seed Production*, 3rd edn. CABI International, Wallingford, UK.
Harries, A. and Cortes, J.	2005	*Procedures Manual for Phytosanitary Accreditation for Seed Export*. Seed Center, Iowa State University.
Kelly, A.F.	1988	*Seed Production of Agricultural Crops*. Wiley, Hoboken, New Jersey.
McDonald, M.B. and Copeland, L.O.	1997	*Seed Production: Principles and Practices*. Chapman & Hall, New York.
McDonald, M.B. and Kwong, F.Y.	2004	*Flower Seeds: Biology and Technology*. CAB International, Wallingford, UK.
Welbaum, G.	2015	*Vegetable Production and Practices*. CAB International, Wallingford, UK.

efficient to transport and may be stored for long periods at low moisture content and with refrigeration. Seeds are one of the true wonders of nature because of their ability to sense the surrounding environment, germinate rapidly, and show persistence under a variety of conditions. Typically, seeds consist of an embryo enclosed in a protective outer coat, usually with some stored nutrients to sustain early seedling growth until establishment. Seeds have sophisticated dispersal mechanisms and often will not germinate until their environment favors growth. Seeds and seedlings have elaborate defense mechanisms to improve survival (Welbaum, 2006).

Producing Seeds

One of the primary skills required for seed production is managing pollination to ensure seeds have the desired genetic composition. The development of most vegetable seeds requires pollen transfer between male and female flower parts. In some species this is accomplished naturally by flower or air movement. Pollen transfer can be facilitated by movement of pollen to flowers on adjacent plants or very short distances from anthers to pistils within the same flower. Other species have evolved elaborate cross-pollination systems that rely on insects, usually bees, to transfer pollen. Therefore, seed production for some vegetables is reliant on management of bees for pollination. Seed scientists work closely with entomologists to study plant–insect interactions to make sure healthy bee populations are available for pollination. In addition, isolation of seed production fields from other genotypes of the same species is required to assure genetic purity of a seed crop.

Most food and fiber crops are produced from seed. Some vegetables, like Irish potato (*Solanum tuberosum*), and sweetpotato (*Ipomoea batatas*), produce seeds that are genetically diverse and do not produce offspring like their parents. For these crops and a few others, vegetative propagation from tubers or roots, respectively, is used commercially. But for most vegetables, seeds are the starting point for crop production and the basis for food security. Since seeds are alive, they have a finite lifespan and must be produced on a recurring basis to ensure annual food and fiber supplies. Seeds can be preserved in dedicated germplasm storage facilities for extremely long periods, but require elaborate and relatively expensive preservation conditions that are unavailable to much of the world and are not adapted for bulk storage of large quantities of seed. This means that the world's population is dependent on annual seed production for its food supply.

Yet human dependence on vegetable seed production is unappreciated by many people not involved in agriculture. The average person has little knowledge about the seeds used to grow their food. Where seeds come from, how they are produced, or the technologies used to produce them are a mystery to the average person. Public knowledge of the seed industry is clouded by inaccuracies, misinformation, and urban legends. Most seed companies are one or two steps removed from the food supply chain and not in the public eye, which contributes to a lack of public familiarity about the industry and its practices.

Who Produces Seeds?

From ancient times until about 70 years ago, farmers in developed countries produced and stored their own vegetable seeds. There were relatively few cultivars to select from and those that were available were open-pollinated. Regional seed companies often developed new cultivars for their area, but over the years many of these smaller companies consolidated or ceased operation. With increased specialization, development of F1 hybrid seeds, post-harvest seed treatments, and advanced genetics, most farmers today buy all their seed from commercial seed companies that specialize in vegetable seed production. This is not the case in less-developed countries where personal seed saving is still common.

Vegetable cultivars tend to have specific climatic requirements for best performance. Each region has unique vegetable pest problems, so access to disease-resistant cultivars is a local concern. Therefore, developing cultivars for specific regions of the world is important for optimizing productivity, controlling pests, and adapting to climate change. However, large international vegetable seed companies tend to concentrate plant breeding and research and development activity in fewer locations due to consolidation. Cultivar development tends to focus on widely adapted cultivars that will do well over a wide range of environments. Prior to globalization, breeders developed more region-specific cultivars. Critics of consolidation and concentration of vegetable seed companies argue that more regional seed companies are needed to better develop cultivars to both quality and environmental needs for specific regions of the world (Contreras, 2002). On the other hand, advances in climate science and prediction are enabling seed companies to identify "climate clones", or regions around the world with similar current or projected climates in which existing or new cultivars may be adapted (for more information, see https://climate.ai/).

The number of companies in the developed world that develop new cultivars and produce vegetable seeds has steadily decreased over the past 60 years. This consolidation of seed companies has paralleled the decrease in the number of vegetable farms in the USA and other countries. Seed company mergers and buyouts occur to achieve economies of size and expand markets. Like other industries, the seed industry has become more specialized. Some seed companies specialize in seed production and cultivar development for the wholesale market and large commercial growers. Other companies are focused on producing cultivars for smaller growers and home gardeners. Some smaller regional seed companies have quit producing seeds and become retailers, distributing seeds from larger companies.

Seeds are the delivery system for advanced crop genetics. Modern seed production and crop genetics requires large investments by companies to keep pace with their competitors. The vegetable seed divisions of large international companies spend as much as 40% of their profits each year on research and development (Malhotra, 2021). Leading companies use molecular techniques such as DNA sequencing, genetic fingerprinting, DNA-marker-based breeding, and gene-editing technology and other emerging technologies, develop new traits, and expedite cultivar development. Small companies often lack resources to invest in ever-changing advanced technologies. Since seeds are in constant demand because of their importance, well-managed seed companies are usually profitable. Because of their profitably and recession resistance, larger seed companies tend to become subsidiaries of even larger multinational companies that acquire them to diversify their business portfolios. In other cases, smaller seed companies continue to grow and buy out their competition until they have an international presence.

This should not imply that small seed companies no longer exist or cannot be successful under the current market conditions. Recent trends have encouraged the formation of new regional or specialized seed companies fulfilling a need for biological seed treatments, open-pollinated heirloom cultivars, home gardening, and organic vegetable seeds. According to the Organic Produce Network in the USA, organic fruit and vegetable sales rose by 22% in 2020 (Malhotra, 2021). The same trend occurred in Europe, where organic/natural sales dramatically increased. In 2020, some organic seed companies experienced a tripling in sales compared to the previous year (Malhotra, 2021). In addition to seed, biological and other natural seed treatments for certified organic production have provided new opportunities for smaller companies to support increased demand.

The Covid-19 pandemic helped stimulate interest in organics as well. The pandemic also increased demand for vegetable seeds for urban agriculture as well as home and community gardening, providing an opportunity for mail order and smaller regional companies (Lal, 2020; Mullins *et al.*, 2021). Recent increased interest and demand for heirloom non-hybrid seed allowed some smaller seed companies to supply this market. Consumer dissatisfaction with large multinational companies, for various reasons, is helping local seed businesses as well.

The European Commission's recent mandate requiring increased use of organically produced seed has boosted this sector. The Commission now requires use of organically grown seed for early sprout production and has a goal of growing all crops certified organic using only organically produced seeds by 2035 (Kowalska and Bieniek, 2022). As plant protection options are more limited, growth of the organic market has increased pressure on seed companies to breed disease resistance into new cultivars. In spinach, for example, the high fraction of organic production results in breakdown of disease resistance almost immediately as new cultivars are released, since the pathogens (e.g. downy mildew) are also genetically variable (and sexual) and overcome resistances quickly if not also controlled with fungicides. Producing disease-free seeds and new cultivars organically is difficult, so the availability and high cost of organic seeds remain challenging for both seed companies and vegetable producers. Rijk Zwaan, Enza Zaden, and Bejo Zaden, along with most global seed companies, are strengthening their organic seed capabilities. Lettuce, spinach, carrot, tomato, pepper, eggplant (aubergine), and cucumber are the most important crops in the organic segment (Malhotra, 2021).

The global vegetable seed market for propagation is large and projected to grow from US$7.2 billion in 2021 to $11.4 billion by 2028, at a compound annual growth rate of 6.8% over that period (Fortune Business Insights, 2022). In the world commercial seed industry, nearly 70% of total seed sales are accounted for by just 22 companies (Malhotra, 2021). Vegetable seeds form the third-largest segment of the seed industry after maize and soybean with a market share of almost 17% in global seed sales for 2020. These numbers do not include the important local seed trade within countries or seed production/saving by individual growers or grower organizations, which supplies up to 80% of demand in some developing countries (for more information, see https://seedsystem.org/). In developed countries, seed-saving organizations play an important role in preserving and providing germplasm and heirloom vegetable seed not available from large commercial seed companies (for more information, see https://www.seedsavers.org/).

Hybrid cultivars dominate the world market, accounting for 75% of the total vegetable seed market in 2021. Open-pollinated vegetable cultivars comprise 25% of the global market (Global Vegetable Seed Market, 2023). The Asia-Pacific market accounted for 45% of the global hybrid vegetable seed market in terms of value in 2021. China and India are the leading countries in Asia, accounting for approximately 56% of that regional market (Global Vegetable Seed Market, 2023).

Compared with other large seed segments such as agronomic crops, which include extensively grown major crops like wheat, maize, soybean, rice, and cotton, the vegetable seed market is more diversified. Using the year 2020 as an example, the top three companies in the international vegetable seed industry accounted for almost 30% of global vegetable seed trade. In contrast, the top three companies accounted for 66%, 58%, and 53% of market shares for global agronomic crops (maize, soybean, and cottonseed), respectively (Malhotra, 2021). The top ten international seed companies account for 64% of global vegetable seed trade (Table 1.2; Malhotra, 2021).

The companies in Table 1.2 are active in the global vegetable market, with many having sales

Table 1.2. Top ten companies in vegetable seed sales in 2020. (Modified from Malhotra, 2021.)

Company	Sales (US$ million)	Global share (%)
Vilmorin	780	10
Bayer	729	10
Syngenta	653	9
Rijk Zwaan	527	7
BASF	511	7
Takii	458	6
Sakata	379	5
Enza	373	5
Bejo Zaden	250	3
East-West	168	2
Total	4828	64
Global vegetable seed sales total	7573	100

in more than 120 countries. Some global companies are active in both seeds and agrochemicals, focus on specific regions or continents, or specialize in seeds for certain crops.

Today, relatively inexpensive, high-quality seeds are marketed through retail channels, from dealers, via catalog, or over the web. Other companies not listed in Table 1.2 focus on treating seeds from other companies. Such seed treatments are covered in a later chapter in this book. Modern seed companies tend to specialize in specific seed crops. Rijk Zwaan, from the Netherlands, is widely recognized for expertise in developing cultivars of lettuce and other salad crops for Europe and North America. Sakata, although a Japanese-based company, is noted for production of Brassica seed worldwide. Bejo Zaden, also from the Netherlands, has long focused on organic vegetable seed. East-West Seeds, headquartered in Thailand, specializes in tropical Asian vegetable seed production for small farmers. The leading vegetable seed company by sales, Vilmorin, a French company, is famous for its diverse selection of tomato, carrot, Cucurbits, cauliflower, and squash seeds.

Although there is tremendous diversity in the vegetables that are consumed around the world, vegetable seed trade is dominated by vegetables from relatively few families: Cucurbitaceae (melon, cucumber, squash), Solanaceae (tomato, pepper), Brassicaceae (various crops), Allioideae (onions), and Apiaceae (carrots) (Table 1.3).

In the last 40 years, genetic improvement technologies have become more sophisticated. Our knowledge of seed physiology and genetics has improved dramatically through research. Much of the success of seed companies lies with plant geneticists who are constantly developing vegetable cultivars with improved traits for disease resistance, controlled environment production, horticultural characteristics, nutritional composition, and production efficiency. However, developing new vegetable cultivars is no longer enough to remain competitive in today's dynamic vegetable seed market. Most vegetable seed companies no longer compete with one another based on seed germinability (whether their seeds germinate) but rather on the vigor of seeds (how well they germinate, particularly under adverse conditions), advanced genetics, and post-harvest seed treatment technologies. To remain competitive, companies such as Syngenta, Bejo Zaden, Rijk Zwaan, and Takii are developing sub-brands for non-traditional markets that appeal to home gardeners, organic growers, and hobbyists. These new brands are in addition to standard brands that continue to target traditional commercial farmers (Malhotra, 2021).

Table 1.3. World vegetable seed market by crop. (Adapted from Welbaum, 2015; data provided by Hazera seed company, Haifa, Israel.)

Total seed traded (%)	Vegetable crop
21	Melon
14	Brassicas
10	Cucumber
8	Long-day onions
8	Field tomato
8	Carrot
7	Pepper
4	Short-day onion
4	Processing tomato
4	Other vegetable crops
3	Small-fruited tomato
3	Greenhouse tomato
3	Squash
3	Leafy greens

References

Contreras, S. (2002) The international seed industry. In: McDonald, M. and Contreras, S. (eds.) *Proceedings International Seed Seminar: Trade, Production and Technology.* Pontificia Universidad Católica de Chile, Facultad de Agronomía e Ingeniería Forestal, Departamento de Ciencias Vegetales, pp. 1–9.

Fortune Business Insights (2022) Seed & Seed Technology/Vegetable Seed Market. Vegetable Seeds Market Size, Share & COVID-19 Impact Analysis, 2021–2028. Report ID: FBI103066. https://www.fortunebusinessinsights.com/vegetable-seed-market-103066 (accessed 14 March 2023).

Global Vegetable Seed Market. (2023) Size, Share, Covid-19 Impact & Forecasts up to 2028. Mordor Intelligence LLP. PDF ID: 6381363. https://www.reportlinker.com/p06381363/Global-Vegetable-Seed-Market-SIZE-SHARE-COVID-19-IMPACT-FORECASTS-UP-TO.html?utm_Source=GNW (accessed 8 September 2023).

Kowalska, A. and Bieniek, M. (2022) Meeting the European green deal objective of expanding organic farming. *Equilibrium. Quarterly Journal of Economics and Economic Policy* 17, 607–633.
Lal, R. (2020) Home gardening and urban agriculture for advancing food and nutritional security in response to the COVID-19 pandemic. *Food Security* 12, 871–876.
Malhotra, B. (2021) Global vegetable seeds market is increasingly fragmented and diversified. Blog Nov. 3, 2021, *S&P Global Commodity Insight*. https://www.spglobal.com/commodityinsights/en/ci/research-analysis/globalvegetable-seeds-market.html
Mullins, L., Charlebois, S., Finch, E. and Music, J. (2021) Home food gardening in Canada in response to the COVID-19 pandemic. *Sustainability* 13, 3056.
Welbaum, G.E. (2006) Natural defense mechanisms in seeds. In: Basra, A.S. (ed.) *Handbook of Seed Science and Technology*. Scientific Publishers, pp. 451–464.
Welbaum, G.E. (2015) *Vegetable Production and Practices*. CAB International, Wallingford, UK.

2

Vegetable Reproductive Biology

Induction of Flowering

Plants progress through a series of developmental stages during their life cycle. Each phase has distinctive morphological traits and/or the development of new organs (Huijser and Schmid, 2011; Jin *et al.*, 2013). Flowering is an essential part of sexual reproduction in angiosperms. Flowering has two distinct developmental transitional phases. The first transition is from an immature juvenile vegetative stage, when the plant is unable to flower, to the mature adult vegetative state when flowering is possible (Poethig, 2003; Sgamma *et al.*, 2014). A second transition occurs at floral induction when the plant switches from vegetative growth to reproductive growth and produces flowers. The initiation of flower development is complex and depends on a number of factors, including plant age, the environmental history (primarily light and temperature), the photosynthetic resources available, and other factors that vary among species. A complex gene regulatory network controls exactly when the transition from vegetative to reproductive growth occurs.

Regulation of flowering is essential for successful reproduction because a plant must flower at an optimum time for pollination or seed dispersal. For out-crossing vegetables, flowering needs to coincide with other plants of the same species close by. For insect-pollinated vegetables, flowering time must be linked with pollinator activity (Huijser and Schmid, 2011; Srikanth and Schmid, 2011). The timing of floral initiation is controlled by several different mechanisms. In many species, plants must complete the juvenile phase and reach a specific size or age before flowering. In biennials like onion or carrot, plants must attain a minimum size before flowering can be induced. In other vegetables, for example tomato, flowering begins at a specific node regardless of environment.

Many plants require specific environmental cues to initiate flowering. The most common environmental signals are temperature (including diurnal and seasonal variations) and light quality and duration. Plants integrate this information with endogenous physiological mechanisms such as plant hormones, circadian rhythms, carbohydrate levels, or vernalization state to regulate flower formation and opening (anthesis) to the appropriate time of the year, and even time of the day (Jackson, 2009; Amasino, 2010; Fornara *et al.*, 2010; Srikanth and Schmid, 2011; Kim *et al.*, 2012; Matsoukas *et al.*, 2012). Most biennial vegetables, for example, require a period of chilling or vernalization for the transition to reproductive development to begin. The most effective temperature is generally around 5°C (45°F) and time periods vary from a few weeks to months. Knowing the vernalization requirements of specific cultivars and parent lines is often critical to the success of seed production. Seed producers must understand when flowering of two parent lines occurs to synchronize

DOI: 10.1079/9781789243260.0002

hybrid seed production. Another large group of plants require specific day lengths to induce flowering. These are generally temperate species from latitudes where day length provides reliable information about the growing season. Some plants, such as spinach, require long days (short nights) to flower and are termed long-day plants (LDP). Others (e.g. soybean) require a minimum night length to flower, and are therefore called short-day plants (SDP), while some will flower under both long and short days and are called day-neutral plants (DNP). It is essential to match the day-length requirements of a crop to the latitude of production to obtain timely flowering and successful seed development. For both temperature and light, the requirements can be either absolute (termed qualitative) or relative (termed quantitative). For example, many biennials have an absolute or qualitative vernalization requirement, but also have a quantitative response to long days. That is, once the chilling requirement has been met, the time to flowering is shortened by long days but will still occur under short days. This matches the general ecological habit of biennials, which develop vegetatively in the fall, overwinter, and then flower the following spring as days lengthen. Specific locations where both the chilling requirement can be met and day lengths are appropriate may become specialized for seed production of selected crops. For example, spinach (*Spinacia oleracea*), table beet (*Beta vulgaris*) and *Brassica* spp. (primarily cabbage) seed production are concentrated in the Skagit Valley in Washington state, USA, Denmark, and other northern European countries. Similarly, the north–south extension and diversity of climatic zones in the state of California, USA and Chile provide excellent growing conditions for Brassica and other vegetable seeds. In addition, seed quality and economic factors such as the availability and cost of labor and favorable seed yields also determine where seeds are grown. The physiological and molecular events that control the induction of flowering are complex and multifactorial, as might be expected from the large number of environmental factors that influence flowering. The transmissible signal that induces meristems to shift from the formation of leaves and stems to the formation of floral organs has long been termed florigen. Florigen (or flowering hormone) refers to the hormone-like molecule(s) responsible for controlling and/or triggering flowering in plants that is produced in the leaves, and changes apical meristems of buds and growing points from vegetative to reproductive development. It is known to be graft-transmissible, and can even function between species. The molecular nature of the systemic floral signal, florigen, is a protein product encoded by the *FT* gene, which is highly conserved across flowering plants (Wang *et al.*, 2023).

Considerable progress has been made in identifying the genetic determinants of flowering time, with over 80 loci identified that respond to a variety of environmental factors (Turck *et al.*, 2008; Izawa, 2021). Responses to light clearly involve photoreceptors, including phytochromes and cryptochromes, that interact with other factors to control sensitivity to both light quality and photoperiod (Guo *et al.*, 1998; Liu *et al.*, 2001). The vernalization response to chilling involves down-regulation of the *FLC* gene (which inhibits flowering) and possibly changes in methylation and acetylation of histones, proteins associated with DNA (Bastow *et al.*, 2004; Angel *et al.*, 2015). Two genes (*VRN1* and *VRN2*) in wheat interact to determine whether vernalization is required for flowering (i.e. determining the difference between winter and spring wheats) (Yan *et al.*, 2003, 2004). Gibberellin (GA) and possibly other plant hormones are involved in flowering, although it has been difficult to prove this conclusively. Restriction of cell-to-cell (symplastic) movement of compounds was also correlated with flowering, suggesting that leaves may also export anti-florigenic signals to the meristem (Gisel *et al.*, 2002).

As timing of flowering and floral organogenesis both significantly affect plant fitness and crop yield, a detailed understanding of the regulatory mechanisms governing flowering time is needed for genetic improvement of vegetable seeds as well as their successful production (Huijser and Schmid, 2011; Srikanth and Schmid, 2011). Research has shown that many genes contribute to the different environmental and endogenous regulatory pathways. In *Arabidopsis*, a model plant for studying reproductive mechanisms, the key environmental signals responsible for flower initiation are photoperiod, ambient temperature, and vernalization pathways, which respond to day length, surrounding temperature, and prolonged cold exposure, respectively (Jackson, 2009; Fornara *et al.*, 2010). Other endogenous factors such as phytohormones and carbohydrate status also regulate flowering through the autonomous, gibberellic acid (GA_3), nutrient-responsive, and aging pathways (Srikanth and Schmid, 2011; Kim *et al.*, 2012; Matsoukas *et al.*, 2012; Yamaguchi and Abe, 2012). All these different regulatory pathways converge on a set of floral pathway integrator genes, namely *FLOWERING LOCUS T* (*FT*) and its paralogue *TWIN SISTER OF FT* (*TSF*), as well as *SUPPRESSOR OF CONSTANS 1* (*SOC1*) and *AGAMOUS-LIKE 24* (*AGL24*). These genes control the expression of a small set of meristem identity genes at the shoot apical and lateral meristems including *LEAFY* (*LFY*), *APETALA 1* (*AP1*) and *FRUITFUL* (*FUL*). Once the expression of these genes reaches a certain level, it induces the expression of floral organ identity genes and flowering occurs (Amasino, 2010; Huijser and Schmid, 2011; Srikanth and Schmid, 2011; Pose *et al.*, 2012). Over the past decades, there have been numerous reports demonstrating that many different microRNA (miRNA) families are involved in regulating flowering time and floral development, by targeting and down-regulating transcription factors involved with flowering (Jones-Rhoades *et al.*, 2006; Chuck *et al.*, 2009; Luo *et al.*, 2013; Spanudakis and Jackson, 2014). Much of this evidence has been obtained from *Arabidopsis*, although a high level of conservation of miRNAs and their regulatory pathways/target genes has been demonstrated in tomato, potato, maize, rice, and other species (Willmann and Poethig, 2007; Sunkar and Jagadeeswaran, 2008; Luo *et al.*, 2013). There are at least 11 different miRNA families that affect different stages of flowering induction and flower development in plants: miR156, miR159, miR160, miR164, miR166/165, miR167, miR169, miR172, miR319, miR390, and miR399.

The main miRNA families regulating timing of flower initiation are the miR156, miR172, and miR390 families, which are involved in the juvenile-to-adult vegetative phase transition and the miR159, miR169, and miR399 families that, along with miR172, affect the vegetative-to-reproductive phase transition (Jones-Rhoades *et al.*, 2006; Huijser and Schmid, 2011; Rubio-Somoza and Weigel, 2011; Zhu and Helliwell, 2011; Kim *et al.*, 2012; Jin *et al.*, 2013; Wang, 2014; Huo *et al.*, 2016; Dong *et al.*, 2022).

Differentiation of Floral Meristems

Once flowering has been induced, the meristems shift from production of vegetative organs (leaves) to reproductive organs (flowers). Meristem changes include an increase in apical height and the initiation of floral organ differentiation, which begins at the margin of the meristem. The floral organs form sequentially in whorls from the outside inward: sepals first, then petals in the next whorl, followed by a whorl of stamens, and finally by the carpels or ovary in the center (Fig. 2.1). This pattern is relatively invariant at the early stages of initiation, although subsequent organ development varies widely, giving rise to the diversity of flowering shapes and types.

Flowers can be broadly categorized into perfect flowers, which contain both the male or pollen-forming organs (stamens) and the female or fruit- and seed-forming organs (pistils or ovaries), and imperfect flowers that are missing either

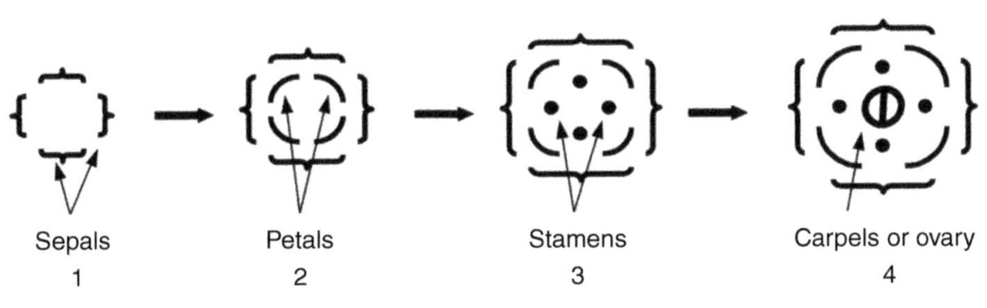

Fig. 2.1. Floral meristems develop inward from the outer whorl. (Courtesy of K.J. Bradford, personal communication.)

stamens or ovary (Fig. 2.1). In some species, such as spinach and asparagus, an individual plant produces either male flowers (with a non-functional ovary) or female flowers (with non-functional stamens). Such species are termed dioecious, and the presence of both male and female plants is required for seed production. In monoecious species, notably the Cucurbits and corn, distinct male and female flowers develop on the same plant. In corn, the tassels at the top of the plant are the pollen-producing male inflorescence, while the ear formed lower on the stem is the female inflorescence. In the Cucurbits, all possibilities can occur, including gynoecious (only female flowers), androecious (only male flowers), andromonoecious (male and perfect flowers), gynomonoecious (female and perfect flowers), and other variants depending upon the genetic composition and the environmental conditions. Altering the presence of or responsiveness to ethylene, a natural plant growth regulator, can also manipulate the sex of flowers in Cucurbits (Fig. 2.2). This is used extensively in the production of hybrid seeds of cucumbers and squashes (Murray *et al.*, 1997).

The "ABC" Model of Flower Development

In recent years, the genetic control of floral differentiation has been elucidated. Based primarily on investigations of mutations in *Arabidopsis* and *Antirrhinum*, the ABC model of floral organogenesis was developed (Fig. 2.3) (Coen and Meyerowitz, 1991; Coen and Carpenter, 1993).

This model proposes that floral differentiation is controlled by a series of related genes known as MADS-box genes (Theißen *et al.*, 2000). These are homeotic genes, or genes that can direct the formation of entire organs. For example, mutation of similar genes in fruit flies (*Drosophila*) can result in a leg growing where an antenna should appear. In plants, three categories of these genes have been identified that are involved in floral differentiation, termed the A, B, and C groups. The model proposes that each of the four whorls of the flower is specified by a unique combination of these genes (Yanofsky, 1995). "A" genes (such as *apetala2*) are expressed in the first whorl and specify sepal formation. In the second whorl, both A and B genes (such as *apetala3* or *pistallata*) are expressed, which results in petal formation. In the third whorl, B genes are still expressed, but C genes (such as *agamous*) turn on and the A genes turn off. In fact, A and C genes are mutually repressive, and normally are not expressed at the same time. Expression of the B and C genes together in the third whorl results in the formation of stamens. Finally, only the C genes are expressed in the fourth whorl, leading to the formation of the carpels or ovary. Further studies have identified additional genes, such as three *sepallata* genes, that cooperate with the ABC genes to result in floral development (Jack, 2001).

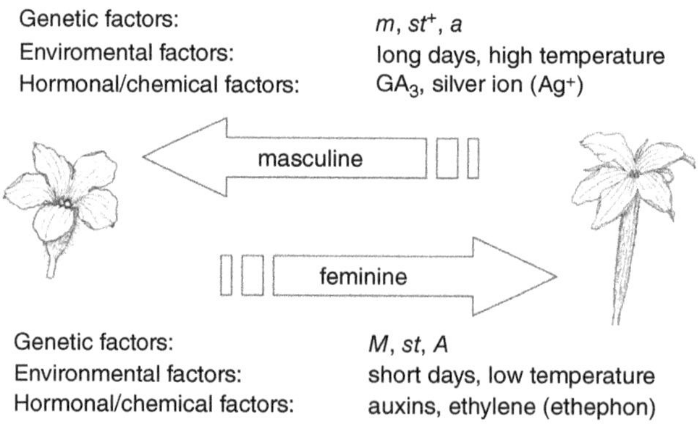

Fig. 2.2. Genetic, environmental, and hormonal factors influencing sex expression of cucumber flowers. The factors shown at the top promote the formation of male flowers (left), while the factors at the bottom promote the formation of female flowers (right). (Courtesy of K.J. Bradford, personal communication.)

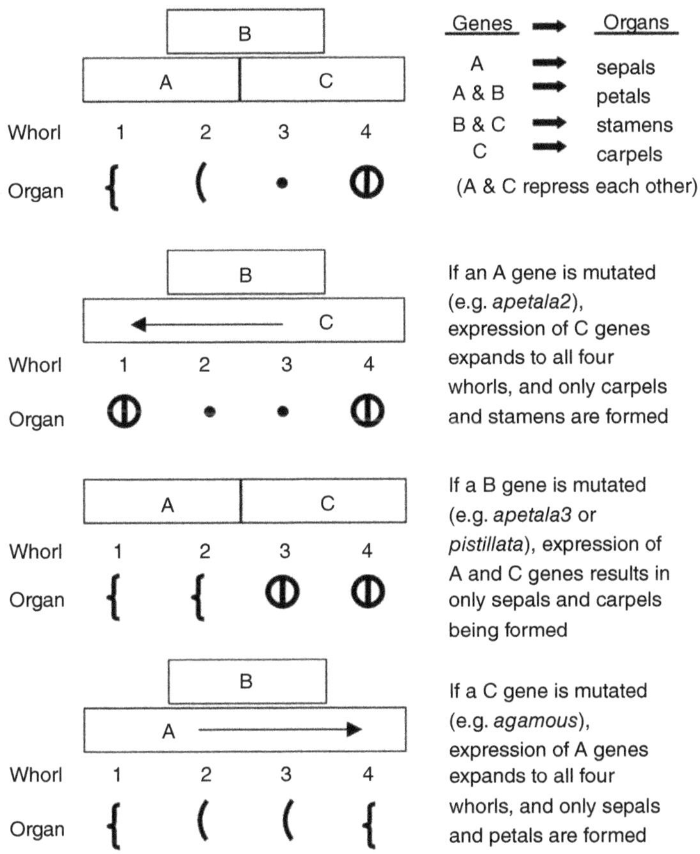

Fig. 2.3. According to the ABC model, three types of MADS-box genes control flower development. (Courtesy of K.J. Bradford, personal communication.)

These functions of the ABC genes have been deduced largely based on the effects of mutations in these genes on floral development. Mutation of either the A or C genes results in the expansion of expression of the other type of gene throughout all four whorls. Altered flowers are formed in the mutated plants as predicted based on the types of ABC genes that would be expressed in each whorl (Okamuro *et al.*, 1993). Interestingly, mutations in all three types of genes simultaneously still results in whorls, but of leaves instead of floral organs. This confirms the long-standing hypothesis that floral organs are modified leaves.

Microsporogenesis (pollen development)

The male gametes are formed within the stamens, which are composed of a filament and an anther (Fig. 2.4). The anther forms locules where pollen mother cells develop (Goldberg *et al.*, 1993). The pollen mother cells undergo meiosis (cell divisions which reduce the number of chromosomes by half) to form a tetrad of microspores, each of which contains half the normal number of chromosomes (haploid). Usually, all four pollen grains in a tetrad are viable. The haploid nucleus in each microspore divides again to form the vegetative and generative nuclei. In some species, the pollen grains are shed at this two-nuclei stage, while in others the generative nucleus divides again to result in three nuclei (a tube nucleus and two sperm nuclei) in the mature pollen grain. In the former type, division of the generative nucleus to form the two gametes occurs following pollen germination. The inner layer of cells in the anther locule is called

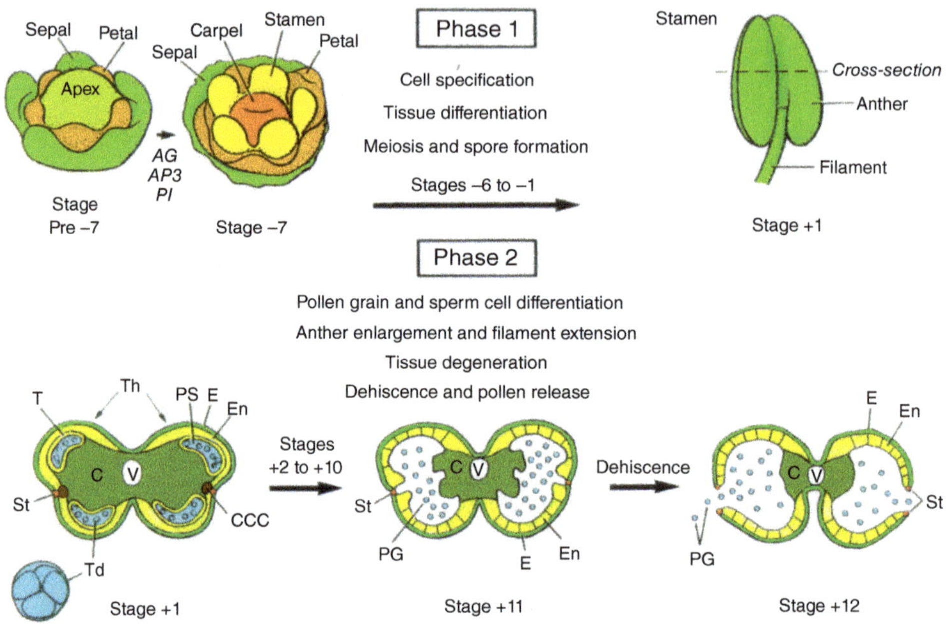

Fig. 2.4. Anther development. E, epidermis; En, endothecium; CCC, circular cell cluster; C, connective; PG, pollen grain; PS, pollen sac; St, stomium; T, tapetum; Td, tetrads; Th, theca; V, vascular bundle. (From Goldberg *et al.*, 1993.) Used with permission of American Society of Plant Biologists; permission conveyed through Copyright Clearance Center, Inc.

the tapetum, which is essential for pollen development and to produce proteins utilized in recognition reactions in some types of self-incompatibility (Goldberg *et al.*, 1993). The tapetum provides nourishment to developing microspores. It also secretes callase enzyme which dissolves the callose that holds the pollen tetrad together, freeing individual microspores or pollen grains. At pollen maturation, the inner tangential and radial walls of tapetal cells secrete small acellular structures of sporopollenin (< 1 μm to 15 μm) which coat pollen grains and increase the thickness of the exine, the outermost layer. At maturity, the anther locules open, releasing pollen grains (Fig. 2.4; Goldberg *et al.*, 1993).

Megasporogenesis (ovule development)

The female gametes develop within the carpel or ovary (Fig. 2.5). The megaspore mother cell undergoes meiosis to yield four megaspores, of which three generally degenerate (Goldberg *et al.*, 1994). Within the remaining megaspore, three cycles of nuclear divisions occur, creating eight haploid nuclei. Cell walls then form to give seven cells in the mature megagametophyte or embryo sac: one egg cell, two synergids, one central cell containing the two polar nuclei, and three antipodals. While this is occurring, the integuments are also developing around the embryo sac to form the ovule (Goldberg *et al.*, 1994). The integuments are derived from maternal tissue and will develop into the seed coat or testa. The funiculus connects the ovule to the ovary wall or placenta. The carpel develops into the fruit, which encloses the seeds of angiosperms (the term means "hidden seed"; Goldberg *et al.*, 1994).

Pollination and Fertilization

Pollination is the transfer of pollen grains from anthers to stigmas and is a prerequisite for fertilization, the fusion of gametes. There are three major pollen transfer strategies utilized by vegetable crops: self-pollination, insect pollination (primarily by bees), or wind pollination (see Table 2.1). Some flower structures promote selfing. Some families like Fabaceae contain male

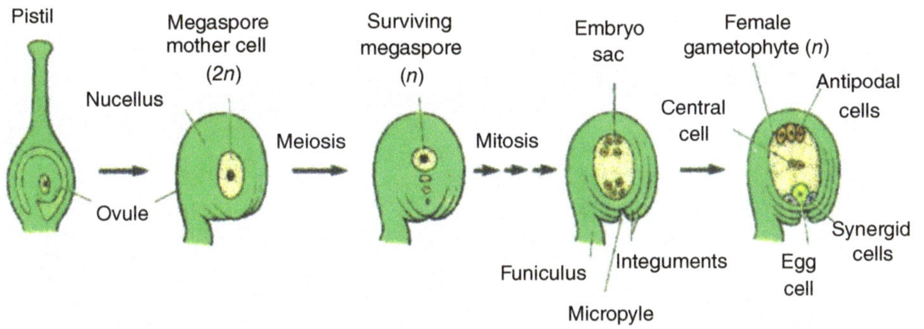

Fig. 2.5. Ovule development. (From Goldberg *et al.*, 1994.) Used with permission of American Association for the Advancement of Science; permission conveyed through Copyright Clearance Center, Inc.

and female flower parts enclosed together in individual flowers to facilitate self-pollination. In practical terms, this means that the pollen grains are light and mobile, and anthers and stigmas are in close proximity only a few millimeters apart often enclosed in "perfect" flowers. Self-pollinated vegetables include lettuce, tomatoes, eggplant (aubergine), peas, and beans. In these crops, the filaments are very thin so the slightest flower movement by wind or mechanical action causes pollen to slough off the anther to pollinate the nearby stigma. In tomatoes, fused anthers form a cone around the stigma. Wind moves the flower, releasing pollen into the cone with some landing on the stigma inside. It is interesting to consider how domestication has affected the pollination of tomato. In wild tomatoes, the stigma exerts beyond the edge of the fused conical anther tube so that pollen released inside cannot successfully fertilize the stigma because it is outside the pollen "tube". During domestication, plant breeders selected genotypes with shorter styles, such that the stigma remains inside the pollen tube for more reliable self-pollination, which is advantageous for cultivated plants. However, when some modern tomato cultivars are grown under environmental stress, the exerted stigma characteristic will recur, reducing pollination. In the case of tomato, domestication caused a transition from an occasional or sometimes frequently cross-pollinated to a more reliable self-pollinated plant due to decreased style length.

Pollen can be carried through the air from male to female flowers in some vegetable species. Important vegetables that are wind-pollinated include beets, chard, sweet corn, and spinach. Vegetables that are wind-pollinated have small inconspicuous flowers without petals and bright colors and lack fragrance and nectar. Their stamens and long prominent stigmas are exposed to air movements to shed and capture pollen, respectively. Pollen of wind-pollinated crops is smooth, light, and easily air-borne. These plants release billions of pollen grains into the air to improve chances for pollination.

Many vegetables such as the Cucurbits, *Brassica* spp., onion, carrot, and celery are insect pollinated, most commonly by bees. Bee-pollinated flowers have large showy, brightly colored petals, large sticky pollen grains, textured stigmas, and anthers and nectaries attached to the receptacle at the flower base inside the petals. Many insects forage among flowers to collect pollen and nectar as food. In the process, pollen grains are spread onto stigmas of many flowers. Among insect pollinators, bees are especially efficient because they consume pollen and nectar exclusively, visit many flowers of the same species during a single trip, and have hairy bodies that collect pollen grains. The honeybee is the most common and effective pollinator used for seed production. Honeybee activity closely coincides with the period when the flower is open from shortly after sunrise until mid-afternoon. For many crops, multiple bee visits are required for adequate pollination. Misshaped or undersized fruit often result from poor or incomplete pollination. Cucurbits have tacky pollen that isn't easily dislodged from the anther. On a perfect flower of a Cucurbit, the stigma and anthers may only be a few millimeters apart, but self-pollination doesn't occur without the assistance of bees because pollen tightly adheres to the anther. Honeybee hives are routinely placed in or around seed fields to

Table 2.1. Reproductive characteristics of some common crops.

Family*	Crop	Genus and species	Flowering type[1]	Selfed or crossed[2]	By[3]	Life cycle[4]	Variety type[5]	Remarks
Asparagaceae (Amaryllidaceae)	Asparagus	*Asparagus officinalis*	Dioecious	C	I	P	OP or H	Propagated by seed
Amaryllidaceae subfamily Allioideae (Alliaceae)	Onion	*Allium cepa*	Perfect	C	I	B	OP or H	
	Garlic	*Allium sativum*	Perfect			P	C	Rarely produces seeds
	Welsh onion	*Allium fistulosum*	Perfect	C	I	B	OP	
	Leek	*Allium ampeloprasum*	Perfect	C	I	B	OP or H	
Apiaceae (Umbelliferae)	Carrot	*Daucus carota*	Perfect	C	I	B	OP or H	Crosses readily with wild carrot
	Celery	*Apium graveolens*	Perfect	C	I	B	OP or H	
	Celeriac	*Apium graveolens*	Perfect	C	I	B	OP or H	
	Parsley	*Petroselinum hortense*	Perfect	C	I	B	OP	
Subfamily Chenopodioideae	Beet	*Beta vulgaris*	Perfect	C	W	B	OP or H	All forms cross, including sugarbeet
	Swiss chard	*Beta vulgaris*	Perfect	C	W	B	OP or H	Crosses with other beets
	Spinach	*Spinacia oleracea*	Dioecious	C	W	A	OP or H	Also contains perfect-flowered plants
Asteraceae (Compositae)	Sunflower	*Helianthus annuus*	Perfect	C	I	A	OP or H	Primarily F1 hybrids
	Lettuce	*Lactuca sativa*	Perfect	S		A	PL	Seed may carry lettuce mosaic virus
	Globe artichoke	*Cynara scolymus*	Perfect	C	I	P	C or OP	Some OP cultivars
	Jerusalem artichoke	*Helianthus tuberosus*	Perfect	C	I	P	C	Seedlings variable
Convolvulacae	Sweetpotato	*Ipomea batatas*	Perfect	C	I	P	C	Propagate by rooted shoots from tubers
Brassicaceae (Cruciferae)	Cabbage	*Brassica oleracea*	Perfect	C	I	B	OP or H	All vegetables in this group cross readily and are often self-incompatible; most commercial varieties are F1 hybrids, often now produced using CMS rather than self-incompatibility
	Cauliflower	*Brassica oleracea*	Perfect	C	I	B	OP or H	
	Broccoli	*Brassica oleracea*	Perfect	C	I	A	OP or H	
	Brussels sprouts	*Brassica oleracea*	Perfect	C	I	B	OP or H	
	Kohlrabi	*Brassica oleracea*	Perfect	C	I	B	OP or H	
	Kale	*Brassica oleracea*	Perfect	C	I	B	OP or H	
	Collards	*Brassica oleracea*	Perfect	C	I	B	OP or H	
	Turnip	*Brassica rapa*	Perfect	C	I	B	OP or H	
	Chinese cabbage	*Brassica rapa*	Perfect	C	I	A	OP or H	
	Rutabaga	*Brassica napus*	Perfect	C	I	B	OP or H	

Continued

Family*	Crop	Genus and species	Flowering type[1]	Selfed or crossed[2]	By[3]	Life cycle[4]	Variety type[5]	Remarks
	Rapeseed	*Brassica napus*	Perfect	C	I	A	OP or H	
	Radish	*Raphanus sativus*	Perfect	C	I	A	OP or H	Crosses with wild forms
Cucurbitaceae	Cucumber	*Cucumis sativus*	Monoecious (various forms)	C	I	A	OP or H	Does not cross with muskmelon
	Muskmelon	*Cucumis melo*	Andromonoecious or monoecious	C	I	A	OP or H	Includes: cantaloupe, honeydew, casaba, etc.
	Squash	*Cucurbita pepo*	Monoecious	C	I	A	OP or H	Some intercrossing can occur
	and	*Cucurbita maxima*	Monoecious	C	I	A	OP or H	among these species
	Pumpkins	*Cucurbita moschata*	Monoecious	C	I	A	OP or H	
	Watermelon	*Citrullus lanatus*	Monoecious	C	I	A	OP or H	Seedless cultivars are sterile triploids
Poaceae (Gramineae)	Corn	*Zea mays*	Monoecious	C	W	A	OP or H	All corn types (pop, flint, dent, flour, sweet) cross
	Wheat	*Triticum aestivum*	Perfect	S		A	PL or H	Winter wheat requires vernalization
	Rice	*Oryza sativa*	Perfect	S		A	PL or H	Hybrids used in China
	Barley	*Hordeum vulgare*	Perfect	S		A	PL	
Fabaceae	Bean	*Phaseolus vulgaris*	Perfect	SS		AA	PL	Seed can carry serious
(Leguminosae)	Lima	*Phaseolus lunatus*	Perfect				PL	diseases
	Pea	*Pisum sativum*	Perfect	S		A	PL	
	Cowpea	*Vigna unguiculata*	Perfect	S		A	PL	Includes blackeye pea
	Soybean	*Glycine max*	Perfect	S		A	PL	
	Alfalfa	*Medicago sativa*	Perfect	C	I	P	OP	"Synthetic" OP cultivars
Malvaceae	Cotton	*Gossypium hirsutum*	Perfect	CS	(I)	P	PL	Primarily self-pollinated
	Okra	*Abelmoschus esculentus*	Perfect	CS	(I)	A	PL	Primarily self-pollinated
Solanaceae	Tomato	*Solanum lycopersicon*	Perfect	S		A	PL or H	
	Pepper	*Capsicum annuum*	Perfect	CS	I	A	OP or H	
	Eggplant	*Solanum melongena*	Perfect	CS	I	A	OP or H	
	Potato	*Solanum tuberosum*	Perfect	S		P	C, OP, or H	Propagated by tubers; some seed propagation

*Archaic names in parenthesis. [1]Perfect = male and female in the same flower; dioecious = male and female flowers on separate plants; monoecious = separate male and female flowers on the same plant; andromonoecious = male and perfect flowers on the same plant. [2]S = self-pollinated; C = cross-pollinated; CS = both crossed and selfed. [3]Pollinated by: I = insects with some selfing; W = wind. [4]A = annual; B = biennial; P = perennial. [5]PL = pure line; C = clone; OP = open-pollinated; H = F1 hybrid.

ensure enough bees are present for adequate pollination. In addition, as bees can fly up to several miles from their hives, bee-pollinated seed crops must be isolated by distance from other varieties or compatible species.

In some crops, flower structures do not develop simultaneously to encourage out-crossing with other flowers. In onion, for example, flowers are protandrous, with the anthers opening first to release pollen and then senescing before the style extends and the stigma becomes receptive. Out-crossing of individual flowers must occur since there is no source of pollen on the same flower when the stigma is fully developed. However, onion flowers within an umbel open at different times, so individual flowers in a given umbel can cross, resulting in self-pollination. Although not a vegetable, avocado flower development is protogynous, illustrating a case in which the female flower parts develop before the male. In protogynous species the stigma and style develop before the anthers to ensure cross-pollination as well.

The number of ovules (and therefore seeds) in an ovary varies widely depending on the crop. *Brassica* spp. may have 25 ovules in each fruit; in contrast, a cantaloupe fruit contains up to 500 or 600 ovules inside each fruit. Each ovule must be fertilized by a single pollen grain for seed development to occur.

Controlling Fertility

For seed production, and specifically for F1 hybrid production, fertility must be controlled. Three major ways that fertility is managed include (Kaul, 2012):

1. Emasculation: physical removal (usually by hand) of the anthers prior to anthesis to prevent selfing. Emasculation is generally followed by hand pollination in crops like tomato and pepper. In maize, the entire male inflorescence (tassel) is removed from the female parent so that natural wind pollination can only occur from the desired male parent.
2. Male sterility: plants in which the anthers or pollen fail to develop cannot self-pollinate and can be used as female parents for F1 hybrid seed production.
3. Self-incompatibility: some plants have physiological mechanisms to prevent self-fertilization, and these can be used for production of F1 hybrid seeds.

There are three different types of male sterility used for seed production: genic sterility, cytoplasmic sterility, and cytoplasmic-genic sterility.

Genic male sterility, or failure to produce viable pollen due to a genetic defect, has been discovered in many crops (Kaul, 2012). Genic male sterility must be maintained in a heterozygous parent, which is selfed to produce 50% heterozygous line (the maintaining line) and 50% male-sterile progeny. Thus, the fertile segregants must be rogued from the female plantings prior to anthesis. Therefore, genic sterility is not used widely for vegetable seed production due to the segregation of the female parent and the need for rogueing out fertile plants. In some self-pollinating crops, genic male sterility offers the benefit of eliminating the need for emasculation prior to hand pollination.

Cytoplasmic male sterility (CMS) occurs when the defect in pollen production is based in the maternal cytoplasm, often as a mutation in mitochondrial DNA (Kaul, 2012). All progeny from a CMS plant are also sterile, so CMS was initially used for commercial development of hybrids of several crops in which organs other than seeds are marketed, including onion, carrot, radish, table beet, and cauliflower. The parental CMS lines are maintained by pollinating with a sister line having normal cytoplasm.

In many cases, nuclear restorer genes have been identified that can be supplied through the male parent to restore fertility in progeny from CMS maternal parents (Kaul, 2012). Such cytoplasmic-genic male sterility is widely used in production of sunflowers and was used in maize (field and sweet corn) hybrid production until 1970, when it was discovered that the Texas or "T" cytoplasm used resulted in sensitivity to the Southern corn leaf blight disease. CMS initially from radish has been transferred into most commercial Brassica crops via protoplast fusion (which transfers the cytoplasm), allowing hybrids of these crops to be produced using cytoplasmic-genic sterility methods in addition to self-incompatibility (see below). Thermosensitive and photoperiod-sensitive CMS has been used for rice breeding, allowing the CMS parent to self-pollinate when produced in the conducive environment and simplifying propagation of the female parent. However, the appropriate lines have not been identified or developed for use in vegetables.

Once pollen lands on the stigma of a plant from the same species with compatible recognition factors, the pollen grain will germinate, sending a pollen tube growing into the stigma. However, many plants exhibit self-incompatibility, or the failure of the same plant (or plants carrying the same incompatibility factors) to successfully fertilize the ovule. This system exists to ensure that the plants out-cross rather than self-pollinate. Self-incompatibility doesn't occur in all plant families and is most common amongst vegetables of the family Brassicaceae. If genetically controlled recognition factors are incompatible (i.e. too similar between the pollen and the stigma or style), pollen germination or pollen tube growth is blocked, depending upon the type of self-incompatibility. There are two major self-incompatibility (SI) systems distinguished by sporophytic or gametophytic determination of the self-incompatible reaction (Fig. 2.6; Newbigin *et al.*, 1993).

Both systems are controlled by the S-locus, a single, multiallelic (over 60 are known in *Brassica* spp.) locus that gives pollen a specific phenotype recognized by the stigma and style of potential receptor flowers. In the gametophytic self-incompatibility system, fertilization does *not* occur any time the S-allele of the gamete matches an active S-allele of the plant to be fertilized. Typically, with gametophytic incompatibility, pollen completes germination on the stigma but, at some point the pollen tube dies during growth in the transmitting tissues, probably due to uptake of RNases present on the stigma and in the style (Newbigin *et al.*, 1993). The sporophytic self-incompatibility system uses the ability of the stigma to detect pollen produced by plants sharing the same S-locus alleles (Nasrallah and Nasrallah, 1993). The phenotype of each pollen grain is determined sporophytically by the diploid genome of the parent plant. Protein products of the S-alleles are deposited in the exine of the pollen during development, and are detected by sensors in the style. Typically, inhibition is rapid when genetically related pollen lands on the stigma because the protruding pollen tube is blocked by a layer of callose that forms on the stigmatic surface. Typically, the stigma develops incompatibility to its own pollen only one day prior to anthesis, enabling bud-pollination to be used to maintain inbred parental lines. If pollen lands on the stigma of a plant from a different species, germination won't occur because of incompatible biochemical interactions preventing interspecific crossing.

If the pollen and receiving flower are compatible, pollen tube growth will continue

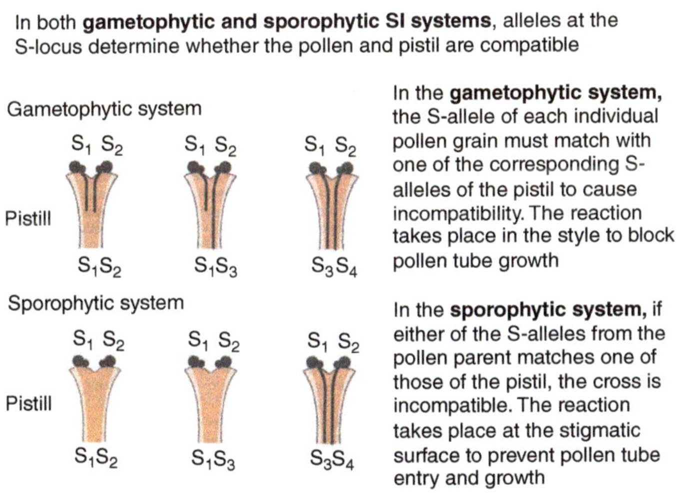

Fig. 2.6. Gameophytic and sporophytic self-incompatibility systems. (From Newbigin *et al.*, 1993.) Used with permission of American Society of Plant Biologists; permission conveyed through Copyright Clearance Center, Inc.

through the style to the ovary and on to an ovule where fertilization occurs. The pollen tube is not hollow but at least partially filled with cytoplasm including vesicles, endoplasmic reticulum, and mitochondria from the pollen grain as well as the tube nucleus and two sperm nuclei cells (Fig. 2.7).

The pollen tube grows through the style from its tip, pulling along the cytoplasm and the nuclei. The void space behind the advancing cytoplasm and nuclei is walled off with callose plugs to ensure migration is toward the ovule and to limit the volume of cytoplasm to the active growth region. Corn is a good crop to conceptualize pollen tube growth leading to fertilization. Silks on an ear of corn (*Zea mays*) are very long styles, each connected to an individual ovule on the cob. Each pollen tube must grow the length of the silk to fertilize an ovule which is required for successful kernel development. This remarkable, complex process is repeated millions of times in a large corn field.

The elongating pollen tube must grow into the small micropyle opening between the integuments in order to penetrate the embryo sac and fertilize the ovule (Fig. 2.8). Growth of the pollen tube is guided by recognition and growth factors in the style and funiculus and help the pollen tube find its way (Higashiyama and Yang, 2017).

These recognition factors attract pollen tubes to the micropylar region of the ovule and direct growth into the micropyle. Pollen tube growth is not a random act but a carefully orchestrated process under genetic control to ensure each individual ovule receives one pollen tube (Higashiyama *et al.*, 2003; Mizuta and Higashiyama, 2018). If genes directing pollen tube growth are mutated, the pollen tubes lose their way, growing randomly and failing to find their micropyle, so fertilization cannot occur (Wang *et al.*, 2008). When the pollen tube reaches the ovule's micropyle, it penetrates the embryo sac and releases the two sperm nuclei inside. During fertilization the tube nucleus, which guided pollen tube growth, disintegrates. Double fertilization joins one haploid sperm nucleus with the haploid egg to form the diploid zygote which rapidly divides to form the embryo (Bewley *et al.*, 2000). The second haploid sperm nucleus fuses with the two central cells' haploid nuclei in the ovule to form the triploid endosperm.

Embryogenesis

Embryogenesis refers to development from the time of fertilization until the embryo is fully developed and germinable or some cases dormant. After fertilization, a period of rapid cell division begins as embryos develop in stages.

Pollen tube growth

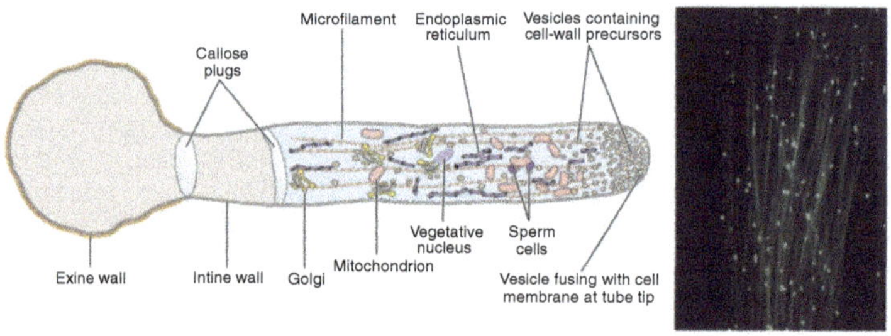

The pollen tube grows from the tip, pulling the cytoplasm along. Callose plugs form behind to limit the volume of the cytoplasm. These can be seen in the micrograph of pollen tubes shown at right

Fig. 2.7. Pollen tube growth through the style. (From Bewley *et al.*, 2000.) Used with permission of American Society of Plant Biologists; permission conveyed through Copyright Clearance Center, Inc.

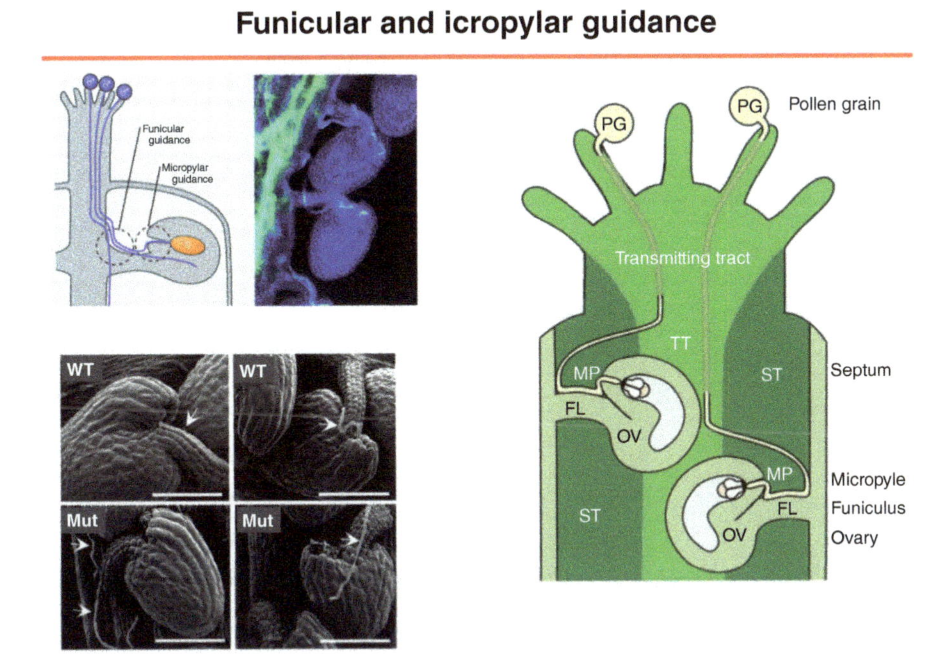

Fig. 2.8. Pollen tube growth is not a random event (right) (PG = pollen grain, TT = transmitting tract, ST = septum, MP = micropyle, FL = funiculus, OV = ovary; from Higashiyama and Yang, 2017[1]). Pollen tubes are guided through the funiculus to find a micropyle (upper left) (from Higashiyama *et al.*, 2003[2]). This system ensures that a single tube grows to each ovule. When genes controlling pollen growth are mutated, tubes fail to find the micropyle and no fertilization occurs (lower left) (WT = wild type, Mut = mutation of the guidance system; from Wang *et al.*, 2008[3]). [1]Used with permission of American Society of Plant Biologists; permission conveyed through Copyright Clearance Center, Inc. [2]Used with permission of Elsevier. [3]Used with permission of American Society of Plant Biologists; permission conveyed through Copyright Clearance Center, Inc.

Embryogenesis in higher plants can be divided conceptually into three overlapping phases. The first phase is morphogenesis, when the polar axis of the embryo is defined. Polarity is established in the first cell division following fertilization. These two cells are very different, and give rise to different structures, establishing polarity in the embryo. The small apical cell is on the top and contains most of the cytoplasm from the original zygote. Embryo development is frequently characterized by names that describe changes in appearance and include: 4-, 8-, and 16-cell, globular, heart, torpedo, and proembryo stages of development (Goldberg *et al.*, 1994). Important morphological changes during the transition from the globular stage to the heart stage are the first visible sign of the formation of the cotyledons and the axis, two embryonic organ systems. The apical cell will further divide and develop into the shoot apical meristem, hypocotyl, and cotyledons. The larger basal cell (with a large vacuole) is on the bottom and will give rise to the radicle, the root cap, and suspensor. The emergence of the cotyledons from a radially symmetrical globular embryo indicates cells in the apical region are induced to proliferate. The embryo is fully developed after the suspensor disappears at the torpedo stage due to programmed cell death (Goldberg *et al.*, 1994).

At the end of the morphogenesis stage in angiosperms, the fully developed embryo includes the shoot apical meristem, hypocotyl, root meristem, root cap, and cotyledons. Seed water content reaches its maximum when morphogenesis and expansion are completed, and then declines as water is exchanged for dry

Double fertilization

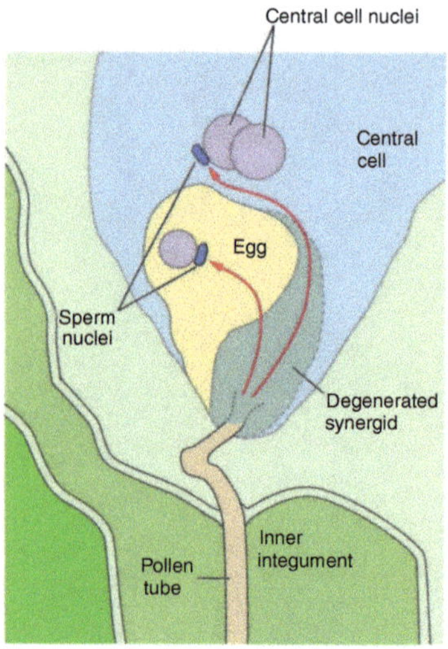

The synergid cells are involved in releasing pollen tube attractants. The pollen tube penetrates one of the synergids and releases the sperm nuclei. One then fuses with the egg nucleus to form the diploid zygote, while the other fuses with the central cell nuclei to form the triploid endosperm

Fig. 2.9. Double fertilization creates the zygote and the triploid endosperm. (From Bewley *et al.*, 2000.) Used with permission of American Society of Plant Biologists; permission conveyed through Copyright Clearance Center, Inc.

Late dicot embryogenesis

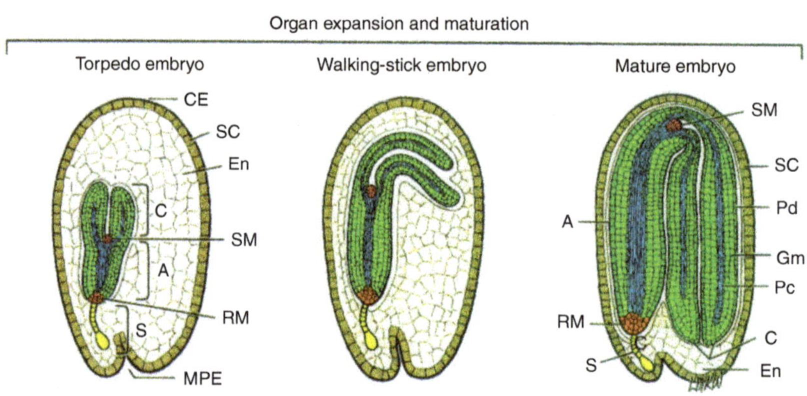

Fig. 2.10. Late embryogensis of a dicot embryo. Abbreviations: S = suspensor; Pd = protoderm; Pc = procambium; Gm = ground meristem; C = cotyledon; A = axis; MPE = micropylar end; CE = chalazal end; SC = seed coat; En = endosperm; SM = shoot meristem; RM = root meristem. (From Goldberg *et al.*, 1994.) Used with permission of American Association for the Advancement of Science; permission conveyed through Copyright Clearance Center, Inc.

matter during the seed filling and maturation stages. Plant embryogenesis differs from animals because immature plants that develop in seeds generally lack many common tissues like leaves, stems, and reproductive organs found in mature plants (Suarez and Bozhkov, 2008).

Endosperm development is coincident with embryo development. Three different types of endosperm occur in angiosperms based on the formation of cell walls. Endosperm is classified as nuclear, cellular, and helobial, an intermediate type between nuclear and cellular (Goldberg *et al.*, 1994). Helobial endosperm is most common in monocotyledons. If repeated free-nuclear divisions occur before cell wall formation, the endosperm is considered nuclear. In cellular endosperm formation, cell-wall formation is coincident with nuclear divisions. In many plants the liquid endosperm that forms early in development is absorbed and stored mostly in the cotyledons of the embryo, or the remaining nucellus that forms the perisperm (a maternally derived storage tissue), which is prominent in beet, or even the integuments that mature into the seed coat (Bewley *et al.*, 2000). In other species, such as those in the family Fabaceae, much of the endosperm is absorbed during embryo development and the function of storage tissue is performed by enlarged cotyledons or seed leaves; these seeds are called "exalbuminous" or "cotyledonous" (Bewley *et al.*, 2000). In some crops, such as corn, the endosperm persists in mature seeds as storage tissue. Mature endosperm tissue stores fats or starch depending on the species to nourish the young seedling following germination. Such seeds are called "albuminous" or "endospermous". The amount and type of carbohydrates stored in the helobial endosperm of sweet corn (*Zea mays* var. *saccharata*) depends on which naturally occurring recessive mutation controls conversion of sugar to starch. In certain species (e.g. tomato), the storage function is distributed between both endosperm and the embryo. Ancestral flowering plants have seeds with small, relatively immature embryos and abundant endosperm. Celery and carrot seeds have this type of endosperm which is considered evolutionarily less advanced and results in more variable germination and seedling emergence. The evolution of flowering plants trends toward mature seeds with little or no endosperm. In more evolved plants the embryo occupies most of the seed and the endosperm is non-developed or consumed before seed maturation (Friedman, 1998).

In some species, endosperm tissue is not a significant source of storage reserves and has evolved a protective role. In lettuce, melon, and cucumber, seeds have storage cotyledons and only a single-cell endosperm surrounding the embryo. This seemingly insignificant endosperm layer has evolved into a formidable barrier between the seed coat and embryo that allows water to pass but not larger molecular weight solutes. In melon, for example, the endosperm is a single suberized cell layer covered with a deposit of callose that forms during development around 30 to 35 days after anthesis to osmotically isolate the seed from fruit pericarp tissue (Welbaum, 1993; Yim and Bradford, 1998). Resilient protective endosperm tissues of this type are enzymatically broken down to allow radicle protrusion during germination (Welbaum *et al.*, 1998; Müller *et al.*, 2006).

The second phase in embryo maturation occurs when plants achieve maximum dry mass. During maturation, cell growth continues as water continues to be exchanged for dry matter as cells accumulate storage reserves such as oils, starches, and proteins required to nourish seedling growth following radicle emergence (Suarez and Bozhkov, 2008). Biological compounds formed in leaves, sometimes called "source" tissues, are translocated through the phloem to fruit and seed, often called "sink" tissues. Compounds translocated through the phloem are disaccharides such as sucrose and sometimes energy-storing monosaccharides produced by photosynthesis. These compounds are not delivered directly to the embryo but must pass through extracellular space (the apoplast) separating the two generations, prior to uptake into the embryo and endosperm. The osmotic environment of seed tissues affects transport and uptake of assimilates. Class I and Class II heat shock proteins are produced during later stages of embryogenesis (Sun *et al.*, 2002). Late embryogenesis abundant proteins (LEA proteins), as the name implies, accumulate late in maturation and protect against protein aggregation due to desiccation and osmotic stresses. LEA proteins are particularly protective of mitochondrial membranes against dehydration damage (Suarez and Bozhkov, 2008). The maternally derived integuments surrounding the embryo fully develop into the seed coat during seed maturation in parallel with embryo development.

Most major vegetable crop seeds are desiccation tolerant and termed "orthodox" in contrast to "recalcitrant" seeds that do not survive desiccation. During final maturation, the embryo prepares for dehydration, desiccates, and enters a period of developmental arrest. The most obvious developmental difference between zygotic and somatic embryos (the latter produced in tissue culture) is perhaps that somatic embryos lack a quiescent resting phase. Many seeds have primary dormancy that develops during maturity to protect from precocious germination prior to dehydration. The depth of primary dormancy varies among species of vegetable crops, cultivars within species, and the growing environment. Seeds that developed in fleshy fruits like tomato and the Cucurbits have a prolonged period of hydration lasting as long as several weeks before fruits dehisce, and seeds are shed. In melon seeds, the osmotic environment in fruit due to solute accumulation during later maturation helps prevent precocious germination before seeds are released (Welbaum and Bradford, 1988).

The biological function of seeds is to ensure continuation of the species for another generation. Seed dispersal mechanisms are often divided into five categories: wind, gravity, ballistics, water, and by animals. Modes of dispersal are typically inferred based on their structures, such as wings that help seeds move through air, shattering of seed pods that spread seeds, or fleshy fruit that animals consume and spread. For example, wild legume fruits explode and disperse their seeds ballistically over a wider area (Ceccarelli, 2009). However, such dispersal is not an advantage in agriculture, so domesticated legumes produce pods that resist splitting and stay intact much longer. Similarly, grain crops, like wheat, rice, and maize that were domesticated long ago and harvested for their seeds are quite resistant to shattering compared to their wild progenitors because of breeding and selection. Loss of shattering is among the first of the domesticated traits (Ceccarelli, 2009). However, since many vegetables are consumed in the vegetative stage before flowering, selecting against shattering in the reproductive stage has not been a priority for most breeders. Therefore, the reproductive stage of crops like lettuce, carrot, and onion remains primitive, similar to wild plants which flower over long periods and often shatter before seeds fully mature (Oliva *et al.*, 1988). Vegetable horticultural traits like yield, shelf life, color, flavor, disease resistance, uniformity, and others are the focus of genetic improvement programs and take precedence over reproductive traits. Greater resistance to shattering would allow seeds to fully mature before harvest. Selecting vegetables with favorable reproductive characteristics like seeds that remain on the mother plant and do not shatter and Brassica siliques that do not split open would go a long way toward improving seed yield and the quality of many vegetables (Jalink *et al.*, 1998; Still and Bradford, 1998).

When seeds develop in fleshy fruit, selection of melons and tomatoes that are highly crack and burst resistant adversely affects seed quality as well. Because fruit are selected for maximum shelf life and to resist cracking, seed dispersal is delayed compared to undomesticated relatives. Often seeds are held at near full hydration, in the harsh osmotic environment, and at high temperatures, conditions that can quickly age seeds before harvest (Welbaum, 1993). Horticultural characteristics that are favorable for consumers, but complicate seed production, will be discussed in later chapters. Suffice to say, plant breeding and selection of beneficial flower and seed characteristics are currently under-utilized tools that could improve seed quality and put less of a burden on the management skills of seed producers.

References

Amasino, R. (2010) Seasonal and developmental timing of flowering. *Plant Journal* 61, 1001–1013.

Angel, A., Song, J., Yang, H., Questa, J.I., Dean, C. and Howard, M. (2015) Vernalizing cold is registered digitally at *FLC*. *Proceedings of the National Academy of Sciences USA* 112, 4146–4151. doi:10.1073/pnas.1503100112 (accessed 15 September 2023).

Bastow, R., Mylne, J.S., Lister, C., Lippman, Z., Martienssen, R.A. and Dean, C. (2004) Vernalization requires epigenetic silencing of FLC by histone methylation. *Nature* 427, 164–167.

Bewley, J.D., Hempel, F.D., McCormick, S. and Zambryski, P. (2000) Reproductive development. In: Buchanan, B.B., Gruissem, W. and Jones, R.L. (eds) *Biochemistry and Molecular Biology of Plants*. American Society of Plant Physiologists, Rockville, Maryland, pp. 988–1043.

Ceccarelli, S. (2009) Evolution, plant breeding and biodiversity. *Journal of Agriculture and Environment for International Development* 103, 131–145.

Chuck, G., Candela, H. and Hake, S. (2009) Big impacts by small RNAs in development. *Current Opinions in Plant Biology* 12, 81–86.
Coen, E.S. and Carpenter, R. (1993) The metamorphosis of flowers. *The Plant Cell* 5, 1175.
Coen, E.S. and Meyerowitz, E.M. (1991) The war of the whorls: genetic interactions controlling flower development. *Nature* 353, 31–37.
Dong, Q., Hu, B. and Zhang, C. (2022) MicroRNAs and their roles in plant development. *Frontiers in Plant Science* 13, 824240. doi:10.3389/fpls.2022.824240 (accessed 15 September 2023).
Fornara, F., de Montaigu, A. and Coupland, G. (2010) SnapShot: control of flowering in *Arabidopsis*. *Cell* 141, 550.e1– 550.e2.
Friedman, W.E. 1998. The evolution of double fertilization and endosperm: an "historical perspective". *Sexual Plant Reproduction* 11, 6–16.
Gisel, A., Hempel, F.D., Barella, S. and Zambryski, P. (2002) Leaf-to-shoot apex movement of symplastic tracer is restricted coincident with flowering in *Arabidopsis*. *Proceedings of the National Academy of Sciences USA* 99, 1713–1717.
Goldberg, R.B., Beals, T.P. and Sanders, P.M. (1993) Anther development: basic principles and practical applications. *The Plant Cell* 5, 1217–1229.
Goldberg, R.B., De Paiva, G. and Yadegari, R. (1994) Plant embryogenesis: zygote to seed. *Science* 266, 605–614.
Guo, H., Yang, H., Mockler, T.C. and Lin, C. (1998) Regulation of flowering time by *Arabidopsis* photoreceptors. *Science* 279, 1360–1363.
Higashiyama, T. and Yang, W.C. (2017) Gametophytic pollen tube guidance: attractant peptides, gametic controls, and receptors. *Plant Physiology* 173, 112–121.
Higashiyama, T., Kuroiwa, H. and Kuroiwa, T. (2003) Pollen-tube guidance: beacons from the female gametophyte. *Current Opinion in Plant Biology* 6, 36–41.
Huijser, P. and Schmid, P. (2011) The control of developmental phase transitions in plants. *Development* 138, 4117–4129.
Huo, H., Wei, S. and Bradford, K.J. (2016) *DELAY OF GERMINATION1* (*DOG1*) regulates both seed dormancy and flowering time through microRNA pathways. *Proceedings of the National Academy of Sciences USA* 115, E2199–E2206. doi/10.1073/pnas.1600558113 (accessed 21 September 2023).
Izawa, T. (2021) What is going on with the hormonal control of flowering in plants? *The Plant Journal* 105, 431–445.
Jack, T. (2001) Relearning our ABCs: new twists on an old model. *Trends in Plant Science* 6, 310–316.
Jackson, S.D. (2009) Plant responses to photoperiod. *New Phytologist* 181, 517–531.
Jalink, H., Van der Schoor, R., Frandas, A., Van Pijlen, J.G. and Bino, R.J. (1998) Chlorophyll fluorescence of *Brassica oleracea* seeds as a non-destructive marker for seed maturity and seed performance. *Seed Science Research* 8, 437–443.
Jin, D., Wang, Y., Zhao, Y. and Chen, M. (2013) MicroRNAs and their cross-talks in plant development. *Journal of Genetics and Genomics* 40, 161–170.
Jones-Rhoades, M., Bartel, D.P. and Bartel, B. (2006) MicroRNAs and their regulatory roles in plants. *Annual Review Plant Biology* 57, 19–53.
Kaul, M.L. (2012) *Male Sterility in Higher Plants*. Springer Berlin, Heidelberg, Germany.
Kim, J.J., Lee, J.H., Kim, W., Jung, H.S., Huijser, P. and Ahn, J.H. (2012) The microRNA156-SQUAMOSA PROMOTER BINDING PROTEIN-LIKE3 module regulates ambient temperature-responsive flowering via FLOWERING LOCUS T in *Arabidopsis*. *Plant Physiology* 159, 461–478.
Liu, X.L., Covington, M.F., Fankhauser, C., Chory, J. and Wagner, D.R. (2001) ELF3 encodes a circadian clock-regulated nuclear protein that functions in an *Arabidopsis* PHYB signal transduction pathway. *The Plant Cell* 13, 1293–1304.
Luo, Y., Guo, Z. and Li, L. (2013) Evolutionary conservation of microRNA regulatory programs in plant flower development. *Developmental Biology* 380, 133–144.
Matsoukas, I.G., Massiah, A.J. and Thomas, B. (2012) Florigenic and antiflorigenic signaling in plants. *Plant Cell Physiology* 53, 1827–1842.
Mizuta, Y. and Higashiyama, T. (2018) Chemical signaling for pollen tube guidance at a glance. *Journal of Cell Science* 131, jcs208447. doi:10.1242/jcs.208447 (accessed 15 September 2023).
Müller, K., Tintelnot, S. and Leubner-Metzger, G. (2006) Endosperm-limited Brassicaceae seed germination: abscisic acid inhibits embryo-induced endosperm weakening of *Lepidium sativum* (cress) and endosperm rupture of cress and *Arabidopsis thaliana*. *Plant and Cell Physiology* 47, 864–877.

Murray, M., Hartz, T. and Bradford, K.J. (1997) *Cucurbit seed production in California*. UC Cooperative Extension Publication 7229. University of California, Davis.

Nasrallah, J.B. and Nasrallah, M.E. (1993) Pollen stigma signaling in the sporophytic self-incompatibility response. *The Plant Cell* 5, 1325–1335.

Newbigin, E., Anderson, M.A. and Clarke, A.E. (1993) Gametophytic self-incompatibility systems. *The Plant Cell* 5, 1315–1324.

Okamuro, J.K., Den Boer, B.G. and Jofuku, K.D. (1993) Regulation of *Arabidopsis* flower development. *The Plant Cell* 5, 1183.

Oliva, R.N., Tissaoui, T. and Bradford, K.J. (1988) Relationships of plant density and harvest index to seed yield and quality in carrot. *Journal of the American Society for Horticultural Science* 113, 532–537.

Poethig, R.S. (2003) Phase change and the regulation of developmental timing in plants. *Science* 301, 334–336.

Pose, D., Yant, L. and Schmid, M. (2012) The end of innocence: flowering networks explode in complexity. *Current Opinions in Plant Biology* 15, 45–50.

Rubio-Somoza, I. and Weigel, D. (2011) MicroRNA networks and developmental plasticity in plants. *Trends in Plant Science* 16, 258–264.

Sgamma, T., Jackson, A., Muleo, R., Thomas, B. and Massiah, A. (2014) TEMPRANILLO is a regulator of juvenility in plants. *Science Reports* 4, 3704. doi:10.1038/srep03704 (accessed 21 September 2023).

Spanudakis, E. and Jackson, S. (2014) The role of microRNAs in the control of flowering time. *Journal of Experimental Botany* 65, 365–380.

Srikanth, A. and Schmid, M. (2011) Regulation of flowering time: all roads lead to Rome. *Cellular and Molecular Life Science* 68, 2013–2037.

Still, D.W. and Bradford, K.J. (1998) Using hydrotime and ABA-time models to quantify seed quality of Brassicas during development. *Journal of the American Society for Horticultural Science* 123, 692–699.

Suarez, M.F. and Bozhkov, P.V. (eds) (2008) *Plant Embryogenesis* (Vol. 427). Springer, New York, p.184.

Sun, W., Van Montagu, M. and Verbruggen, N. (2002) Small heat shock proteins and stress tolerance in plants. *Biochimica and Biophysica Acta* 1577, 1–9.

Sunkar, R. and Jagadeeswaran, G. (2008) In silico identification of conserved microRNAs in large number of diverse plant species. *BMC Plant Biology* 8, 37.

Theiβen, G., Becker, A., Di Rosa, A., Kanno, A., Kim, J.T., *et al.* (2000) A short history of MADS-box genes in plants. *Plant Molecular Biology* 42, 115–149. doi:10.1023/A:1006332105728 (accessed 15 September 2023).

Turck, F., Fornara, F. and Coupland, G. (2008) Regulation and identity of florigen: FLOWERING LOCUS T moves center stage. *Annual Review of Plant Biology* 59, 573–594.

Wang, F., Li, S., Kong, F., Lin, X. and Lu, S. (2023) Altered regulation of flowering expands growth ranges and maximizes yields in major crops. *Frontiers in Plant Science* 14. doi.org/10.3389/fpls.2023.1094411 (accessed 15 September 2023).

Wang, H., Boavida, L.C., Ron, M. and McCormick, S. (2008) Truncation of a protein disulfide isomerase, PDIL2-1, delays embryo sac maturation and disrupts pollen tube guidance in *Arabidopsis thaliana*. *The Plant Cell* 20, 3300–3311.

Wang, J.W. (2014) Regulation of flowering time by the miR156-mediated age pathway. *Journal of Experimental Botany* 65, 4723–4730.

Welbaum, G.E. (1993) Water relations of seed development and germination in muskmelon (*Cucumis melo* L.): VIII. Development of osmotically distended seeds. *Journal of Experimental Botany* 44, 1245–1252.

Welbaum, G.E. (1999) Cucurbit seed development and production. *HortTechnology* 9, 41–48.

Welbaum, G.E. and Bradford, K.J. (1988) Water relations of seed development and germination in muskmelon (*Cucumis melo* L.): I. Water relations of seed and fruit development. *Plant Physiology* 86, 406–411.

Welbaum, G.E., Bradford, K.J., Yim, K.O., Booth, D.T. and Oluoch, M.O. (1998) Biophysical, physiological and biochemical processes regulating seed germination. *Seed Science Research* 8, 161–172.

Willmann, M.R. and Poethig, R.S. (2007) Conservation and evolution of miRNA regulatory programs in plant development. *Current Opinion in Plant Biology* 10, 503–511.

Yan, L., Loukoianov, A., Blechl, A., Tranquilli, G., Ramakrishna, W., *et al.* (2004) The wheat VRN2 gene is a flowering repressor down-regulated by vernalization. *Science* 303, 1640–1644.

Yan, L., Loukoianov, A., Tranquilli, G., Helguera, M., Fahima, T. and Dubcovsky, J. (2003) Positional cloning of the wheat vernalization gene VRN1. *Proceedings of the National Academy of Sciences USA* 100, 6263–6268.

Yamaguchi, A. and Abe, M. (2012) Regulation of reproductive development by non-coding RNA in *Arabidopsis*: to flower or not to flower. *Journal of Plant Research* 125, 693–704.

Yanofsky, M.F. (1995) Floral meristems to floral organs: genes controlling early events in *Arabidopsis* flower development. *Annual Review of Plant Biology* 46, 167–188.

Yim, K.O. and Bradford, K.J. (1998) Callose deposition is responsible for apoplastic semipermeability of the endosperm envelope of muskmelon seeds. *Plant Physiology* 118, 83–90.

Zhu, Q.H. and Helliwell, C.A. (2011) Regulation of flowering time and floral patterning by miR172. *Journal of Experimental Botany* 62, 487–495.

3

Genetic Improvement Methods for Vegetable Seeds

Outside the seed industry there is a common misconception that plant breeders produce the seeds that are sold by a seed company, ignoring the critical roles of the seed production and technology specialists and contracted farmers who are responsible for producing and preparing seed for commercial sale.

Plant breeders develop new cultivars using a wide array of techniques from traditional plant breeding to genomics and molecular biology. After new cultivars are developed and released, the responsibility for mass-producing seeds for sale falls to seed production personnel. Seed production is a separate discipline from plant breeding/genetic improvement. However, breeding and seed production are closely related and both are essential for commercial success, so both components of a company must work closely together to successfully produce and market seeds. Information about cultivar traits and crossing systems must be communicated and coordinated between breeders and seed producers for the successful operation of a seed company. In order to have an appreciation about how commercial vegetable cultivars are developed, and the implications for seed production, we will review some of the tools that breeders use to develop new cultivars.

Sources of Genetic Diversity

Cultivar development involves collaboration from several different departments and specialties within a seed company. It involves input and cooperation among: marketing, sales, breeding, production, production research, and quality control departments in an organization (Fig. 3.1).

New vegetable cultivars are created by combining traits in novel combinations. The objectives of genetic manipulation for cultivar development involve the integration of several traits, such as resistance to diseases, nutritional composition, changing plant stature for improved harvest efficiency, increasing yield, or matching consumer preferences. Some common sources of genetic diversity include old "heirloom" cultivars, plant introduction lines preserved in gene banks, wild germplasm, synthetic genes, edited genes, and novel genes created through induced mutation.

Gene banks are extremely valuable sources of diversity for genetic improvement programs. Wild relatives of crop plants are often collected and stored in gene banks for the valuable genes that they contain. Plant gene banks preserve genes (responsible for the observed traits, or phenotypes) that contain the information "blueprint" for vegetable cultivars in viable seeds or sometimes other tissues (Plucknett and Smith, 2014). A genome is the distinct and unique combination of genes in an organism and gene banks are currently the only place where plant genomes are systematically stored. Plant germplasm collections are assemblies of genotypes representing primitive botanical varieties or landraces of indigenous plants, agriculturally

DOI: 10.1079/9781789243260.0003

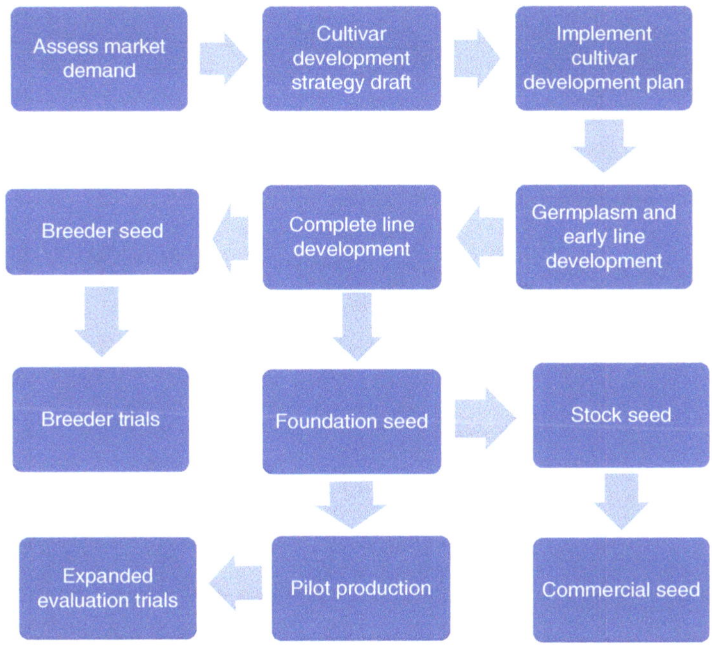

Fig. 3.1. Key steps in the development of new cultivars for commercial sale.

obsolete cultivars, and modern cultivars that are the product of scientific breeding and wild plants, including weedy relatives of crops.

Gene banks exist at different levels. Long-term storage gene banks are operated primarily by national programs in both industrial and developing countries, such as the Kew Millennium Seed Bank in Wakehurst, Sussex, UK, the US National Plant Germplasm System (NPGS) in Fort Collins, Colorado, USA, as well as international agricultural research centers, including the World Vegetable Center Gene Bank in Taiwan, the International Maize and Wheat Improvement Center (CIMMYT) in Mexico, the International Rice Research Institute (IRRI) in the Philippines, and the Svalbard Global Seed Vault in Norway. Developing countries house a substantial portion of the world's crop germplasm (Plucknett and Smith, 2014). Private companies concentrate on assembling working collections for their breeding programs rather than on operating long-term preservation of germplasm for posterity. It is important to note that seeds stored in even the best conditions (dry and cold) will eventually lose viability. Thus, an important function of germplasm storage organizations is to monitor the viability of the seeds in their accessions and also regrow them periodically to replenish seed stocks, doing this in a way that preserves their inherent genetic variation (Hay and Whitehouse, 2017).

Consolidation of vegetable seed companies in the developed world combined with loss of genetic resources because of land clearing has placed renewed emphasis on government and public gene bank storage and evaluation. For example, if Dr Charles Rick had not collected wild tomato relatives throughout the Andes in South America in the 1940s and 1950s and established the Tomato Genetics Resource Center (TGRC; see https://tgrc.ucdavis.edu/), much of those genetic resources would have been lost to subsequent land development. Most government and public gene banks not only store seeds but share seeds with researchers and breeding programs. Upon request, and if sufficient seeds are available, many gene banks will share samples of requested genotypes sometimes for a small fee. This world germplasm network provides an invaluable genetic resource by preserving and providing genes for crop improvement (Plucknett and Smith, 2014). To be useful to breeders, accessions collected for genetic archiving must be screened for resistance to plant diseases and other useful traits before they can be used as breeding lines for cultivar development. Some organizations, such as the Collaboration for Plant Pathogen Strain Identification (CPPSI; see https://cppsi.ucdavis.edu/), focus on developing collections of germplasm by identifying specific pathogen variants and their corresponding resistance genes.

Traditional Plant Breeding

Once desirable traits have been identified in existing germplasm, they must be transferred to a genetic background expressing the right horticultural traits to make a successful new cultivar. There are many technologies available today to accomplish this. Breeding systems used for cultivar development are important to understand because they may dictate subsequent seed production procedures.

Plant breeding began with the discovery of genetics by Gregor Mendel (Bateson and Mendel, 2013). Mendel's work (first published in 1866) was ignored for several decades, but constitutes the starting point for the development of plant breeding as a systematic discipline. Plant breeding is the oldest technique for creating new vegetable cultivars. Traditional plant breeding involves making sexual crosses between two plants by manually pollinating the stigma of one plant with pollen from an anther of another while excluding unwanted pollen. The progeny seeds from such crosses are then phenotyped to identify those that have inherited desirable traits, and this process is repeated for multiple generations to incorporate new genes (and the traits that they encode) into a cultivar. Vegetable breeding is thus the intentional genetic manipulation of plants through sexual means to create desired genotypes and phenotypes for enhancing crop production.

Plant breeding is the most common form of genetic enhancement practiced in the world today. Vegetable breeding programs are carried out worldwide by government institutions and private companies to develop new cultivars. Classical plant breeding uses deliberate cross-breeding of plants within the same species and occasionally different related species to produce novel plants with desirable properties. Thus, plants are cross-bred to transfer traits/genes from one line into a new genetic background. For example, a blight-resistant tomato may be crossed with a high-yielding but blight-susceptible tomato, and progeny carrying both traits are selected. The goal of such breeding is to introduce blight resistance while retaining the high-yielding characteristics. Progeny from the initial cross exhibiting blight resistance would be crossed again (backcrossed) with the high-yielding parent to transfer additional genes responsible for yield. This backcrossing process is repeated until the new parent line is stable and the cultivar has blight resistance along with high yields similar the original parents.

Sexual crosses combine complete genomes containing many genes in addition to the gene of interest, which complicates and slows selection of specific traits. Only a few individuals from a population of progeny may contain the gene combinations of interest. The selection process is time consuming and inefficient because most of the population derived from a cross is discarded to select a few desirable plants. To improve plant-breeding efficiency, marker-selected breeding was developed to rapidly screen plants for the trait of interest using genetic markers for specific locations in the genome. There are a variety of tests available, depending upon the trait, from simple seed germination tests in the presence of an herbicide to polymerase chain reaction (PCR) tests that detect the specific gene(s) that have been introduced into the cultivar. Advantages of classic plant breeding are that it is relatively inexpensive and is non-controversial compared to some of the molecular techniques described below.

Other methods of genetic improvement used for cultivar development involve selection of spontaneous mutations that occur in interbreeding populations. While many mutations are lethal, very rarely spontaneous mutations result in desirable traits conferring disease resistance, changes in plant architecture, color, nutritional composition, edible qualities, or size. For example, standard *su* gene sweet corn arose from a simple mutation that resulted in kernel accumulation of sugar rather than starch, while an orange cauliflower mutant was discovered in a field in Canada in the 1970s. Seeds of plants expressing stable mutations can pass on traits to future generations and be used in breeding programs. Mutations also can be induced using radiation or chemical mutagens (Spencer-Lopes *et al.*, 2018). Mutational breeding is very much a "shotgun" approach with a low percentage of favorable stable mutations, and therefore requires screening of large populations. Mutagens can be difficult to work with and require many safety precautions.

As our understanding of plant genetics and reproductive biology has advanced, vegetable breeding has become much more sophisticated. Modern molecular biology, genetics, and information technology have enhanced the capabilities of plant breeders through technologies such as marker-assisted selection and gene sequencing to track gene segregation across generations and to complement traditional techniques for manipulating genotypes through sexual crossing. Today, many different breeding techniques ranging from simple selection to direct gene editing are used in cultivar development. Many of the more advanced techniques make selection more efficient and help to reduce the time required to develop new cultivars. This is especially helpful to overcome the rapid development and spread of new races of several plant diseases, such *Bremia lactucae* in lettuce.

A major limitation to traditional plant breeding is that routine sexual crosses are only possible between plants within the same species, although occasionally wider crosses among species or even genera are possible. Somatic fusion, also called protoplast fusion, is a type of genetic modification that combines genetic material from different species. Somatic protoplasts of two distinct plant species are fused together using electrofusion or chemical treatment to form a new somatic hybrid plant with characteristics of both. The resulting hybrid has the chromosomes of both plants and is thus similar to polyploid plants (Torrence, 2008). Successfully fused cells can be regenerated into plantlets using tissue culture. Protoplast fusion has been used successfully to improve disease resistance in potato (Helgeson *et al.*, 1986).

Another technique to simplify complex genetic backgrounds as part of a genetic improvement strategy for vegetables is anther culture. Anther culture refers to plant regeneration from the haploid microspore cells with the aim of creating haploid plants containing only one set of chromosomes. These haploid plants can be treated with colchicine to duplicate those chromosomes and create dihaploid plants, which are completely "inbred", as both copies of each chromosome are identical. This greatly shortens the time required to create inbreds, which otherwise require repeated selfing for six to ten generations. Dihaploids have several advantages, such as a shortened breeding cycle by immediate fixation of homozygosity, high selection efficiency, wider genetic variability through the production of gametoclonal variants, and earlier expression of recessive genes suitable for breeding (Devaux and Pickering, 2005).

Modern Crop Improvement Methods

The past 40 years have seen the rapid evolution of molecular techniques for plant genetic improvement. These precise techniques allow direct transfer of genes from one organism to another regardless of their biological relationships. These technologies are based on the discovery and utilization of restriction endonucleases, enzymes that make direct manipulation of nucleic acids possible. It is now possible to identify a gene of interest in one organism and transfer and express it in another organism without sexual crossing. Such gene technologies are not restricted to inter-generic and inter-species transfers like traditional plant breeding, opening new opportunities for novel genetic combinations. The ability to directly manipulate nucleic acids, introduce them into cells, and regenerate those genetically transformed cells into whole plants are the basis of molecular breeding which has revolutionized the field of plant genetic improvement.

Agrobacterium tumefaciens is a naturally occurring bacterium that lives in soil. It infects susceptible plants, causing crown gall disease. *A. tumefaciens* infects plants by transferring its own DNA into cells using transfer DNA via its T-1 plasmid. The DNA fragment contributed by *A. tumefaciens* becomes stably integrated and expressed in the infected plant genome in a series of stages (Fig. 3.2). Plant biologists can replace the *A. tumefaciens* T-1 plasmid DNA with a cloned gene expression cassette containing the transgene they wish to incorporate and express

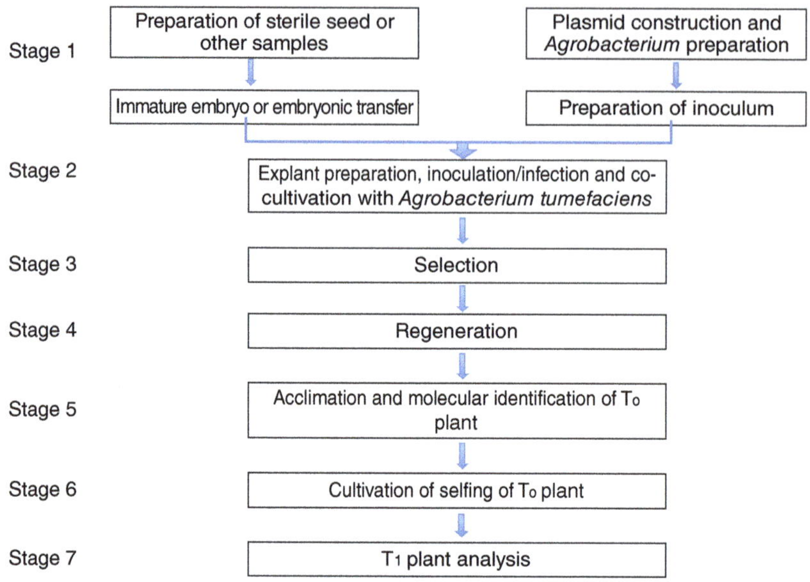

Fig. 3.2. Steps required for creating transgenic plants using *Agrobacterium tumefaciens*-mediated transfer. (Courtesy of K.J. Bradford, personal communication.)

in plant cells. Although a wide range of nascent plants cells may be used for *A. tumefaciens* transformation, immature embryo or embryogenic tissues are most common (Fig. 3.2). After co-incubation of *A. tumefaciens* containing the expression cassette with the transformation target plant tissue, a small percentage of the plants will be stably transformed if the system is optimized. Biolistics, also called ballistics, is another genetic transformation method in addition to *A. tumefaciens*-mediated transfer that has been used for the delivery of nucleic acid to cells by high-speed particle bombardment. Traditionally used on monocots, this technique uses nucleic acid-coated particles propelled by a pressurized gun (gene gun) to transform cells or organelles. A very small percentage of cells incorporate and express genetic material after treatment.

Genetically transformed cells can be identified by selectable markers (Fig. 3.2). A selectable marker helps to identify and delete non-transformants and allows transformants to expand selectively. A selectable marker is a gene inserted into a cell, in particular a bacterium or a cultured cell, which confers a trait appropriate for artificial selection. Selection may be positive or negative. For example, positive selection can involve targeting the desired cell population with an antibody specific to a surface marker. Targeted cells giving a positive response are then retained for downstream analysis. Negative selection is when several cell types are removed, such as by selection for introduced herbicide resistance, leaving the cell type of interest untouched. One objection to early transgenic crops was use of selectable markers that conveyed antibiotic resistance that were expressed in the crop plant. Modern transformation protocols no longer rely on antibiotic resistance as a selectable marker, so transgenic crops can be developed without expressing antibiotic resistance genes. Once transformed cells have been identified, they can be regenerated into whole plants using tissue culture (Fig. 3.2). Regenerated plants must be tested for expression to see if the transgene is stably expressed after regeneration. Stably transformed regenerated plants can be propagated by seed or sexually crossed with other breeding lines (Fig. 3.2).

Plant molecular biology has its own unique terminology that can be very confusing to outsiders (Table 3.1). The rapid evolution of molecular biology techniques complicates fluency in this field. The term "biotechnology" has become

Table 3.1. Some important terms describing genetic improvement technologies used for vegetable seeds.

Term	Definition
Biotechnology	Use of restriction endonucleases and related technologies to directly manipulate genes in living organisms. It may be used to describe any modern technology used in biology from tissue culture to enzyme-linked immunosorbent assay (ELISA)
Molecular biology	The study of life processes, especially gene expression, on a molecular level
Genetically modified	Manipulating the genes of a living organism by any means, including traditional plant breeding. Often misused in the popular press to describe organisms that were genetically modified using *Agrobacterium*
Genetically engineered	Genes are moved directly from one organism and expressed in another using one of several asexual gene transfer technologies e.g. *Agrobacterium*
Transgenic crops	Containing genes from another organism that were not transferred via sexual crossing
Genetically transformed	An organism that contains genes from another organism that were not transferred via sexual crossing
Plant breeding	Plant breeding is the science of changing plant traits to enhance desired characteristics. This is often accomplished by controlled sexual crossing of selected individuals and selecting progeny with desirable traits
Molecular plant breeding	Genetic manipulation or analysis at the DNA level to improve plant traits. It may include genetic engineering or gene manipulation, molecular marker-assisted selection, and genomic selection
Marker-assisted breeding	Marker-assisted breeding uses DNA marker-associated traits of interest to select plants for inclusion in a breeding program early in their development. This dramatically reduces the time needed to identify individuals expressing desired traits

a cliché meaning different things to different people. The term was first widely used in the 1980s to describe technologies that allowed foreign genes to be incorporated and expressed in organisms that were not closely related. Biotechnology in this sense was a significant development in plant science because previously genes could only be incorporated through sexual crossing between relatively close relatives within the same species. Thus, biotechnology enabled genes to be systematically recruited from diverse plant species or even other organisms for expression in crop plants. Since the initial use of biotechnology was coined, some now use the term rather loosely to mean any modern technology used in plant genetic research such as tissue culture and immunohistochemistry.

Some consumers are under the false impression that most vegetables sold commercially are transgenic. While it is technologically possible to create transgenic vegetables using genes from outside a species in conjunction with *Agrobacterium*-mediated transfer, the reality is that essentially none of the commercial vegetable crops sold in the world, including the USA, Canada, Mexico, and Europe, are transgenic (Table 3.1). Transgenic crops in most countries must pass strict evaluation and safety assessment requirements prior to sale or import. Currently grown transgenic crops are mostly field or agronomic crops that are extensively grown and that are not directly consumed by humans like grain soybeans, corn, canola, and cotton. Consumers and governments are more accepting of transgenic crops that are not consumed directly by humans but are more commonly industrial crops that are used for animal feed, fiber, or oil production.

One of the greatest obstacles to the development and use of genetically engineered vegetable crops today is consumer acceptance. Interestingly, the initial introduction of a transgenic tomato in 1994 proceeded, at least in the USA, South America, and Australia, with relatively little public concern. Regulatory procedures for the licensing and testing of transgenic crops were already in place when that product first appeared, and by and large the public had confidence in government agencies responsible for ensuring food safety. Following a series of high-profile events such as the "mad-cow disease" and other food safety issues such as biological contamination of vegetables in England, Europe, and other places in the world (none related to transgenic crops), consumers, particularly in those countries, lost confidence in their governments' ability to ensure food safety. Some groups opposed to genetically engineered foods on moral or philosophical grounds also played on these fears to sensationalize the situation, such as by labeling transgenic crops "Frankenfoods". Other factors for consumer rejection are based on complex sociopolitical dynamics caused by a lack of trust and understanding of the technology, the idea that such technologies are harmful to the environment and human health, DNA technologies are unnatural, and that large multinational corporations have excessive control without sufficient input from consumers.

Scientifically, fears regarding the safety of consuming transgenic crops are largely unfounded, as breeders have many years of experience with developing genetically improved cultivars by classical methods and know the types of factors, such as allergenicity, that need to be tested. Environmental concerns are also raised and, there, the uncertainty may be more justified in some cases. For example, the report that Monarch butterfly larvae were killed after eating pollen from maize plants expressing the *Bt* toxin has been widely publicized (Losey *et al.*, 1999). However, subsequent field trials indicate that while this is true in laboratory tests, exposure of the larvae to such pollen is quite low in the natural environment (Sears *et al.*, 2001). The risk to non-target insects from *Bt* pollen is much less than that from insecticides that would be used if the crop did not contain the *Bt* trait (Pimentel and Raven, 2000). Techniques like tissue-specific promoters limit expression of the *Bt* protein only to the leaves, stem, or roots, eliminating the toxin protein from pollen. Thus, both the benefits and the risks must be analyzed together and compared to current practices to determine the relative value or danger.

Experimentally, for research purposes, most vegetables (both monocots and dicots) can be genetically transformed using ballistics or *Agrobacterium*, but only a handful of vegetable cultivars have been approved by international governments for commercial production. In the 1990s transgenic tomatoes, squash, and potatoes were approved in the USA and brought to market but these vegetables were unsuccessful, largely because of a lack of consumer acceptance due to aggressive negative campaigns by organizations opposed to them. Few are offered for sale today because the lack of consumer trust and confidence in transgenic crops continues. These include some hybrids of sweet corn and squash; however, most cultivars of these crops are non-transgenic. In sweet corn, the *Bt* trait in some cultivars reduces the heavy use of pesticides to control insects such as earworm. From a corporate standing point, development of commercial transgenic vegetable cultivars has in many cases stopped because of high development costs and what is often viewed as excessive government regulation in the face of a lack of consumer confidence. For example, as of 1 January 2022, the National Bioengineered Food Disclosure Law requires all foods produced with bioengineered ingredients, also known as genetically modified organisms (GMOs), be labeled as such (Table 3.1). This law increases food costs by requiring more labeling and government oversight. Interpretation is problematic if transgenes are only in non-edible plant tissues and not in food products. The National Organic Program in the USA and other countries prohibits transgenics, so consumers could buy organic without additional government oversight.

The approval to create and sell transgenic vegetables is controlled by governments in most countries. Currently there is no unified policy on use of transgenic technologies; every country or political entity has their own approval procedure. At this time, there are many more transgenic agronomic crops grown around the world than vegetable crops. What then is limiting acceptance and use of transgenic vegetables? Two major factors can be identified. One is consumer mistrust. As consumers become more comfortable with transgenic vegetables, attitudes and buying preferences may change. Another major limiting factor is the exorbitant costs of getting a transgenic vegetable through the regulatory process for the large number of cultivars sold. For vegetables, no single cultivar can cover costs of government approval through sales. Many companies are unwilling to incur high development and regulatory costs of transgenic vegetables until consumers are receptive and willing to buy them. In addition, many vegetable seed companies operate in more than 100 countries, so until there is greater homogenization of regulatory systems, it is just beyond the ability of vegetable seed companies to gain the approvals required.

It is unclear when current consumer attitudes toward transgenics and the regulatory process may change. Public education could improve acceptance. A reduction in the bureaucracy without sacrificing public safety could help stimulate more transgenic crop development. Recently more scientists have spoken publicly in favor of a transgenic approach to crop improvement. Since inherent genetic diversity with crop species has nearly been exhausted, future advances will be reliant on trans/synthetic genes to provide novel traits. To date, many transgenes incorporated into crop plants have dealt with herbicide, insect, and disease resistances, extended shelf life, stress tolerance, and traits that do not directly benefit consumers. Transgenic vegetables with enhanced nutrition or flavor may appeal to consumers, such as the high-anthocyanin purple tomato recently approved by the US Department of Agriculture (USDA). Some predict that transgenic crop development and usage will only accelerate once a major world government, e.g. India, the European Union (EU), China, the USA, etc., endorses and supports their use. Currently acceptance of genetically engineered crops varies with country and EU countries are among the most conservative, while China, Brazil, Argentina, and Australia are among the more liberal.

One important development in genetic improvement is CRISPR (clustered regularly interspaced short palindromic repeats) technology to edit genes without transforming them with *A. tumefaciens* or other transgenic technologies. Crops whose nucleic acids have edited with CRISPR (pronounced "crisper") are not considered transgenic and do not require special labeling in most markets. This means that specific gene sequences can directly altered with surgical precision. CRISPR is derived from a natural bacterial defense system against viruses that can identify and cut DNA at specified locations. CRISPR/Cas9 gene editing is a technique that rapidly edits genomic DNA at a very high efficiency and specificity. Cas9 is an RNA-guided DNA endonuclease enzyme associated with CRISPR. CRISPR/Cas9 gene editing will likely be more widely used by crop genetic improvement in the future. In the USA, USDA regulations have already been modified to reduce the requirements for release of CRISPR-edited products if they are similar to the changes that can occur naturally or due to mutation. Although CRISPR can also be used to introduce genes in specific locations, if those genes are not from the same species, the product is still considered to be transgenic and subject to higher regulatory restrictions.

Controlling Pollination for Seed Production

In Chapter 2 we discussed the pollination biology of vegetables. Selfing and out-crossing systems are utilized by seed producers to mass-produce seed. We can think of seed production as controlled sexual crossing of plants. Understanding pollination systems is especially important for production of F1 hybrid seed, where inbred parental lines must be crossed in mass to produce the seeds of commerce. The following is an introduction to terminology used in the vegetable seed industry to describe the ways that various seeds are produced.

Pure line

Self-pollinated crops that do not out-cross are often referred to as a pure-line cultivar because all the seeds are derived from individual flowers pollinating themselves, assuring selfing and leading to genetic homozygosity. Lettuce, many common beans, and peas fall within this category. Pure-line seeds are usually uniform and will grow true-to-type from seed. Factors impacting the integrity of a pure-line cultivar may include physical contamination from other cultivars. In some production areas, a crop like beans can be subject to a low percentage of out-crossing when there is a high population of insect pollinators present during flowering. Maintaining genetic purity of seed crops of these species focuses primarily on rogueing off-types and preventing mechanical mixtures.

Open-pollinated

As the name implies, plants in an open-pollinated seed production field freely cross-pollinate with one another. The emphasis during seed production is on rogueing and removing off-types and assuring physical isolation from cross-compatible plants. This term is often used rather loosely to describe any cultivar that is a non-hybrid. A stricter definition is: “a heterogeneous cultivar of a cross-pollinated crop allowed to inter-pollinate freely during seed production” (Chahal and Gosal, 2002). Open-pollinated cultivars will produce plants reasonably true-to-type but by nature there is more natural variation in an open-pollinated than a self-pollinated crop. Open-pollinated cultivars are more subject to genetic drift if parental (or basic or foundation) seed stocks are not properly maintained or out-crossing occurs when proper isolation distances between fields are not adequate. By the same token, an F1 hybrid will be more uniform than an open-pollinated cultivar of the same crop.

Self- versus cross-pollinated crops

Some crops, such as lettuce, have perfect flowers that contain both male (anthers) and female (pistils) flower parts that develop simultaneously without barriers and therefore self-pollinate (Chahal and Gosal, 2002), while other crops must cross-pollinate by transferring pollen grains from the anther of one flower to the stigma of another plant of the same species. Some species have perfect flowers but cross-pollinate because flower parts do not develop at the same time, the pollen is not compatible with stigmas on the same plant, or the pollen is sticky and must be moved by bees or other insects to adjacent flowers. Other crops cross-pollinate because they are monoecious and have separate male and female flowers on the same plant. Dioecious crops like asparagus and spinach produce separate male and female plants and therefore must cross-pollinate.

Cross-pollination promotes genetic diversity because traits from many different plants are combined. We often describe cross-pollinated crops as being more genetically heterozygous than ones that are self-pollinated. This higher degree of heterozygosity produces offspring that are less true-to-type than self-pollinated plants. True-to-type describes how offspring resemble their parents and express traits that define a cultivar. Pollination of cross-pollinated crops may be less reliable because of dependence on pollen vectors such as insects or wind to spread pollen from plant to plant. Self-pollinating plants set seed more reliably since a pollen vector is not needed and plants use less energy attracting pollinators with colorful petals or odors. Self-pollinating plants tend to produce more uniform or homozygous progeny. However, self-pollinated crops may have difficulty as a whole resisting disease, for example, because of their limited genetic diversity (Chahal and Gosal, 2002).

Production of F1 Hybrid Seeds

While plants are often hybridized to created genetic diversity as part of a breeding program, F1 hybrids refer to a type of hybridization that is the most important commercially. F1 hybrids are created by sexually crossing two inbred lines that are homozygous at all loci (Chahal and Gosal, 2002). Inbred parent lines are developed by repeated self-pollination for six or more generations. In many plants, continuous inbreeding results in a loss of plant vigor and seed yield known as inbreeding depression, which complicates seed production such as in carrot and onion (Fig. 3.3). When the right combination of inbred (genetically homozygous) parents are crossed, the resulting F1 hybrid seeds receive one set of chromosomes from each parent, and this genetic heterozygosity results in hybrid vigor or heterosis (Fig. 3.3; Chahal and Gosal, 2002). In other words, the F1 plants outperform the original out-crossed parental lines. Combining ability describes how well inbred lines respond to hybrid crosses. Some combinations of inbred lines produce superior hybrids to others, so breeders must test combining ability of diverse inbreds as part of cultivar development (Fig. 3.3). This creates offspring that are heterozygous with different alleles at most loci. An F1 means “filial 1” because the seeds are derived from the generation following the parental generation. F1 hybrids are popular because of their greater vigor, uniformity, and increased yield potential (Chahal and Gosal, 2002). Seed companies also like F1 hybrids because seeds harvested from F1 crops will segregate in the next generation and cannot be used successfully as propagules to establish another hybrid field. F1 hybrid seeds are also more expensive than the open-pollinated

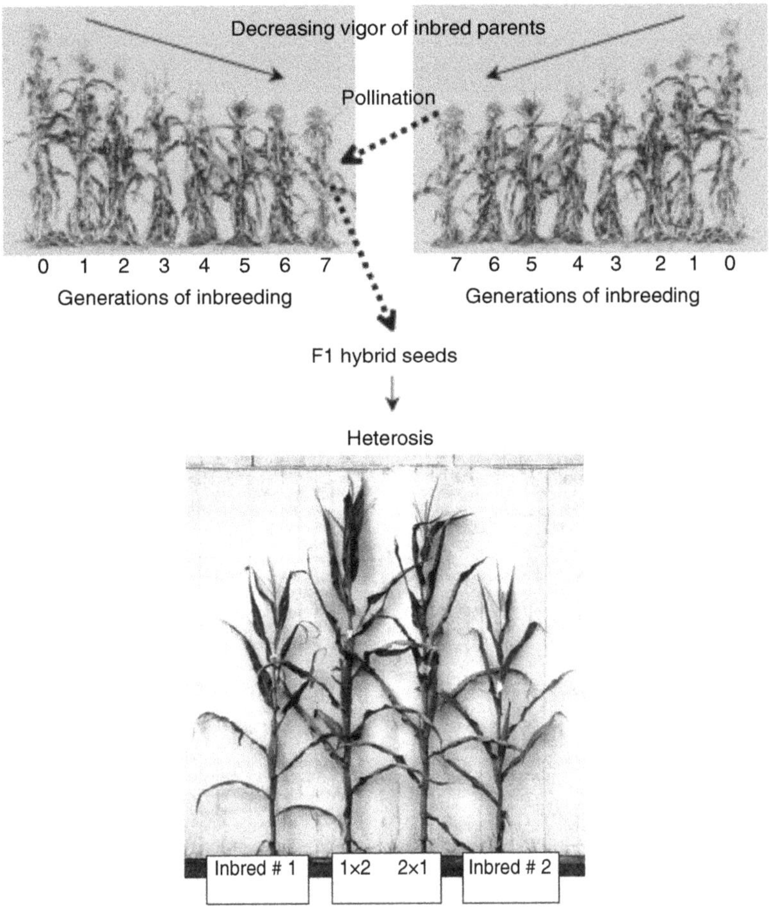

Fig. 3.3. Inbreeding depression and heterosis. The top panels illustrate the reduction in vigor (inbreeding depression) that occurs in corn if plants are continuously self-pollinated. However, if two distinct inbred parents are crossed, the resulting seeds are called F1 hybrids, and the heterosis (vigor) and yield of plants produced from these seeds exceeds both inbred parents (bottom panel). Reciprocal crosses show differences in the vigor of F1 offspring. (Modified from J.A. Birchler *et al.*, 2003 and K.J. Bradford, personal communication.)

seeds but worth the extra cost for vegetable production due to their greater productivity, and even hybrid seeds represent a rather small cost compared to other expenses.

Another advantage of F1 hybrids is a given female parent can be used in combination with several different male parent lines to provide unique hybrids for other market applications. For hybrid seed production, one line is designated the male parent and the other as the female parent. In order to prevent self-pollination of the female line, its pollen fertility must be controlled to permit fertilization only by pollen from the male parent. Where inbreeding depression is a challenge, such as in carrot or onions, three-way hybrids are utilized where two unique or even closely related lines are crossed to produce the female parent. The female parent is then crossed with the male parent to produce higher yields of hybrid seed for the market and lower production costs.

A critical requirement for successful pollination of hybrid seed crops is synchronous flowering of the male and female lines, known as nicking. If the flowering times of the parent lines do not coincide or nick, pollen will not be available when the female flowers are receptive, seed yields will be reduced, and the opportunity for unwanted out-crossing or self-pollination is increased. Depending on the crop, various methods, such as staggered planting times of the parent lines, multiple plantings of the male parent, and manipulating or pruning plants, ensure that the flowering times of parental lines match. Nicking trials in a particular environment to determine the flowering responses of each parent line and to test production techniques are essential for successful hybrid seed production.

For some crops, crossing can be forced by physically removing the male flowers or anthers from individual flowers, so selfing is impossible. For example, the separation of male and female flowers in corn allows plants to be emasculated by detasseling the female lines by hand or machine, which is widely used to produce hybrid corn seed. For tomato, the fused anthers form a pollen tube that can be removed with hand tweezers and pollen collected from male parents can be manually deposited on the exposed stigma of the female parent. Each emasculated and hand-pollinated tomato flower will produce about 250 seeds. Lettuce has perfect flowers, but each floret produces only a single seed so emasculating and hand pollinating each flower is not cost effective. Manipulation of pollination by genetic or hormonal mechanisms is not available, so there are no F1 hybrid lettuce or bean seeds available now. Gametocides that kill developing pollen have been developed, allowing treated rows to serve as a female parent accepting pollen collected from a male parent. However, such treatments are not totally effective and can result in a percentage of selfed plants, so are currently not common for vegetable seed production. New and improved seed crop hybrid systems are being developed all the time, helping reduce costly labor and human error, and will likely replace hand pollination sometime in the future. Seed producers must be aware of new technologies that improve techniques and overall operational efficiency.

F1 hybrids are available for most major vegetable crops where pollination systems enable convenient and economical cross-hybridization. Hybrid seeds are produced by manipulating pollination systems so mass crosses are easily made. For example, genetic and cytoplasmic male sterility, genetic pollen incompatibility, and use of plant-growth-regulating compounds to change floral sex expression are a few of the techniques used to control crosses to produce F1 hybrid seeds (Chahal and Gosal, 2002). Some of these methods may need to be incorporated as traits into the parental lines (e.g. male sterility, S-loci, or sex expression), so breeders are also involved in facilitating the production of hybrid seeds.

Hand emasculation is expensive, time consuming, and can lead to human error. There are several mechanisms to alter sex expression that are widely used by the seed industry to make hybrid production easier and more efficient. Incorporating the male sterility trait results in a line unable to produce viable pollen. This male-sterile line cannot self-pollinate and serves as the female parent in an F1 hybrid cross so seed formation is dependent upon pollen from the male parent. Due to the importance of controlling male fertility in the production of hybrid seeds, anther and pollen development and fertilization processes have been rigorously investigated (Goldberg *et al.*, 1993). Many nuclear genes involved in pollen development have been identified as mutants that result in pollen abortion and male sterility. This genic male sterility is used in hybrid seed production but has limitations due to the need to maintain female parent lines as heterozygotes and the segregation of fertile and sterile plants each generation (Fig. 3.4A). Cytoplasmic male sterility (CMS) is also used extensively in hybrid seed production. In this case, the sterility is transmitted only through the female and all progenies will be sterile (Fig. 3.4B). Sterile progeny is not a problem for crops such as onions or carrots where the crop produced in the F1 generation is vegetative. These CMS lines must be maintained by repeated crossing to a sister line (known as the maintainer line) that is genetically identical except that it possesses normal cytoplasm and is therefore male fertile. Cytoplasmic male sterility is due to alterations in the mitochondrial genome, as the mitochondria and plastids in the embryo are derived from the egg cell (Zubko, 2004). In some cases, nuclear restorer genes have been identified to overcome the effect of the sterile cytoplasm and restore fertility to the next generation. This is known as cytoplasmic-genic male sterility (Fig. 3.4C). Restoration of fertility is essential in crops where the harvested commodity from the F1 generation is a seed, such as sweet corn.

Male sterility has also been engineered into plants using the tissue-specific expression of a ribonuclease (RNase) called barnase (see Fig. 3.5; Mariani *et al.*, 1990, 1992). A promoter that is active only in the cells of the tapetum was

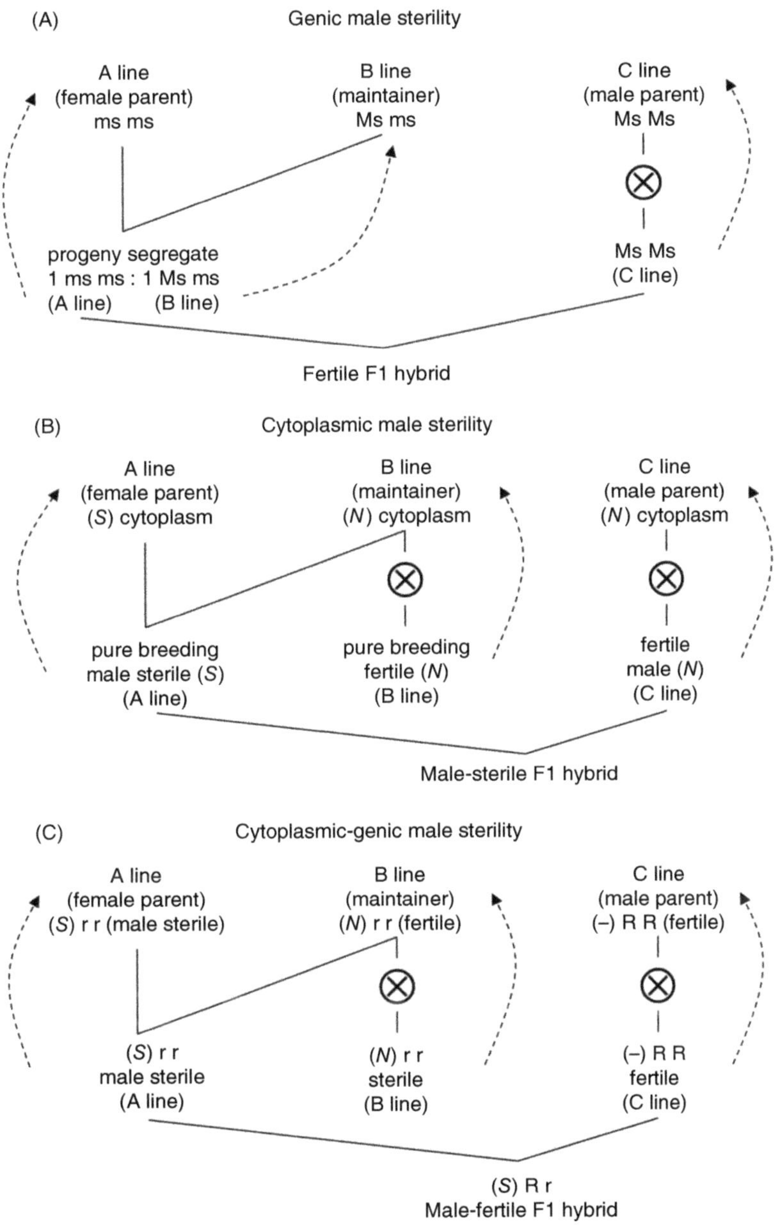

Fig. 3.4. Schemes for maintenance of parent lines and hybrid seed production utilizing (A) genic male sterility, (B) cytoplasmic male sterility, and (C) cytoplasmic-genic male sterility. Abbreviations: ms = recessive nuclear male sterility gene; MS = dominant nuclear male fertility gene; (*S*) = male-sterile cytoplasm; (*N*) = male-fertile cytoplasm; r = recessive fertility restorer gene (male sterile); R = dominant fertility restorer gene (overcomes effect of *S* cytoplasm) (male fertile); ⊗ = self-pollination. Solid lines indicate sexual crosses; dashed lines indicate seed production of the male sterile parent. (Courtesy of K.J. Bradford, personal communication.)

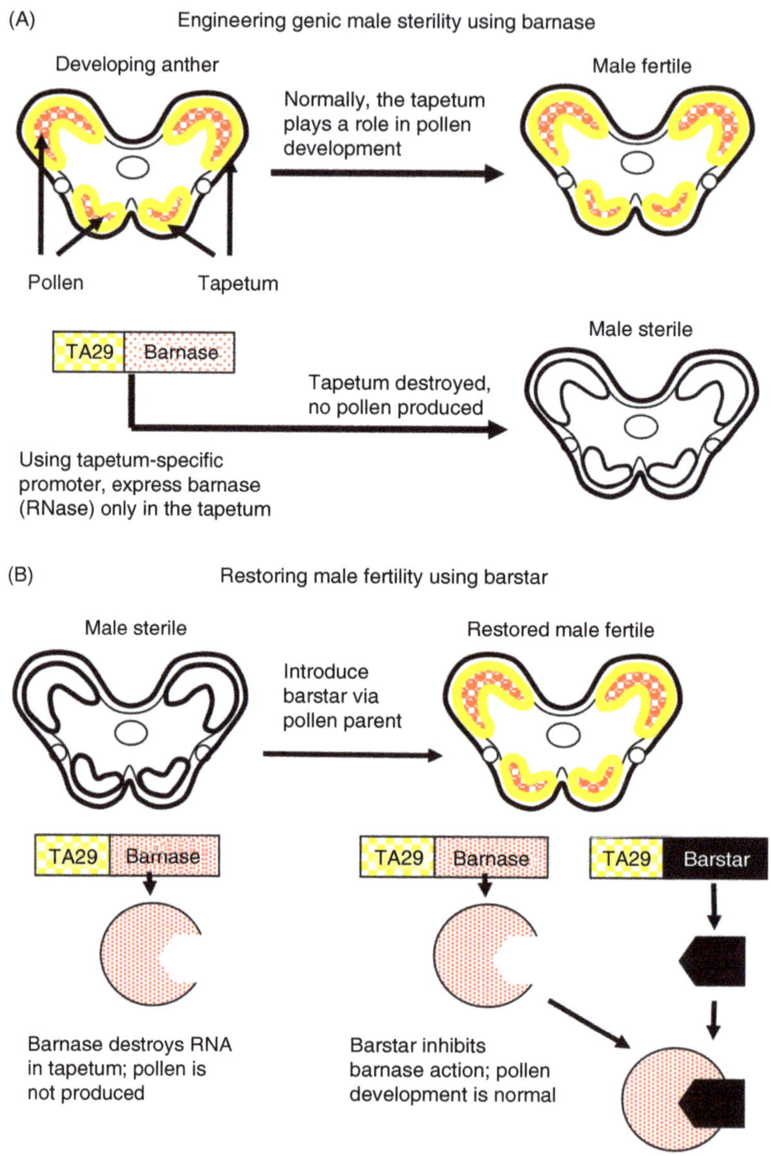

Fig. 3.5. Engineering genic male sterility using barnase (A) and restoring fertility to F1 plants using barstar (B) (from Goldberg *et al.*, 1993). Barnase is an RNase that will destroy the RNA of tapetal cells when expressed under the TA29 promoter, which is only expressed in the developing tapetum. This prevents the formation of viable pollen (A). To restore fertility to F1 plants (B), a gene coding for barstar, an inhibitor of barnase, can be introduced via the male parent under the control of the same TA29 promoter. In this case both barnase and barstar will be produced, the RNase activity will be inhibited, and normal pollen development will occur. (B) Used with permission of American Society of Plant Biologists; permission conveyed through Copyright Clearance Center, Inc.

isolated and used to cause expression of barnase during anther development. The barnase destroys the RNA in tapetal cells, stopping their development and preventing pollen grain formation. In this case, fertility can be restored by expressing an inhibitor of barnase, called barstar, using the same promoter. Thus, when the barstar gene is delivered through the pollen, it is also expressed in the next generation, blocking the action of barnase and allowing normal tapetum and pollen development. This system has been shown to work in numerous species and is being used commercially in hybrid seed production of corn and canola. Since it is a genic type of male sterility, the fertile and sterile plants will segregate. However, the barnase gene can be linked with an herbicide resistance gene, so fertile plants can be easily removed (Fig. 3.6). In all male-sterility systems, procedures for propagating the female parent lines are an important consideration (Perez-Prat and van Lookeren Campagne, 2002).

Protoplast fusion, described above, has been used to transfer CMS among species. For example, a sterile cytoplasm identified in radish (*Raphanus sativa*) has been transferred to *Brassica oleracea* crops. While this was done by repeated backcrossing into the radish cytoplasm background, the resulting plants where chlorotic due to incompatibility between the Brassica nucleus and radish chloroplasts. By fusing entire cells of the two species, cells were created that had all combinations of nuclei, chloroplasts, and mitochondria from the donor and recipient plants (Fig. 3.7). From this mixture, cells could be selected and regenerated into plants that contained the Brassica nuclei and chloroplasts, as well as the radish mitochondria responsible for the male sterility trait. This CMS is more desirable for hybrid seed production than the self-incompatibility system described below. It has been introduced into Brassica hybrid seed production.

Another mechanism for controlling fertility is self-incompatibility. Incompatibility is used in some crops, particularly Brassicas, for F1 hybrid seed production of crops like cabbage, broccoli, cauliflower, etc. Two general types of self-incompatibility (SI) exist, sporophytic and gametophytic (described in Chapter 2). These terms describe whether the maternal (sporophytic or 2N) genotype or the paternal (gametophytic or 1N) genotype specifies the self-recognition interaction. In addition, there are differences in pollen and stigma characteristics between the two systems (Table 3.2). In particular, the hydration requirements of the pollen and sites of the recognition reaction distinguish each system. Hydration of the pollen grain is a necessity following transmission to the stigma. In some cases, such as the Cucurbits, the stigma is wet and its secretions quickly engulf pollen grains. In other cases, as in the Brassicas, the surface of the stigma is dry and waxy. Hydration of each pollen grain is critical to the fertilization process and acts as a control point. In species that exhibit the sporophytic type of self-incompatibility, like the Brassicas, controlled hydration of pollen is often required and recognition reactions between pollen and stigmatic surface (Table 3.2). In contrast, pollen of species exhibiting

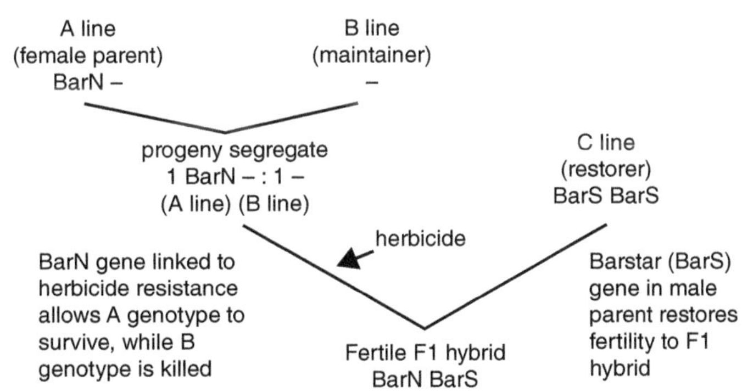

Fig. 3.6. Using engineered male sterility in F1 hybrid seed production. Engineered genic male sterility is a dominant trait (BarN = barnase). Female lines are maintained by crossing to a homozygous maintainer line. The BarN gene can be linked to an herbicide resistance gene to allow fertile plants to be removed easily from the segregating population. The remaining sterile (and herbicide-resistant) plants can be used as the female parent to produce F1 hybrid seeds. If the male parent line (C) carries the BarStar (BarS) gene, its product will inhibit barnase activity and restore fertility to the F1 progeny. (Courtesy of K.J. Bradford, personal communication.)

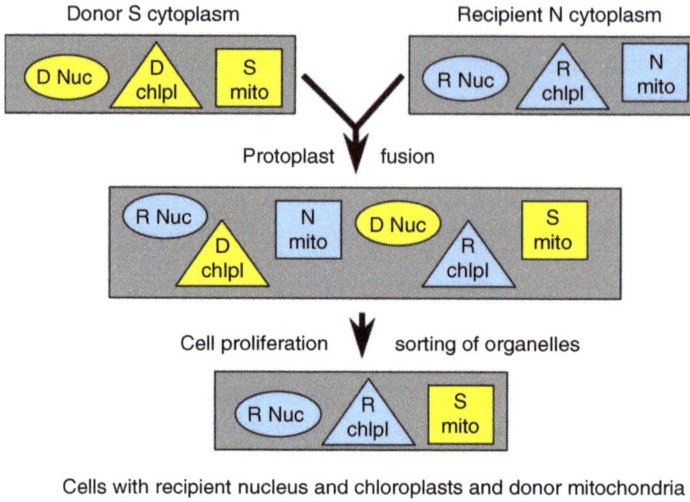

Fig. 3.7. Protoplast fusion to transfer CMS between species. Cells are fused to mix the nuclei (Nuc), chloroplasts (chlpl) and mitochondria (mito) of the donor (D) and recipient (R) cells. During cell proliferation, these organelles sort into different cells, and some cells containing the desired combination can be selected and regenerated into male-sterile plants. (Courtesy of K.J. Bradford, personal communication.)

Table 3.2. Relationship of self-incompatibility type with pollen and pistil characteristics. (Courtesy of K.J. Bradford, personal communication.)

	Gametophytic type	Sporophytic type
Pollen	Bicellular	Tricellular
	Low respiratory rate	High respiratory rate
	Long-lived	Short-lived
	Germinates readily in liquid media	Controlled hydration required for germination *in vitro*
	Exine with little protein storage	Exine with abundant tapetal protein storage
Stigma	Fluid secretion on surface when receptive	Dry surface; papillae with extracuticular receptor layer
Site of incompatibility recognition	Tissue of transmission (style)	Stigma surface

the gametophytic type of self-incompatibility generally germinate readily and the recognition reactions occur inside the style during pollen tube growth (Brugière *et al.*, 2000; Hiscock and McInnis, 2003).

In the Brassicas, over 70 S-locus genes have been identified that specify self-incompatibility variants. When the S-locus genes in the pollen and the stigma match, the incompatibility reaction is triggered (Fig. 3.8). The sporophytic SI system of *Brassica oleracea* (broccoli, cauliflower, cabbage, etc.) is used commercially to produce F1 hybrid seeds. The two parent lines are selected to be cross-compatible, while the SI system prevents selfing. This system is not always successful, as it can break down under some extreme environmental conditions, such as high temperatures, to allow selfing. In addition, if there is not a good nick between the two lines, plants will eventually begin to set selfed seeds, resulting in unmarketable seed. Developing inbred parental lines is laborious, time consuming and expensive, as selfing must be accomplished by bud pollination, where the flowers are

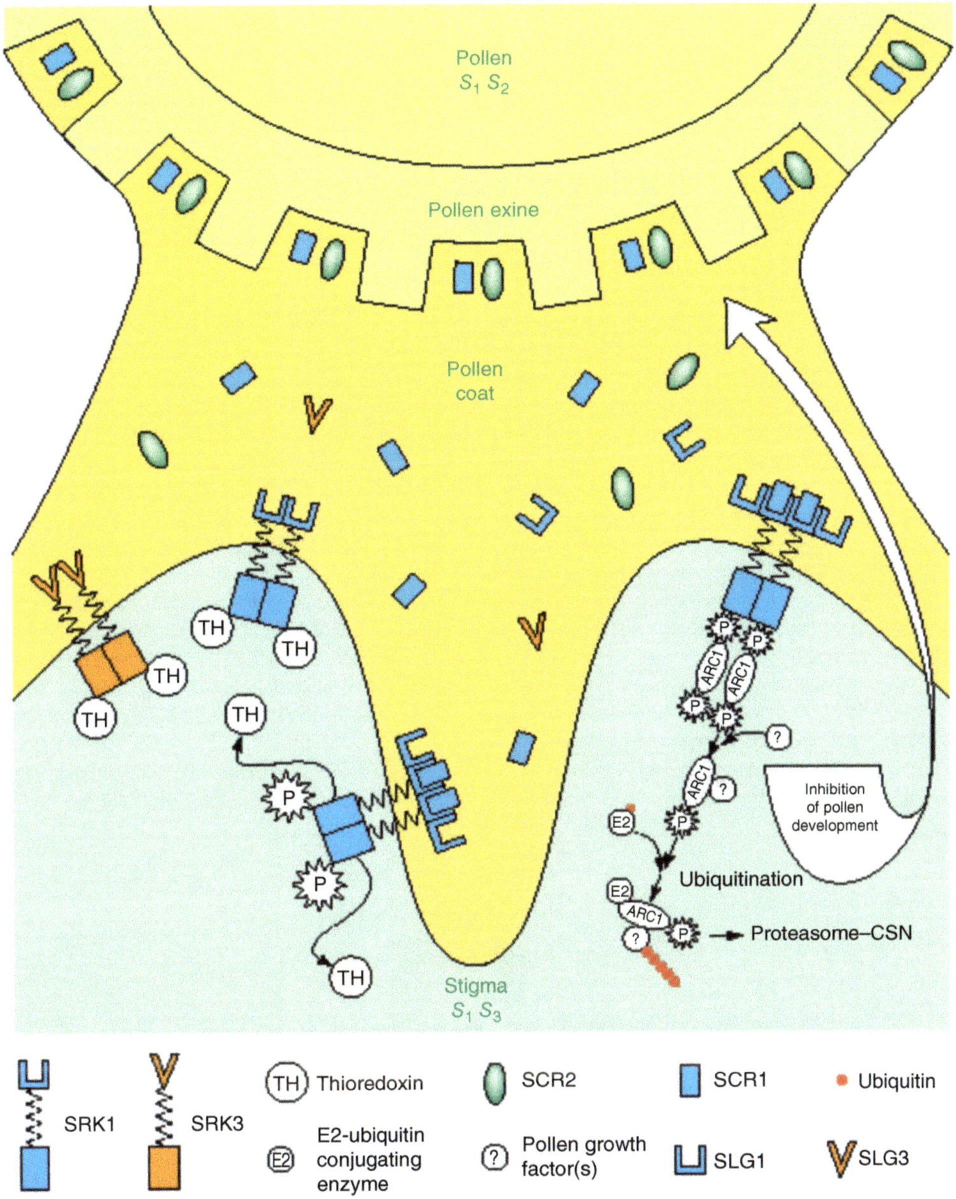

Fig. 3.8. Pollen–stigma signaling in sporophytic self-incompatibility. The exine of the pollen grain contains SCR (S-locus cysteine rich) proteins that can be recognized by corresponding SLG (S-locus glycoproteins) present in the papillar cell walls of the stigma (but which may not be essential for the SI recognition). The SCR proteins bind to the extracellular domain of the corresponding SRK (S-locus receptor-like kinase). This activates the intracellular kinase domain of SRK, which phosphorylates other proteins (ARC1) to initiate the self-incompatibility response, including the synthesis of callose at the point of contact with the pollen grain and inhibition of pollen development. Thioredoxin may prevent the kinase from autophosphorylating, and ubiquitin is involved in regulation of ARC1 action via the proteasome pathway. COP9 signalasome (CSN). (From Hiscock and McInnis, 2003.)

pollinated by hand before they open and before the SI system has fully developed. Exposure to high levels of CO_2 after pollination or other environmental stresses may also permit self-fertilization and reduce the labor and cost of producing parent seed. Alternatively, inbred parents may be propagated by meristem or tissue culture.

In the gametophytic self-incompatibility system, growth of pollen tubes sharing a common S-locus with the stylar tissue is inhibited before they reach the ovule, preventing fertilization. Some of the S-locus genes expressed in styles code for ribonucleases (Newbigin *et al.*, 1993). Both self and non-self RNase molecules can enter the pollen grain, but RNases from a non-selfed plant are inhibited and degraded, while those with a matching S-allele are not inhibited and therefore degrade RNA of the pollen tube, blocking its growth (Fig. 3.9; Franklin-Tong and Franklin, 2003). Other mechanisms of gametophytic SI are also known, including one which triggers increases in cytosolic calcium levels and programmed cell death (Fig. 3.10; Franklin-Tong and Franklin, 2003).

Breakdown of incompatibility systems particularly under environmental stressors like heat and moisture may increase unwanted selfing in F1 hybrid seed production. To reduce the level of selfed-seed due to the breakdown of incompatibility system, seed companies will utilize a 1:2 female-to-male planting ratio rather than the normal 2:2 ratio. This results in lower overall seed yields but helps meet legal and market requirements for hybrid seed labeling. Considering this environmental plasticity, many seed companies now prefer male sterility systems for F1 hybrid seed production of Brassica crops instead of incompatibility.

F2 Generation Seeds

Progeny of F1 seeds is the F2 generation. F2 means "filial 2" because the seeds are derived from the second generation following the crossing of the original inbred parents. One of the identifying characteristics of F1 generation plants is their uniformity. If F1 seeds are self-pollinated, however, their offspring, the F2 generation, is a generation of maximum variability showing all possible combinations of the original two inbreds' characteristics due to random chromosomal segregation during gamete formation. Heterosis is also absent in the F2 generation. Such variation is unacceptable for quality commercial vegetable production, so F2 seeds are not used commercially. F1 hybrid seeds can only be produced by crossing the original inbred lines.

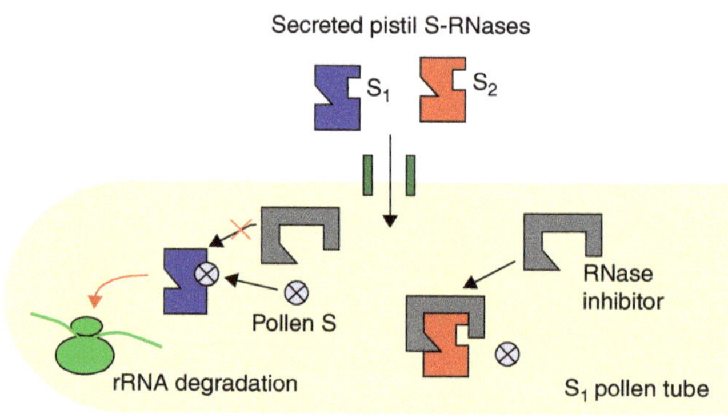

Fig. 3.9. Possible mechanism of action of gametophytic self-incompatibility in Solanaceae species. SI-allele-specific RNases secreted by the pistil enter the pollen tube. Those that are recognized as non-self are inhibited and degraded. Those from self-styles are recognized by a pollen component (unidentified), which prevents the RNase inhibitor from binding, and allows the RNase to degrade pollen RNA. (From Franklin-Tong and Franklin, 2003.)

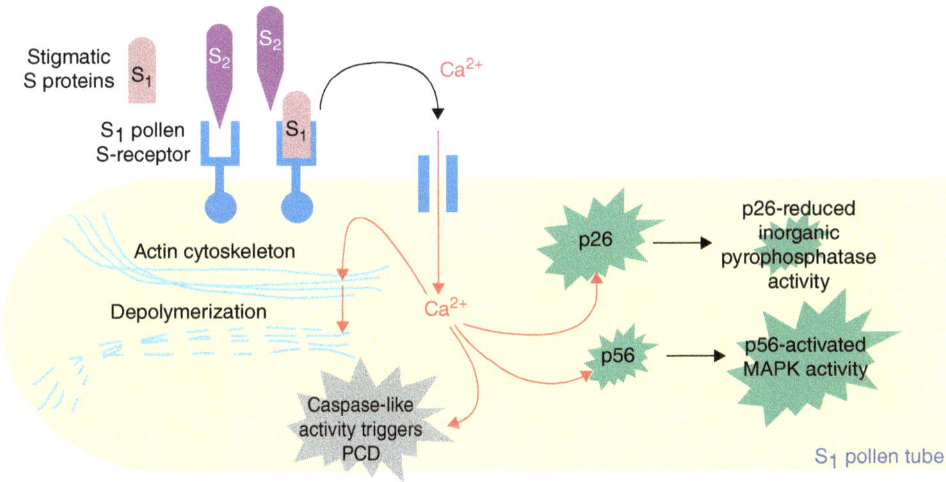

Fig. 3.10. Possible mechanism of action of gametophytic self-incompatibility in *Papaver rhoeas*. SI-allele-specific proteins secreted by the stigma interact with membrane receptors on the pollen tube which trigger an influx of calcium. This leads to several cellular responses that result in programmed cell death (PCD). (From Franklin-Tong and Franklin, 2003.)

References

Bateson, W. and Mendel, G. (2013) *Mendel's Principles of Heredity*. Dover, Mineola, New York.

Birchler, J.A., Auger, D.L. and Riddle, N.C. (2003) In search of the molecular basis of heterosis. *The Plant Cell* 15, 2236–2239.

Brugière, N., Rothstein, S.J. and Cui, Y. (2000) Molecular mechanisms of self-recognition in *Brassica* self-incompatibility. *Trends in Plant Science* 5, 432–438.

Chahal, G.S. and Gosal, S.S. (2002) *Principles and Procedures of Plant Breeding: Biotechnological and Conventional Approaches*. Alpha Science International Publishing, Harrow, UK.

Devaux, P. and Pickering, R. (2005) Haploids in the improvement of Poaceae. In: Palmer, C.E.D, Keller, W.A. and Kasha, K.J. (eds) *Haploids in Crop Improvement II*. Springer Berlin, Heidelberg, Germany, pp. 215–242.

Franklin-Tong, N.V. and Franklin, F.C.H. (2003) Gametophytic self-incompatibility inhibits pollen tube growth using different mechanisms. *Trends in Plant Science* 8, 598–605.

Goldberg, R.B., Beals, T.P. and Sanders, P.M. (1993) Anther development: basic principles and practical applications. *The Plant Cell* 5, 1217–1229.

Hay, F.R. and Whitehouse, K.J. (2017) Rethinking the approach to viability monitoring in seed genebanks. *Conservation Physiology* 5, cox009. doi.org/10.1093/conphys/cox009 (accessed 4 September 2023).

Helgeson, J.P., Hunt, G.J., Haberlach, G.T. and Austin, S. (1986) Somatic hybrids between *Solanum brevidens* and *Solanum tuberosum*: expression of a late blight resistance gene and potato leaf roll resistance. *Plant Cell Reports* 5, 212–214.

Hiscock, S.J. and McInnis, S.M. (2003) Pollen recognition and rejection during the sporophytic self-incompatibility response: Brassica and beyond. *Trends in Plant Science* 8, 606–613.

Losey, J.E., Rayor, L.S. and Carter, M.E. (1999) Transgenic pollen harms monarch larvae. *Nature* 399, 214–214.

Mariani, C., De Beuckeleer, M., Truettner, J., Leemans, J. and Goldberg, R.B. (1990) Induction of male sterility in plants by a chimaeric ribonuclease gene. *Nature* 347, 737–741.

Mariani, C., Gossele, V., De Beuckeleer, M., De Block, M., Goldberg, R.B., *et al.* (1992) A chimaeric ribonuclease-inhibitor gene restores fertility to male sterile plants. *Nature* 357, 384–387.

Newbigin, E., Anderson, M.A. and Clarke, A.E. (1993) Gametophytic self-incompatibility systems. *The Plant Cell* 5, 1315–1324.

Perez-Prat, E. and van Lookeren Campagne, M.M. (2002) Hybrid seed production and the challenge of propagating male-sterile plants. *Trends in Plant Science* 7, 199–203.

Pimentel, D.S. and Raven, P.H. (2000) *Bt* corn pollen impacts on nontarget Lepidoptera: assessment of effects in nature. *Proceedings of the National Academy of Sciences USA* 97, 8198–8199.

Plucknett, D.L. and Smith, N.J. (2014) *Gene Banks and the World's Food*. Princeton University Press, Princeton, New Jersey.

Sears, M.K., Hellmich, R.L., Stanley-Horn, D.E., Oberhauser, K.S., Pleasants, J.M., *et al.* (2001) Impact of *Bt* corn pollen on monarch butterfly populations: a risk assessment. *Proceedings of the National Academy of Sciences USA* 98, 11937–11942.

Spencer-Lopes, M.M., Forster, B.P. and Jankuloski, L. (2018) *Manual on Mutation Breeding*, 3rd edn. International Atomic Energy Agency, Vienna, pp. 251–266.

Torrence, J. (2008) *Higher Biology*. 2nd edn. Hodder Education Publishing, UK.

Zubko, M.K. (2004) Mitochondrial tuning fork in nuclear homeotic functions. *Trends in Plant Science* 9, 61–64.

4

Producing High-Quality Vegetable Seeds

Importance and Appeal of Seed Production

Seed production is the catalyst and starting point for world food and fiber production. Because of our reliance on them, the importance of seeds to our society should not be underestimated or trivialized. Most world food and fiber crops, such as many vegetables, rice, wheat, maize, beans, cotton, and flax, for example, are all produced from seed. Therefore, our lives depend on a continuous supply of high-quality viable seeds every year. Seed production is truly an international activity because seeds are needed in all parts of the world.

Seed production is a noble profession that brings genetics to life. Seeds are the delivery system for modern advanced crop genetics. Why are people attracted to work in the seed industry? Seed production requires creativity and relies on personal intuition and intellectual capacity. Many people find seed production very stimulating because the parameters are challenging and rewarding. The skill set requires knowledge of plant and seed physiology, genetics, and plant pathology with strong human relations assets and the ability to apply these skills on a practical basis with a business mindset. Decisions made during the growing season can have a direct impact on the quality, quantity, cost, and profit margins of the sellable seeds. Responding to problems with creative solutions through experimentation can be invigorating, and many seed companies have their own seed production research programs to develop techniques specific to their cultivars and germplasm. In the seed industry, you're always interacting with living things in changeable environments and responding to their needs daily by providing inputs such as irrigation, fertilization, biologicals, or pesticides to protect and maximize growth. Recent advancements and innovations related to precision agriculture, reducing our carbon footprint, robotics, and drone technologies are becoming increasing important.

Where Seed Production Occurs

Commercial seed production tends to be clustered in certain areas of the world that are well suited environmentally and proven to deliver adequate quantities of high-quality seed (see Fig. 4.1). The best environments for growing seed crops differ from those for vegetable crops grown as food. Vegetable crops for food are harvested for their vegetative (celery, carrot) or reproductive (tomato, broccoli) organs for human consumption rather than mature seeds. Seed production for most crops requires the growing cycle to be extended beyond the human consumption stage, requiring further exposure to plant diseases, insects, and the environment that can directly impact seed quality and quantity. In most cases, the

DOI: 10.1079/9781789243260.0004

harvest of the food crop occurs under different environmental conditions than those optimal for vegetable seed production. For example, temperature, photoperiod, length of growing season, and precipitation may vary between vegetable food and seed crops. Those vegetables harvested for market in the vegetative stage, such as carrot, onion, radish, Brassicas, and lettuce, are typically grown in areas with lower temperatures than those required for a seed crop in order to maintain quality growth, tenderness, and minimize bolting. Seed production often requires higher temperatures to encourage bolting although excessively high temperatures also hinder pollen development and fertilization of onion, tomato, and lettuce. In contrast, beet and carrot have been selected to avoid bolting under mild conditions for market use but require sufficient exposure to cool temperatures (vernalization) to induce flowering and thus seed production.

Some vegetables require specific photoperiods to properly develop. For example, onion cultivars exhibit a wide range of photoperiod responses to trigger bulb formation. This influences where seed production occurs because there is a need to examine bulb development and rogue off-type plants from seed fields. Seed production of long-day onions generally occurs at latitudes greater than 45° such as Idaho and Washington, USA in the northern hemisphere and short-day onions at latitudes no greater than 35° like Mendoza, Argentina in the southern hemisphere. Since onions are biennial, bulbs developing during the first season are often stored over the summer and replanted in the fall to allow vernalization and flowering the next spring and summer. Even when appropriate photoperiods are present to check bulb development, all other environmental factors and especially temperatures must be suitable for the completion of the entire life cycle including seed development.

Precipitation amount and distribution also vary for food and seed crops. Many vegetable crops, such as leafy greens, harvested for their vegetative tissues, require significant early-season rainfall or irrigation for rapid succulent development. Seed crops, in contrast, demand uniform water supply during vegetative and reproductive development followed by drier conditions during seed maturation and harvest. Dry conditions in the late season for seed production result in fewer weather-related harvest delays, reduces shattering and precocious germination, and minimizes pathogen infestations.

Normally, good climates for vegetable seed production are relatively dry (with irrigation available) with significant diurnal temperature fluctuations appropriate for the crop during flowering, development, and maturation. A suitable production area should have a growing season that is at least a month longer than needed with suitable climatic conditions to produce mature seed. This extra time provides flexibility in planting and harvesting in case delays are encountered and allows seed companies to utilize different planting windows to spread their risks. Seed production requires high light intensity since vegetables are sun-loving plants and sunlight reduces damp conditions that may promote certain diseases (Welbaum, 2015).

Vegetable seed production is a critical and integral part of a business entity. Decisions to produce are deliberate to meet customer needs with a focus to supply the right quantities and quality of seed on time. The vegetable seed business is global in nature and seeds of a given species for market use can be planted essentially every day somewhere across the world. The typical model for vegetable seed production includes multiple growing areas followed by precleaning prior to exporting seeds to a central processing facility in another country near the major market. After final upgrading, seeds are shipped to customers across multiple countries. Thus, the planning process and inventory management are critical to the success of a seed company. A business consideration to produce vegetable seeds often involves the utilization of more than one seed production area for a given species to spread risks, preferably in both the northern and southern hemispheres, to ensure a continuous supply for customers. Crop planning and placement are critical processes and require extensive data analysis to understand the strengths and weaknesses of each production area and to avoid under- or overproduction as much as possible. Typically, each cultivar or hybrid is assigned an expected base yield rate (kg/ha or lbs/acre), which can vary by production area. This becomes the basis for determining the surface area required for producing a target quantity of seeds. Considering the various production areas around the world, the annual or biennial nature of the seed crop, and seed processing and shipping time requirements, the crop planning for a given cultivar or hybrid can extend several years in advance. The criteria for selection of production areas also include several quality attributes (germination, vigor, genetics, seed health, etc.) to meet market requirements, historical reliability of the area, labor availability, cost, and the ability to meet phytosanitary requirements of the destination countries where the seed will be processed and eventually sold. Once a production area has been decided, the local production personnel work with seed growers in the area to select fields or protected environments and agree on the requirements of proper crop rotation, isolation distances, and pricing before finalizing the production contract. Open fields of cross-pollinated

Fig. 4.1. Major commercial vegetable seed production regions of the world. (Personal communication.)

Fig. 4.1. *Continued.*

Reference	Country	Areas	Major seed crops[1, 2, 3]
1	Argentina	Mendoza, San Juan	Carrot, onion
2A	Australia	New South Wales, Victoria, South Australia	Beet, broccoli, cabbage, carrot, cauliflower, chard, lettuce, misc. Brassicas, onion, radish
2B	Australia	Tasmania	Beet, cabbage, carrot, cauliflower, chard, misc. Brassicas
3	Chile	Central and Southern	Beans, broccoli, cabbage, carrot, cauliflower, cucumber, lettuce, melon, misc. Brassicas, onion, peas, pepper, spinach, squash, sweet corn, tomato, watermelon
4A	China	North (Gansu, Xinxiang)	Aubergine (eggplant) cucumber, melon, pepper, pumpkin, squash, tomato, watermelon
4B	China	East (Henan, Liaoning)	Cabbage, Chinese cabbage, cucumber, pepper, radish, tomato
4C	China	South (Yunnan, Sichuan)	Broccoli, cabbage, cauliflower, Chinese cabbage, misc. Brassicas, radish
5	Denmark	All areas	Chinese cabbage, misc. Brassicas, radish, spinach
6A	France	North and Central	Beans, beet, broccoli, cabbage, carrot, cauliflower, lettuce, onion, pea, radish, spinach
6B	France	South and Southwest	Beet, carrot, chard, cucumber, melon, onion, pepper, squash, tomato, watermelon
7	Hungary	All areas	Peas, sweet corn
8	India	All areas	Aubergine (eggplant) beans, cabbage, cauliflower, gourd, melon, okra, onion, pepper, tomato, watermelon
9	Israel	All areas	Melon, pepper, squash, tomato, watermelon
10	Italy	All areas	Beans, beet, broccoli, cabbage, carrot, cauliflower, cucumber, lettuce, onion, pea, pumpkin, spinach, squash
11	Korea	All areas	Chinese cabbage, misc. Brassicas, onion, radish
12	Mexico	West Coast	Broccoli, cauliflower, cucumber, lettuce, pepper, tomato
13	Netherlands	All areas	Cucumber, pepper, spinach, tomato
14	New Zealand	South Island	Beet, carrot, misc. Brassicas, onion, pea, radish, spinach, sweet corn
15	Peru	North and South	Cucumber, lettuce, melon, pepper, sweet corn, tomato, watermelon
16	South Africa	Oudtshoorn	Bean, beet, broccoli, cabbage, carrot, cauliflower, chard, cucumber, lettuce, pea, onion, pumpkin, squash, tomato
17	Thailand	Central and North	Cucumber, gourds, melon, pepper, squash, sweet corn, tomato, watermelon
18	Turkey	All areas	Aubergine (eggplant), bean, cabbage, cucumber, melon, onion, pea, pepper, spinach, tomato, watermelon
19A	USA	Eastern Oregon	Carrot, onion
19B	USA	Eastern Washington	Bean, carrot, misc. Brassicas, onion, peas, radish, sweet corn
19C	USA	Idaho (Treasure, Magic Valleys)	Beans, carrot, misc. Brassicas, onion, pea, radish, sweet corn
19D	USA	Northern California	Cucumber, onion, pumpkins, squash, watermelon
19E	USA	Coastal and Central California	Broccoli, cauliflower, cucumber, lettuce, onion, squash
19F	USA	Southern California (Imperial Valley)	Broccoli, cauliflower, onion

Continued

Fig. 4.1. *Continued.*

Reference	Country	Areas	Major seed crops[1, 2, 3]
19G	USA	Southwest Arizona	Broccoli, cauliflower, onion
19H	USA	Western Oregon	Cabbage, radish, spinach
19I	USA	Western Washington	Beet, cabbage, chard, Chinese cabbage, spinach

[1]Minor vegetable seed crops within various families are often grown in similar areas as those shown. For example, parsley seed production is typically grown in the same area as carrots.
[2]Major seed crops listed include export seed production in addition to domestic production. Additional domestic crops may also be produced within a given country due to local tradition or country seed laws.
[3]Minor countries include Ethiopia, Kenya, Tanzania, Vietnam, among others and seed production for export is often tied a single company or a limited number of crops with low volumes.

crops by either wind or insects require certain isolation distances between seed production fields. Field personnel must work with other seed companies and sometimes local organizations or associations to register and pin each field location prior to planting.

Many seed production areas are irrigated maritime or medium-altitude deserts. These areas are scattered around the world and their numbers are limited. The utilization of drier climates for seed production is especially important for the management and control of seed-borne diseases. In fact, some areas where critical seed-borne diseases are present or cannot be adequately controlled are not considered for seed production for certain crops like tomato or watermelon. This is the principal reason much of the vegetable seed production occurs in dry climates such as western North America rather than wetter regions in the eastern portion of the continent. The maritime-like climate of the Pacific Northwest is unique and assures cool, dry summers and mild, wet winters for optimum development of many biennial seed crops such as cabbage and table beets. The mild winters of inland medium-altitude deserts of central Oregon, Washington, and Idaho, USA, and Mendoza, Argentina, are ideal for carrot and onion seed production. Southern California and Yuma, Arizona, USA, are hot and dry during the summer but ideal for broccoli and cauliflower seed production during the winter months. The Central Valley of California and several valleys in central Chile are ideal for many of the Cucurbits as well as lettuce, onion, and other seed crops. Dry conditions at these locations require that seed crops be irrigated with adequate moisture at the proper time for optimum seed quality. Thus, most vegetable seeds are produced under irrigation, primarily using drip or furrow methods. Overhead irrigation is generally avoided because of greater disease potential and low water use efficiency but is used for larger-volume crops like garden beans and peas in areas like the Columbia Basin of Washington state in the USA.

There are dangers of genetic shifts that occur when producing seed crops in climates markedly different from where the cultivar was developed (Rao and Hodgkin, 2002). For example, when seed crops are produced in warmer drier climates, plants in the population that express alleles that confer greater fitness to these conditions may be favored. Several generations of seed production in a warm climate may cause genetic movement away from the original cultivar's characteristics. In general, random drift causes rare alleles, whether adaptive or not, to be lost from small populations, whereas selection tends to eliminate the less adaptive alleles and to increase the frequency of alleles that are better adapted to their generation environment. This loss of alleles due to genetic shifts is complex and related to the effective population size, the growing environment, and the genetic constitution of the crop (Rao and Hodgkin, 2002). Naturally heterozygotic crops that are cross-pollinated like melon, corn (maize), and spinach are more prone to genetic movement compared to relatively homozygous self-pollinated crops like lettuce and endive (Breese, 1989). Evidence of phenotypic changes are well-documented in cross-pollinated populations, especially when multiplication is performed in growing regions that differ from their origins (Breese, 1989). Rogueing may sometimes help remove unwanted phenotypes but shifts in response to environment may be difficult to detect. One solution is to produce seed in its original environment periodically or consistently. Much of sweet corn (maize) seed production occurs in drier regions where the seed is utilized to prevent genetic shifts that would occur if seed were produced in a much different environment.

A focus in new genotype development along with maintaining current cultivars is to establish parent lines and cultivars with stable genetics. This is a requirement for releasing new cultivars and F1 hybrids for registration and federal approval in most countries. Using modern molecular technologies, plant breeders work with their quality programs to establish the baseline genetics of each parent line or cultivar. This provides an ability to detect when and if genetic drift occurs in later generations. The original source of each cultivar or parent line is called breeder seed and small quantities are maintained in cold storage, utilized for parent production and increased only when needed (Chapter 3). The maintenance of each parent line and cultivar requires close inspection, rogueing, and testing during foundation and stock seed production and is the source for producing commercial seeds. This system allows companies to grow seeds in different international production areas as well as to maintain the unique genetics of a given cultivar over its lifetime.

The suitability of some areas can be improved for vegetable seed production by controlling environments. An increasing volume of vegetable seeds are grown under protected environments to enhance the climate for more consistent production and higher yields, to control cross-pollination from insects or wind, or for the exclusion of insects and diseases for phytosanitary reasons. This is especially true for higher value seeds of hybrid tomato, pepper, and melon, and for production of small seed quantities of foundation seed, stock seed, and experimental hybrids. Limited seed production, usually for breeder or foundation seed, also can be accomplished in glasshouses where supplemental light is provided, or light excluded to promote flowering. Examples of protected environments vary from simple to advanced permanent structures including net cages, tunnels, net or plastic greenhouses, or glasshouses. Plastic or fine mesh coverings provide isolation and prevent contamination from cross-pollination by bees and wind-borne pollen. Netted coverings can also reduce insect feeding damage and prevent disease transmission (Acharya *et al.*, 2020). Coverings may also optimize climatic and agronomic conditions for seed production. For example, when temperatures are too cold, enclosures may provide passive solar warming or supplemental heating may be used. If irrigation is available, the crops can be shielded from rainfall by coverings to reduce foliar disease and seed shattering (Welbaum, 2015).

Producing Quality Seeds

Quality requirements for vegetable seed production starts with an understanding that markets vary between and within countries. The important attributes that are critical for successful seed sales include physiological, physical, genetic, and pathological quality. Seed company quality standards must meet customer needs. These standards are the basis for selecting production areas and developing grower contracts.

Differing requirements for commercial open-field production and protected culture seed markets provide a contrast between these segments. Typically, cultivars and hybrids targeted for the protected culture seed market demand higher quality standards. Seed produced under protected culture also bring a much higher selling price compared to open-field produced tomato, pepper, and cucumber. The seed testing requirements can also vary by market segments. For example, for species where the crop is transplanted, a usable transplant test may be conducted in addition to standard germination tests.

Although achieving high yields of seed crops is important for profitability, producing high-quality seed is equally important. Physiological quality includes a standard germination test to determine whether a seed can produce normal seedlings as required by law for each country. Seeds for international trade must meet rules for labeling established by organizations such as the International Seed Testing Association (ISTA). Other physiological tests include seed vigor, usable transplants, and various tests to detect seed dormancy. More advanced technologies to assess seed quality include imaging and measurement of respiration rates during germination. Physical purity is freedom from inert matter, insects, and their reproductive structures, and seeds or plant structures of other species. Crops, such as carrot and onion, produced in open fields must be grown weed-free to avoid costly and challenging additional cleaning steps and special equipment. In contrast, hand-harvested crops produced in protected environments seldom have contamination issues from weed or other crop seeds. Another element of seed quality is genetic purity. Genetically pure seeds reproduce "true-to-type" with the characteristics of their parents in the case of non-hybrid cultivars. For F1 hybrids, this means seeds are the product of crossing of parental line without selfing, mutation, or out-crossing with other seed crops or volunteer plants. Genetic quality can include both molecular tests and phenotype evaluation. Quality seeds must also be tested for seed-borne diseases using various tests ranging from grow-outs to molecular detection assays. Each quality attribute is critical for customer sales. For example, a given seed lot can have high physiological quality, but if it fails a test for seed health, it is unmarketable and a loss for the producing company.

The production process must involve constant monitoring and testing to ensure seeds meet company, state, national, and if sold globally, international standards for germination, physical purity, genetic purity, and phytosanitation. Seed crops must be inspected by phytosanitary agencies and pass field and lab tests during the growing season before they can be exported. Vegetable production for edible consumption is based on specific planting windows (Welbaum, 2015). Seed of a desired cultivar must be available to growers at specific planting times, or they will sourced from a competitor resulting in lost sales and damaged customer relations.

Maintaining target pricing is important to achieve the desired product margin to operate a profitable business. Companies must be competitive against other seed companies that are selling similar products in terms of product quality, service, and price. Being a successful seed company is more than just producing seeds. Business considerations are essential for maintaining profitability, stability, as well as good customer and consumer relationships.

Collaboration among different groups including breeding, marketing, sales, production, planning, inventory, logistics and quality control is essential for a successful seed company. Sales and marketing are essential because potential customers must be informed of existing and new products. Production locations must provide a consistent supply of seed to sales departments at the right time, in the right quantity, and at the right price. Some companies may wholesale most of their production to businesses that specialize in retail sales and do not produce their own seed or supplement their own production with outside sources. Seed companies that specialize in wholesale trade may have smaller marketing departments than operations that specialize in retail sales. The inventory management department is a gatekeeper between production and marketing. Inventory management determines how much of which seeds needs to be produced and informs the marketing department about how much of which seeds are available for sale. Quality management must ensure testing is implemented to assure all seeds are free of disease, physically pure, genetically pure, viable, and vigorous. All test results must be accurately documented with auditing systems in place (George, 2009). Typically, quality management also works with the sales department to address and resolve product complaints from customers. It is critical that companies systematically and efficiently receive, analyze, and resolve complaints. Working in a seed company requires strong interpersonal skills because of the continuous interaction needed among sales personnel, breeders, planning, production, quality control, logistics, and inventory specialists.

Seed production is very challenging. It is not a static industry, and the dynamics change from year-to-year based on forecasted supply needed, current available inventory, and changes in product portfolio that impact production plans. Each production problem tends to be somewhat different, and solutions are not textbook or "cookie cutter" and require patient analysis and critical thinking. A system such as International Organization for Standardization (ISO) certification helps systematically address problems and improve quality and service (Lo and Chang, 2007). The vegetable seed industry favors flexible people who can easily change directions to solve problems from season-to-season. Seed production is very dynamic, and procedures must adapt and change very quickly and respond to circumstances. A successful seed production supply chain favors people who are very proactive. Reacting to problems after they occur is difficult and can have a negative financial impact. Proactively identifying and correcting problems before they exist is a key to success in the seed industry.

Contracting with Seed Growers

It is important to have good personal relationships within and outside one's company. One of the most important relationships is between the seed producer and the contracting company. While some small seed companies may produce their own seed, many companies contract production to farmers who specialize in seed crops. In many cases, contract seed production occurs on private farms under the direct control and supervision of a seed company field representative. Larger seed companies may also work with third-party enterprises that specialize in certain aspects of production like harvest and cleaning, for example.

Companies must select growers carefully to make sure they have the skills to produce a successful crop. International guidelines help seed companies work with other enterprises and seed growers to ensure the integrity of all parties involved (VSPGPG, 2017). The company and the grower must have an open and honest business relationship. A free exchange of information and continuous dialogue is needed to ensure a production plan is carried out and problems are resolved. All seed company expectations should be included in the written contract the grower signs. A seed company should establish multiple areas of leverage for geographic risk, so all production of a particular crop is not reliant on a single grower or location. Seed companies should establish healthy competition among growers in a region to leverage economic risk. This is necessary so a grower who is not doing a good job will fear loss of his/her acreage to another more competent producer in the future. Good seed growers are motivated, curious, and engaged to clarify ambiguities about the seed crop they are producing. Seed producers must be organized, disciplined, and focused with attention to detail, so stages of production are completed on time to a high standard. The seed company must trust their growers to treat their product with discretion and integrity because in many cases, growers have access to advanced genetics and technology developed by their clients and Intellectual Property Rights (IP) must be respected (VSPGPG, 2017). Growers must be effective at labor management and adhere to all laws, safety, and human rights requirements because producing seed crops requires much hired labor. Growers must be good cost managers without cutting corners to save money that may negatively affect the crop or laborers.

There are two common scenarios for harvesting vegetable seed crops. The seed company can harvest the crop using its own equipment or labor, or the grower or a third party conducts the harvest. After seed harvest, either the seed company or the grower will complete the post-harvest handling of the seed crop such as seed cleaning and/or drying for wet-seed crops like tomato, Cucurbits, or watermelon. Eventually, the seed company will take possession of the seed according to the contract. Benchmarks for standard germination percentages, physical purity, and freedom from inert material, and in some cases genetic and seed health standards, are specified in the production contract. A seed company representative will frequently inspect seed fields to make sure production is on schedule in accordance with the contract, good management practices are followed, and that isolation is maintained. Producers expect seed company representatives to be knowledgeable, fair, and honest so both groups benefit.

Much like any business, things go wrong during seed production. When things don't go according to plan, both the company and seed producer need to be professional and flexible. Companies need to conduct ongoing trials of promising new cultivars to optimize production practices prior to scaling up production. Generally, breeding technology is evolving faster than production technology, suggesting that genetic solutions to seed production problems (e.g. shattering) may be available in the future.

Establishing Seed Crops

Planting seeds or transplants to grow a crop is called "stand establishment" (Welbaum, 2015). Non-dormant, viable vegetable seeds germinate most rapidly when fully hydrated (no moisture stress) at their optimum temperature for germination. Seeds should also be planted deep enough to access moisture. If moisture is limiting, seeds can be planted a little deeper, so they will have sufficient moisture to germinate. A common recommendation is to plant to a depth three times the length of the seed, so large seeds are planted deeper than smaller ones. "Plant to moisture" is a rule of thumb often applied for direct seeding, but seeds shouldn't be planted so deep that emergence is inhibited. Good seed-to-soil contact facilitates seed imbibition. Soil contact can be improved by using coated or pelleted seeds and proper soil preparation. Soils with high clay content can "crust" and inhibit emergence. Crusting can be minimized by keeping the soil surface moist using sprinkler irrigation until emergence. Overhead irrigation also helps cool soils, which is important when establishing a cool-season crop in warm soil. When temperatures exceed the optimum for establishment, night transplanting may reduce shock and improve survivability particularly for crops like broccoli grown in the Imperial Valley of California or Yuma, Arizona. Many crops are established from transplants to reduce the time in the field, increase uniformity, improve nicking, and ensure a good stand (Welbaum, 2015). With the high value of the parent lines and the need to assure uniform stand establishment, transplants are used to establish seed crops of Brassicas, Solanums, and certain Cucurbits.

Two planting techniques are common for establishing biennial vegetable seed crops. The "seed-to-seed" method is accomplished by direct seeding seed in late summer or early fall for the first season. The established vegetative plants are overwintered and reproduce for seed harvest the following year (George, 2009). It is essential to use high-quality "true-to-type" stock seed as this method does not permit the selection or rogueing of roots or bulbs for genetic off-type characteristics. Many carrot and onion seed crops are produced by this method, which is less expensive and complex than the "root-to-seed" method described below. One risk of the seed-to-seed method is crop failure caused by loss of stands due to abnormally low temperatures causing winter kill. This sometimes can be mitigated by providing frost-protective covers for the more susceptible parent lines or cultivars. Another risk is the potential increase in insect and disease pressure from overlapping seed crops in nearby fields. For example, if new carrot crops are direct-seeded and emerge before an existing nearby carrot seed crop has been harvested, seed-borne pathogens like *Xanthomonas campestris* pv. *carotae* can be transmitted to the newly planted seed field. In this scenario, the new crop could become contaminated and would require seed sanitation treatment before sale.

Biennial vegetables such as onions, beets, carrots, and cabbage must develop sufficient vegetative growth prior to cool exposure to be vernalized and flower the following season. Planting these crops too early increases chances for winter kill or late-season pest infestations. Planting too late may prevent vernalization, limiting flowering and reducing seed yield (George, 2009).

The "root-to-seed" method requires planting seed the first year in a separate bed or nursery, typically located away from the primary seed production area. The resulting roots or bulbs are removed from the soil, inspected and culled for off-types, and replanted after refrigerated storage to ensure vernalization. The seed crop is harvested after reproductive growth and senescence during the second season. The replanted root, bulb, or expanded hypocotyl is referred to as a "steckling" (George, 2009). This method permits rigorous selection and removal of off-type and diseased plants. Stecklings may be easier to vernalize in a cold room or field because they are a well-developed, vegetative structure. In the USA, carrots are seeded in a plant bed in the Imperial Valley of California in early fall, dug and sorted in mid-winter, and placed in cold storage to vernalize for 6 to 8 weeks before transplanting in the seed production fields in the Pacific Northwest. The root-to-seed method allows seed companies more time to adjust production plans to better manage their inventories. This method has the advantage of ensuring uniform stand establishment and crop growth and synchronizing the timing of flowering between male and female parents for hybrid seed production. The root-to-seed method is used by plant breeders of biennial vegetables because the time needed for cultivar development is decreased by make selections prior to transplanting. Sometimes there are restrictions on use of the root-to-seed method. For example, onions bulbs typically stay close to where they will be used. One exception is when a short-day parent is to be crossed with a long-day parent to produce breeder seed; bulbs are moved to a higher latitude where days for seed production are longer, e.g. California bulbs replanted further north in Oregon or Washington, USA. In some cases, bulbs cannot be moved into certain areas because of inherent disease problems, e.g. because of white rot (*Stromatinia cepivora*), no bulbs can be moved into Idaho. In the case of Argentina, only bulbs produced locally are used in production areas. Carrots are another example of a seed crop that is produced near areas where carrots are grown for human consumption. Root-to-seed is the most expensive method of vegetable seed production due to transportation, handling, storage, and transplanting costs and is used primarily for hybrid seed production. The benefits of checking genetic quality before replanting are important in vegetable seed production to prevent genetic drift and to maintain "true-to-type" cultivars.

Vegetable seed crops must complete their juvenile phase before reproductive growth can begin. Plants need to develop sufficient leaf area and nutrient reserves during the juvenile phase to support later reproductive growth (Welbaum, 2015). In addition, the increasing emphasis on hybrid seed production necessitates establishment of male and female lines at times that assure successful crossing or "nicking". Inbred lines are notoriously weak because of inbreeding depression and must be carefully managed to achieve maximum seed yield for F1 hybrid crops. Different male and female inbred lines may need to be planted at differing times to guarantee synchronous flowering. Sometimes staggered establishment of the male lines (both before and after female lines) is often necessary to ensure proper nicking. Many Brassica crop parents for hybrid production are transplanted at different times to ensure synchronous flowering. Hybrid cucumber and squash are usually direct-seeded but because of the differences in male and female line flowering times, they are planted in a relay fashion where the earlier flowering line is planted first before the later line (George, 2009).

Vegetable seeds vary in their optimal soil temperatures at planting (Welbaum, 2015). Open-pollinated or self-pollinated cultivars of large-seeded crops such as beans, peas, and sweet corn are direct-seeded. Peas, radish, and spinach require cool soil temperatures for optimal emergence and are direct-seeded in early spring for seed production. Beets, carrot, and onion are tolerant of cool soil temperatures, although they do better in warmer soils, and are established by either seed-to-seed or root-to-seed methods. Brassica crops are also tolerant of cool soils and most are transplanted. Cucurbits require warm soil temperatures for optimal emergence. Melons are plug-transplanted and cucumber, squash, and watermelon are either direct-seeded or plug-transplanted depending on the production area. In regions with high summer temperatures, vegetables such as tomato, pepper, aubergine (eggplant), and Cucurbits are planted in the early spring to optimize flowering and seed set in early summer (Welbaum, 2015). Tomato, pepper, and eggplant seed crops are usually established by transplanting.

Row spacings and planting densities of vegetable seed crops differ from those for fresh market production (George, 2009). Sufficient space for flower development, air movement to reduce pathogens, mechanical cultivation, unrestricted access to inflorescences by pollinators, and harvest operations are necessary for seed crops. Insect-pollinated vegetables include Cucurbits, most Brassica crops, carrot, and onion. Sufficient plant spacing is needed for effective rogueing of genetic off-type plants and allow higher yields per plant. In some cases (e.g. eggplant, pepper, tomato, muskmelon, and watermelon), row spacings for seed production are like those for fresh market. For lettuce, plant spacing must be sufficient to properly rogue plants for genetic off-types and expression of lettuce mosaic virus (LMV). Root-to-seed beet and carrot seed fields are planted at much lower densities compared to those grown for food. Higher densities for seed production to promote primary umbels leads to lodging and dense fields that prevent entry for rogueing and inspection (Oliva *et al.*, 1988). Higher planting densities of onions grown by the bulb-to-seed method reduce lodging (George, 2009). The method (wind or insect) of pollination also affects crop spacing. Hybrid beet, sweet corn, and spinach are wind-pollinated and the female-to-male row ratios are established based on male pollen production to ensure optimum crossing. Row spacings and planting densities are also important considerations in the production of vegetable seed hybrids. In order to maximize the amount of hybrid seed produced per unit area, the optimum ratio of female-to-male rows must be determined through production research.

Production Considerations

Seed producers must have an intimate knowledge of the crops they are producing. Production considerations include the diverse areas of plant physiology, agronomy, and pathology. The goal of a seed producer is to grow a plant that is healthy both above and below ground. To produce a good seed crop, plants must have healthy roots, leaves, stems, and shoots. Promoting soil and root health are important considerations for maximizing seed crop productivity.

It is easier to manage healthy soils. Soil health can be improved through proper management. Most vegetables prefer deep, friable, well-drained loams with an abundance of soil organic matter (Welbaum, 2015). Soil aggregation can be improved by adjusting soil pH to the range of 6.5 to 7.0 which is compatible with most vegetables. Soil organic matter can be increased by incorporating cover crops or composted waste materials. Growing deep-rooted cover crops that create micropores in the root profile can improve soil aeration and structure (Welbaum *et al.*, 2004). A healthy and balanced microbiome can be maintained through crop rotation, growing green manure crops, managing irrigation, maintaining soil organic matter, and adding biological amendments to soils (Welbaum *et al.*, 2004).

Proper fertilization is essential for crop development and reproductive growth. Soils should be tested for nutrient levels before a seed crop is established. Soil pH is a major factor in absorption of minerals, so it is essential for pH to be in the optimum range to ensure plants are properly fed and healthy. Nitrogen, potassium, phosphate, calcium, magnesium, sulfur, and micronutrients are often limiting in soils (George, 2009). Adequate N-P-K are needed to support rapid vegetative growth early in development. Mineral nutrition requirements are not static but change during the life cycle. Over-fertilization, particularly with nitrogen, can promote vegetative growth at the expense of reproductive growth. The key for seed production is to manage fertilization to produce a balanced plant with a good vegetative structure and size before entering the reproductive phase. Foliar or petiole analyses can assess the nutrient status of plants during the growing season (Welbaum, 2015). It is important to diagnose and correct deficiencies before symptoms are visible and growth is affected.

An understanding of basic plant physiology helps grow a healthy plant that will produce a high yield of vigorous, highly germinable, and healthy seeds. Photosynthesis is important to seed production. Any factor that limits photosynthesis, e.g. drought stress, insect damage, disease, shading, weed competition, or mechanical damage, will directly reduce plant growth and seed yield. Cultural practices should always be tailored to maximize photosynthesis for best seed production (Welbaum, 2015). Many crops transfer resources to the developing seeds from senescing tissues. Thus, it is important to allow the seed maturation process to proceed to conclusion (i.e. dry down for dry seeds or full fruit maturity for wet seeds) prior to harvesting.

Since commercial seed crops are often grown in desert areas, as discussed previously, irrigation is a necessity. Sprinkler irrigation ensures uniform establishment of a seed crops, but doesn't use water efficiently and wets foliage leading to higher disease incidence. Drip or furrow irrigation is often substituted for overhead irrigation after establishment. There are different levels of drought stress. Slight drought stress reduces expansive growth which reduces leaf growth and canopy size (Hsiao and Bradford, 1983). Smaller canopies fix less atmospheric carbon by photosynthesis and thus have lower yield potential. Seedlings grow vegetatively until they transition from juvenile to mature stages when reproductive growth is possible. Excessive moisture and nitrogen fertilization delay reproductive growth. Proper water and nutrient management helps plants transition to reproductive growth so there are no delays in flowering and fruit set.

Atmospheric CO_2 that's fixed by photosynthesis creates the carbon building blocks that are translocated from leaf to fruit providing the dry matter for seed development (Hsiao and Bradford, 1983). However, more severe moisture stress at the time of flowering reduces fruit set, resulting in stunted growth and smaller seed yield (Hsiao and Bradford, 1983). Soil moisture should be monitored with tensiometers or other accurate soil monitoring devices (Welbaum, 2015). The amount of water lost from a field can be monitored by pan evaporation to tell how much water should be reapplied during irrigation. The amount of water required varies greatly with soil type, crop developmental stage, temperature, wind, and humidity (Hsiao and Bradford, 1983). A simple rule of thumb for estimating water needs is seed crops require a minimum of 2.5 to 3.7 cm (1 to 1.5 in.) of water per week to maintain growth and more under drying conditions and in sandy soils (Welbaum, 2015). Irrigation systems are becoming more automated and efficient. With advancing technologies, water demand, soil moisture amounts, and water consumption use can be measured on a continuous basis to ensure efficient water application and usage. Drone and satellite technologies utilizing predictive analytics are helping use water more efficiently.

Nutritional demands increase after fruit set as rapidly growing fruits and seeds are strong "sinks" for both photosynthates and minerals. As the seed crop nears maturity, nutrient demand declines. Given changing nutrient demands with crop development, fertigation is superior to single preplant broadcast applications. When fertigation is not available, one or more side dressings can increase nutrient availability for key stages of crop development. Nutrient inputs must be carefully managed to avoid waste and excessive vegetative growth. Excessive fertilizer wastes money and results in runoff that can pollute ground and surface water.

Pest Management

It is important to be proactive and take preventative measures before a crop is damaged by pests. Following established phytosanitary practices goes a long way toward controlling disease in seed fields (Maude, 1996). Burying, burning, or removing diseased tissues from a previous crop will help control future outbreaks. For example, asparagus rust (*Puccinia asparagi*) is a complicated fungal disease with several stages. Rust-infected tissue may remain in the field and infect the next year's seed crop since asparagus is perennial. Burning or burying infected asparagus residue can help prevent or slow future infection. When phytosanitary practices are insufficient, biological control measures or pesticide sprays may be necessary (Maude, 1996). Integrated pest management (IPM) is a multifaceted approach and provides an alternative to intensive chemical usage for pest control (Flint and Van den Bosch, 2012). IPM is an ecosystem-based approach for long-term mitigation of pests combining biological control, habitat manipulation, modification of cultural practices, and use of resistant cultivars. With IPM, pesticides are used discriminately when monitoring indicates applications are necessary following established guidelines (Flint and Van den Bosch, 2012). Treatments are made with the goal of removing only the target organism. Pest control materials are selected and applied in a manner that minimizes risks to human health, beneficial and non-target organisms, and the environment (Flint and Van den Bosch, 2012). For example, insecticide sprays should not be applied in morning hours when bees are most active. Biological controls that do not harm bees should be seriously considered as an alternative to insecticides. The number and quality of biologicals are improving all the time and even more will be available in the future. Fungicide sprays should only be applied when conditions occur that could cause disease.

Some seed-borne diseases are devastating and must be eliminated from the production chain. Depending on the crop, it is important to have proper control and management practices at every step in the seed production process. This includes protocols for risk assessment, testing of stock seed, crop rotation, growing season inspections, field and greenhouse entry, equipment disinfection, harvesting, seed extraction, and post-harvest handling. Some crops such as tomato and pepper require seed sanitation after seed extraction prior to drying of the seed crop and export. One should not rely on control measures like seed treatments to prevent transmission of seed-borne diseases. Selling disease-contaminated seeds can easily lead to legal liability action against a seed company. Following best production practices and using proactive measures to prevent seed contamination work best.

Controlling weed contamination in seed crops is vital. Unwanted plants should never be allowed to release seeds in or near production fields. Controlling weeds in fallow areas around seed fields by mowing or removal prior to flowering is simple and effective. Weeds that harbor seed-borne diseases must be given the highest priority for control. Making sure equipment and stock seeds are free of weed seed is a standard practice. When a field is already contaminated with weed seeds, a different strategy is needed. The stale bed technique, where the first flush of weeds after soil preparation is killed by shallow tillage, can be effective without chemical usage (Welbaum, 2015). Hoeing or other mechanical removal can effectively eliminate weeds but requires more labor. Robotic weeding technology offers effective control with minimal human labor (Fennimore and Cutulle, 2019). Plastic film mulches can control weeds, modify soil temperatures, and retain soil moisture. Solarization in areas with intense sunlight may be effective especially in lower latitudes (Welbaum, 2015). Chemical fumigation with chloropicrin or other chemicals may be an option but requires the field to be covered after application. Some pre- and post-emergence herbicides can be used if labeled for seed crops that are not organic. In addition to environmental concerns about chemical use, damage and liability issues from spray drift to adjacent properties makes this a less attractive option. Organic production that relies on natural chemicals or non-chemical approaches to pest control can be effective and are covered in a later chapter.

References

Acharya, T.P., Welbaum, G.E. and Arancibia, R.A. (2020) Low tunnels reduce insect populations, insecticide application, and chewing insect damage on Brussels sprouts. *Journal of Economic Entomology* 113, 2553–2557.

Breese, E.L. (1989) *Regeneration and Multiplication of Germplasm Resources in Seed Gene Banks: The Scientific Background*. International Board for Plant Genetic Resources, Rome.

Fennimore, S.A. and Cutulle, M. (2019) Robotic weeders can improve weed control options for specialty crops. *Pest Management Science* 75, 1767–1774.

Flint, M.L. and Van den Bosch, R. (2012) *Introduction to Integrated Pest Management*. Springer, New York.

George, R.A.T. (2009) *Vegetable Seed Production,* 3rd edn. CAB International, Wallingford, UK.

Gourion, B., Berrabah, F., Ratet, P. and Stacey, G. (2015) *Rhizobium*–legume symbioses: the crucial role of plant immunity. *Trends in Plant Science* 20, 186–194.

Hsiao, T.C. and Bradford, K.J. (1983) Physiological consequences of cellular water deficits. In: Taylor, H.M., Jordan, W.R. and Sinclair T.R. (eds) *Limitations to Efficient Water Use in Crop Production.* ASA-CSSA-SSSA, Madison, Wisconsin, pp. 227–265.

Lo, L.K. and Chang, D.S. (2007) The difference in the perceived benefits between firms that maintain ISO certification and those that do not. *International Journal of Production Research* 45, 1881–1897.

Maude, R.B. (1996) *Seedborne Diseases and Their Control: Principles and Practice*. CAB International, Wallingford, UK.

Oliva, R.N., Tissaoui, T. and Bradford, K.J. (1988) Relationships of plant density and harvest index to seed yield and quality in carrot. *Journal of the American Society for Horticultural Science* 113, 532–537.

Rao, V.R. and Hodgkin, T. (2002) Genetic diversity and conservation and utilization of plant genetic resources. *Plant Cell, Tissue and Organ Culture* 68, 1–19.

VSPGPG (2017) *Vegetable Seed Production Good Practice Guide*. Prepared by APSA's Vegetables and Ornamentals Special Interest Group (SIG) and ISF Working Group Vegetable Seed Production. International Seed Federation, Nyon, Switzerland.

Welbaum, G.E. (2015) *Vegetable Production and Practices*. CAB International, Wallingford, UK.

Welbaum, G.E., Sturz, A.V., Dong, Z. and Nowak, J. (2004) Managing soil microorganisms to improve productivity of agro-ecosystems. *Critical Reviews in Plant Sciences* 23, 175–193.

5

Post-harvest Seed Enhancement Technologies

Seventy-five percent of the world's crops are grown from seed including most vegetable crops. The world vegetable seed and seed treatment industries are a major part of plant agriculture. The global seed treatment market was US$5.3 billion in 2017 and is projected to increase to $13.5 billion in 2025; the vegetable seed treatment industry is a major contributor to this total (Fortune Business Insights, 2019). Conventional commercial vegetable growers in developed countries rarely use "raw" untreated seed for propagation. Many growers pay a premium for treated seeds to ensure the best possible field establishment. This contrasts with home gardeners who often use "raw" untreated seed for propagation because of concerns about chemical seed treatments.

Concerns about chemical treatments has led to increased sale of organic seeds (Chapter 6). Organic seed standards vary from country to country. Many programs require organic crops to be grown with organic seeds. When organically produced seeds of a given cultivar are not available, there is an exception process to allow use of conventional seeds. However, this loophole in the current organic standards within the European Union (EU) is scheduled to close by 2035. A common thread is that seeds for organic production must not be treated with any artificial chemicals or unnatural treatments. Organically accepted seed treatments are one of the fastest growing areas of the seed technology sector. This chapter examines some of the more common organic and non-organic treatments applied to vegetable seeds.

Seeds are the delivery systems for crop biotechnology. In order to utilize the genetic improvements that plant breeders and molecular biologists are continuously incorporating into crops, high-quality vigorous seed is required to successfully establish crops in the field. To aid infield establishment, seed technologies providing both protection and enhancement may be applied. The protective treatments, both synthetics and biologicals, help seeds survive attacks from pests. Enhancements improve physiological quality, vigor, and synchrony of germination, especially when vegetables are planted under less than optimum field conditions. Seed technologists don't produce seeds, but they make seeds germinate better. Seed technology enhancement treatments include: biofertilizers/biostimulants, micronutrients, plant-growth-promoting substances, priming, pelleting, coating, biologicals, chemicals, artificial seeds, and other novel seed treatment methods (Taylor *et al.*, 1998; Halmer, 2000). A new emerging group of biological seed treatments provides enhancement to the plant either below or above ground in conjunction with natural biomass in the soil or microbes.

Often the profitability of a crop is determined by seed performance in a field, which is directly related to stand establishment and crop uniformity. Not only are the final number and spacing of plants important, as is the case for

DOI: 10.1079/9781789243260.0005

lettuce, carrots, or Brassicas, but also the speed and uniformity of emergence can have a major impact on harvest as well. For example, a difference of only two days in emergence of lettuce or broccoli can decrease the synchrony of crop development. This can lead to costly multiple harvests or in extreme cases late-maturing plants are not harvested because multiple harvests are less efficient. Some vegetables are started in greenhouses as transplants. The efficiency of transplant production is also affected by seed quality. All these factors have brought increased attention to the vigor and quality of seeds.

Seed Conditioning/Precision-sized Seeds/Mechanical Upgrading

Vegetable seed conditioning removes undesirable material including inert matter, plant debris, underdeveloped seed, weed seed, diseased and damaged seed from raw harvested seed lots, to create pure uniform lots for use in crop propagation (Gregg and Billups, 2010). Seed conditioning is not the focus of this chapter but deserves mention because it is an important post-harvest enhancement (Gregg and Billups, 2010). Space does not allow description of all the diverse vegetable seed cleaning and sorting equipment described elsewhere (Fig. 5.1; Gregg and Billups, 2010). All vegetable seeds are cleaned and sorted after harvest using specialized machines to improve quality before other post-conditioning enhancements are applied. Seed-processing equipment for vegetable seeds include air screen cleaners, gravity separators, spiral separators, indent cylinder separators, color sorters, and aspirators. Image-based individual seed sorters are also becoming available (Nehoshtan *et al.*, 2021).

After the basic cleaning, usually based on screens and air flow, many vegetable seeds are precision sized and may be followed by additional sorting and imaging processes (Fig. 5.2). Seed sorting using X-ray imaging of embryos is used to remove low-quality seeds (Bruggink and van Duijn, 2017). Brassica and melon seed chlorophyll levels decline with maturity. Chlorophyll fluorescence sorting can separate mature and immature seeds in a lot that otherwise look the same (Wilson *et al.*, 2014; Bello and Bradford, 2021).

Some seeds such as tomato, pepper, or fennel seeds for greenhouse production are sorted by liquid density gradients to improve overall quality. Vegetable seeds are often mechanically sorted and sold based on their size (Fig. 5.2). Larger precision-sized seeds generally perform better and are sold at a premium, because of their higher vigor and uniform emergence. Growers prefer to purchase precision-sized seed for more uniform placement with greenhouse tray, belt, and vacuum seeders.

Biological Seed Treatments

Currently biological seed treatments account for about 16% of total applications and are expected to grow in the future (Fortune Business Insights, 2019). Biological seed treatments in general refer to applying living organisms to seed. Inoculation of legume seed (family Fabaceae) with *Rhizobium* spp. bacteria has long been a seed treatment to facilitate the symbiosis between nitrogen-fixing bacteria and plant roots. The symbiosis results in incorporation of atmospheric nitrogen into amino acids for utilization by plants in exchange for plants providing nutrients that nourish the bacteria. The symbiosis occurs inside root nodules during seedling development after germination. *Rhizobium* seed inoculation is a fine black powder containing the bacteria and spores that are applied as a slurry just before planting or the furrow during planting. Some *Rhizobium* suppliers are using liquid inoculants to replace dry powder. Although free-living *Rhizobium* may survive from season-to-season in soil, *Rhizobium* spp. seed treatments accelerate and maximize nitrogen fixation in legumes. *Rhizobium* are sometimes referred to as "biofertilizers". Biofertilizers contain microbes such as fungi, blue-green algae, and bacteria, which promote plant growth by supplying essential nutrients or growth stimulants. Biological seed treatments include mycorrhizal fungi that help plants take up phosphorus that is unavailable to plants from soil or the bacterium *Bacillus subtilis* that produces plant growth stimulants.

Other biologicals may be antagonistic fungi or bacteria that protect seeds from other harmful disease-causing organisms. For example, *Trichoderma* spp. are free-living fungi common in soil

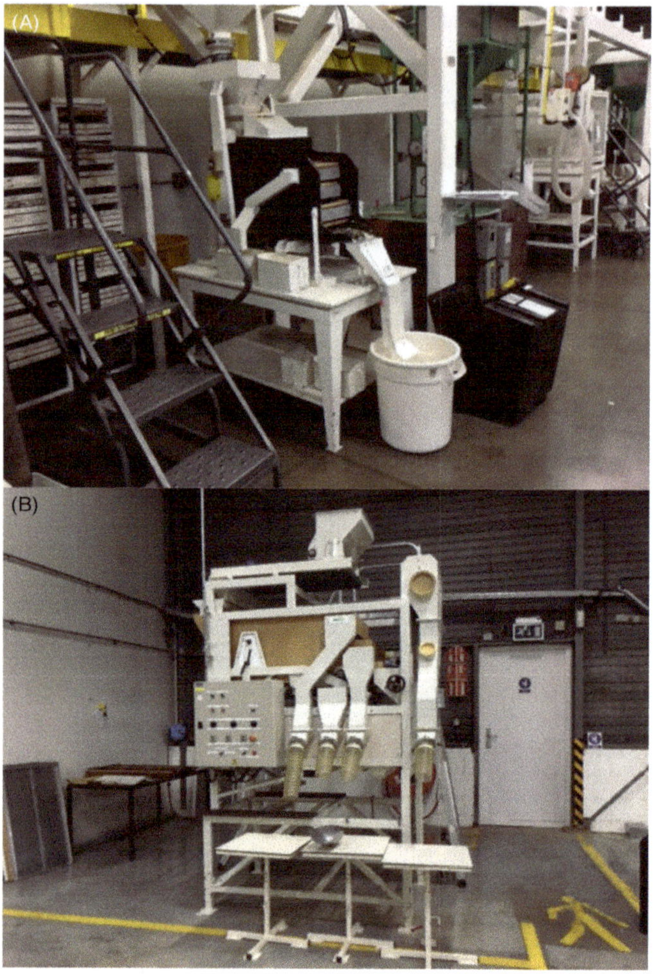

Fig. 5.1. Seed cleaner for small (A) and medium (B) seed lots. (Courtesy of Tim Loeffler.)

and root ecosystems. Recent studies show that Trichoderma are opportunistic, avirulent plant symbionts, as well as parasites of other fungi like ones that cause the seedling disease "damping-off". Organisms delivered as biological seed treatments are designed to multiply in the soil, stimulating plant growth and/or protecting root systems against soil-borne pathogens after germination. A challenge with biologicals is getting enough living organisms on the seed to survive from the time of treatment until planting. Some biologicals may not need to be living to be effective, since stable antimicrobial compounds or plant growth substances may remain in non-living treatments. Currently, it is much easier to register a new biological seed treatment with government agencies than new chemical treatments in many countries.

Chemical Treatments

Pesticides

For conventional production, insect and disease control has traditionally been achieved by application of pesticides to foliage after seedling

Fig. 5.2. Precision seed sizer for smaller lots (A) and large seed lots (B). These seed sizers are most common for small-seeded vegetables. The seed is metered onto a series of vibrating screens progressing from smaller to larger hole sizes. Seeds fall through holes in screens that match the seed size into a container below. This process is also referred to as "seed calibration". (Courtesy of Tim Loeffler.)

emergence. Seed treatments are a superior alternative to protect developing plants against pests because fewer chemicals are needed, and seed treatments are safer and more efficient than field applications of pesticides. One of the most common seed treatments are fungicides applied to control soil-borne diseases such as damping-off. Damping-off is a very destructive disease caused by *Pythium ultima* and other soil-borne fungi that cause rapid seedling collapse after emergence. A light coating of fungicide on the seed surface is sufficient to protect vulnerable seedlings. Thiram (dimethylcarbamothioylsulfanyl N,N-dimethylcarbamodithioate), captan (dimethylcarbamothioylsulfanyl N,N-dimethylcarbamodithioate), and apron (methyl 2-[(2,6-dimethylphenyl)(methoxyacetyl)amino]propanoate) are fungicides often applied as seed treatments. Thiram and captan can no longer be used in some countries because of environmental concerns. Safer products such as Apron Xl (mefenoxam), Maxim (flufioxonil), and Dynasty (azoxystrobin) and others are being used. The availability of these newer seed treatments varies across countries because of differences in seed chemical registration. This makes shipping treated seeds from one country to another challenging because compliance with laws in both the shipping and the destination countries must be met. Systemic insecticides such as Gaucho® (N-{1-[(6-chloro-3-pyridyl)methyl]-4,5-dihydroimidazol-2-yl} nitramide) and Cruiser® (3-[(2-chloro-1,3-thiazol-5-yl)methyl]-5-methyl-N-nitro-1,3,5-oxadiazinan-4-imine) are also applied as vegetable seed treatments. Systemic chemicals applied to seeds are absorbed by young seedlings and are metabolized away as the crop develops, so they only provide protection in the seedling stage and are not present during crop harvest. Each treated seed must contain the correct proper dose of active ingredient according to the label specifications. The "recovery %" of active materials applied to seed should be tested in laboratories as quality control to determine the effectiveness of the treatment. Colored dye is required by law in many countries to show that seeds have been treated with foreign substances like fungicide. However, colored seeds may be treated with materials besides pesticides.

Trace element seed treatments

Small quantities of essential trace minerals added as seed treatments may alleviate deficiencies resulting in healthier crops and higher yields. Such applications are more efficient and precise than soil applications. Seed application is not an efficient way to apply N-P-K or other macronutrients that can damage sensitive young seedlings. Legumes are especially sensitive to salts and are not usually treated with mineral nutrients. However, some seeds are relatively tolerant of salts, so small quantities of micronutrients with low salt index can be effective. Water-soluble and temperature-regulated polymer barriers can protect seeds from nutrient coatings. Similarly, nutrients can also be delivered in pelleting materials that do not directly contact the seed. Matric priming with diatomaceous earth boosted early seedling growth because mineral nutrients were absorbed by the seed (Jett *et al.*, 1996). Applying trace elements as seed treatments continues to be an evolving area of seed technology.

Plant growth regulators

Plant growth regulators are sometimes applied as seed treatments to break dormancy, improve stress responses, or stimulate early seedling root and shoot growth. Since only small amounts are needed, plant growth regulators are often combined with a liquid priming treatment like osmotic priming (described below). Examples of plant growth regulars that can be applied to seeds include indole-3-yl-acetic acid (IAA), indole-3-butyric acid (IBA), gibberellic acid (GA_3), kinetin (KIN), and brassinosteroids (Prajapati *et al.*, 2015). To reduce costs, synthetic versions of these compounds are applied when available.

Seed Coating and Pelleting

Seeds are coated to make them easier to handle, to deliver chemicals, to improve soil contact, or to deliver microbes. Vegetable seeds are often small or irregularly shaped, making them difficult to singulate for planting. Belt seeders with prepunched holes and vacuum planters with a revolving plate with small holes at the circumference have improved precision placement of seeds but still have problems with irregularly shaped, small seeds. Depth control is another problem for vegetable seeding equipment because small seeds may fall between soil particle aggregates to deeper depths than intended. Seed pelleting technology addresses both problems by coating seeds with clay, diatomaceous earth, and/or polymers to create a uniform round shape, larger size, and greater density for precision soil placement (Fig. 5.3).

For example, pelleting is widely applied for lettuce, but small seeds of other species can also be pelleted. Small valuable seeds, such as tomato and celery, are often pelleted to increase their size and facilitate their use in automatic planters in greenhouses for transplant production. Pellets help provide greater contact with the soil to improve imbibition. Modern pelleting treatments either split open or melt upon contact with water, so the coating does not pose a barrier to radicle growth or limit oxygen access to the germinating seeds. Once seeds are coated, they are more difficult to distinguish, so it is critical maintain their identity from treatment to planting. Color coding is often used to identify seeds after pelleting. Colored seeds is also required by law in many countries to indicate chemical treatments (Fig. 5.4).

Coating, encrusting, and pelleting treatments use similar materials (Fig. 5.4). Encrusted and coated seeds have a thin layer of material,

Fig. 5.3. Rotating seed pelletizer. Moist coating material (polymer, clay, and other compounds) with binding agents are sprayed on seeds rotating at high speed. Seed pelleting allows irregularly shaped seeds to be singulated for precision planting. (Courtesy of Tim Loeffler.)

Fig. 5.4. Comparison of seed treatment technologies. Upper left, encrusted seeds; upper right, chemical treatment with fungicide; lower left, polymer film coating; and lower right, pelleted seed.

such as clay, polymer, or diatomaceous earth added, to make the seeds larger without a change in their overall shape (Pedrini *et al.*, 2017). Pelleting and encrusting seeds is a wet process, the pelleted and encrusted seeds are dried immediately afterward under controlled temperature and airflow. With any wet coating process, drying back is needed to ensure the seed stores properly and its longevity is minimally affected. The exact compositions of coating materials are proprietary trade secrets maintained by their developers. An important criterion for coated seeds is to avoid diatomaceous earth and clay materials from sloughing off because of seed-to-seed abrasion. This occurs during shipment creating harmful dust, especially when coatings are combined with pesticides, which can create a health hazard for workers.

Film coating

Until recently, seed treatment chemicals were applied as a slurry that dried on the seed surface. Disadvantages of this method were difficulty obtaining uniform coverage on the seed and "dusting" that occurred as the material rubbed off during transport and handling. Abrasive dust wasted chemicals and was hazardous if breathed or ingested. In recent years, film-coating technology overcame these problems (Fig 5.4). With film coating, the chemicals are applied as a polymer slurry that is sprayed on the seeds as they fall though specialized coating machinery (Fig. 5.3). The polymer dries rapidly, and seeds emerge with a dry smooth polymeric coating that doesn't rub off, reducing waste and worker exposure to dust. Polymer technology was borrowed from the pharmaceutical and candy industries who use similar treatments to coat pills and confections. Film coatings give seeds a smooth surface for improved flowability and singulation and are combined with fungicide, growth promoters, biologicals, or fertilizer treatments. They are water-soluble and are not a barrier to oxygen diffusion, imbibition, or radicle emergence. A wide range of polymers is available for color-coding seed treatments and cultivars (Fig. 5.4). Industrial-scale machinery capable of treating large quantities of seeds is available.

Novel polymer applications have also been developed for specialized applications. For example, polymers have been developed that exhibit temperature-sensitive permeability to water. Permeability to water is low at cold temperatures and increases at warm temperatures. Modifying the composition of the polymer can set the temperature at which the permeability changes. Seeds coated with these polymers will not imbibe water if the temperature is below the set point of the polymer, potentially protecting the seed from imbibitional chilling injury if planted early. The coatings are also being used to delay germination after planting, such as for timing the emergence of male parent lines at different times for hybrid seed production. A starch-based biopolymer is also used in film coating to slow water uptake and alleviate chilling injury.

Seed Priming

A physiological technique for improving germination performance is called seed priming. Seed priming is a controlled imbibition process followed by redrying. Priming involves imbibing seeds to a moisture content below that needed for germination, but sufficiently high to allow embryo metabolism to proceed (Welbaum *et al.*, 1998; McDonald, 2000). During priming, seeds are essentially in Phase II of imbibition, but are prevented from proceeding into Phase III (Fig. 5.5A).

Different techniques control seed moisture during priming. Seeds may be primed in either osmotic solutions (osmoconditioning) (Fig. 5.6), moist solid carrier materials, such as moistened vermiculite or clay (matriconditioning), or water (hydro or drum priming) (Fig. 5.5D). Currently, either osmoconditioning or drum priming is used for most vegetable seeds. After priming, seeds are often pelleted and then dried, packaged, distributed, and planted the same as untreated seeds.

Certain key metabolic steps for germination are completed during priming and need not be completed upon rehydration, allowing seeds to germinate faster (Fig. 5.5B). Priming reduces the average time to germination for many seed lots, effectively improving seed vigor (Fig. 5.5C). Priming may also improve germination under temperature or moisture stress. For example, priming prevents thermodormancy and photodormancy in lettuce (*Lactuca sativa*) seeds. Lettuce seeds may be induced to enter secondary dormancy when planted at temperatures approaching 30°C (86°F) if they are not primed. Priming also alleviates photodormancy in lettuce and is a requirement if the seeds are to be pelleted. Priming dramatically increases the speed of tomato (*Solanum lycopersicum*) and pepper (*Capsicum annuum*) seed germination, thus

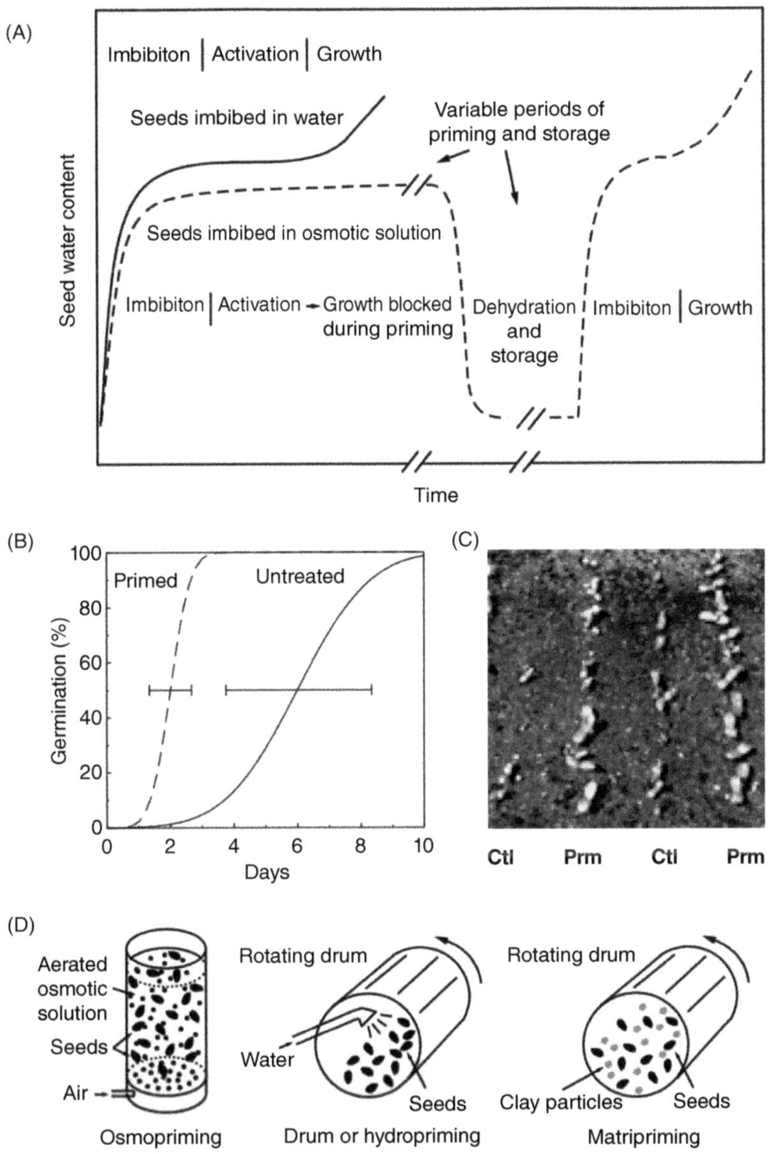

Fig. 5.5. Imbibition kinetics comparing germination with an osmotic priming treatment (A). Germination time course comparing primed and unprimed seeds showing how priming speeds germination while reducing variation in times to radicle emergence (B). For field planting, primed seeds (Prm) emerge faster than unprimed controls (Ctl) (C). Osmopriming, drum or hydropriming, and matric are different priming methods (D). (Courtesy of K.J. Bradford, personal communication.)

Fig. 5.6. Osmotic priming of pepper seeds. Seeds are imbibed in an osmotic solution that controls water uptake by the seeds. (Courtesy of Tim Loeffler.)

reducing the time and improving the uniformity of germination for producing transplants in greenhouses before field planting (Fig. 5.5B, C). Priming improves the germinability of some seeds by removing primary dormancy and substituting for after-ripening. However, priming does not improve the viability of poor-quality, dead, or damaged seeds. In other words, priming cannot bring dead seeds back to life. Priming is usually offered as a value-added seed treatment at an extra charge either by the company that produced the seeds or a third party. In many areas, seed laws do not require primed seeds to be labeled differently than unprimed.

Priming decreases the storage life of seeds, particularly when they are subjected to poor conditions such as high moisture and/or temperature. Prior testing of seed is critical to select those seed lots that will respond favorably to priming. The tests used for determining the best priming recipe include speed of germination, seed vigor, thermogradient table tests, usable transplants, and others. It is best to prime seeds just prior to planting rather than before long-term storage. Refinements to priming technology, particularly the handling of the seeds prior to dehydration, may mitigate reduced longevity (Bruggink *et al.*, 1999; Gurusinghe *et al.*, 2002). After priming, seeds are dried back to their original water content and stored under dry refrigerated conditions to maximize longevity prior to planting.

Biodynamic Seed Soaks and Sprays

Biodynamic preparations are designed to enhance biological activity in soil using seeds as the delivery vehicle. Preparations consist of mineral, plant, or animal manure extracts, often fermented and applied in small proportions to seeds either as a spray treatment or soak treatment. After biodynamic application, seeds are usually dried, packaged, and handled as other seeds.

Seed Disinfection

A simple, non-chemical treatment that is sometimes applied to seeds for disease control is hot water. Briefly soaking certain seeds in hot water can kill bacterial diseases and spores as well as some fungi, oomycetes, and viruses attached to exterior tissues. Tomato (*Solanum lycopersicum*), pepper, carrot, and *Brassica* spp. are seeds that are sometimes treated with hot water. Typically, seeds are treated in water heated to 50°C (122°F) for 20 to 30 min. Longer durations require precise control of temperature and time and may reduce germination percentage and rate. After treatment, seeds are immediately dried back to their original moisture content before further seed processing, seed enhancements, treatment, packaging, and sale. Hot water treatment is viewed as a treatment of last resort because of its potential adverse effects on quality. Hot water treatment is usually applied when other methods of seed disease disinfection do not work. Hot water treatment is also used for organic seeds where other chemicals are not allowed.

Other seed disinfection treatments to remove seed-borne bacterial and viruses are chlorine bleach (NaOCl) and hydrochloric acid (HCl) (Fig. 5.7). Trisodium phosphate dodecahydrate (TSP) can also be used against viruses. For each crop, a specific recipe of chemical concentration, temperature, and time exposure is developed for disinfection. Appropriate controls should be monitored to ensure no loss of seed quality results from treatment (Fig. 5.7).

Pregermination

Pregerminated seeds were developed to increase the uniformity of emergence in species where seed quality was extremely variable. The first generation of the technology in the 1980s was to imbibe seeds in a facility and select ones with protruding radicles for planting. These germinated seeds were intolerant of dehydration and had to be handled while hydrated. This led to the development of various fluid-drilling schemes where germinating seeds were suspended in gels and extruded into the row or transplant tray cell. Although the system worked on an experimental scale, problems with perishability, distribution, and added weight prevented first-generation pregerminated seeds from being a commercial success.

In the 1990s, methods to induce desiccation tolerance in pregerminated seeds were developed (Bruggink and van der Toorn, 1995). This allowed germinated seeds at the radicle stage to be selected for uniformity and slowly dried under controlled conditions prior to packaging, distribution, and planting. Desiccated pregerminated seeds improved longevity and reduced weight compared to hydrated seeds. When planted, pregerminated seeds emerge more quickly and uniformly than conventional ungerminated seeds. Although the technology is available and has been used on vegetable seeds, the widest use is on high-value flower seeds, where the greater cost can be justified.

Synthetic Seeds

"Synthetic" seeds derived from tissue-cultured somatic embryos coated in gelatinous material are a form of clonal propagation. However, compared to sexually derived seeds, they are relatively expensive, difficult to ship, have a short shelf life, and are excessively variable for commercial use. Current research has focused on developing synthetic seeds that are desiccation tolerant and can be handled like conventional seeds. Although progress has been made in mass-producing desiccation-tolerant synthetic seeds, they have not had a major impact on the vegetable seed industry.

Seed Tapes and Plant Tapes

Seeds embedded in water-soluble tape that can be planted at a precise depth and in-row spacing are called seed tapes. This technology may improve uniformity because seeds are precision placed with good seed-to-soil contact. Sufficient water is needed to both dissolve the tape and imbibe the seed, so seed tapes are often used in conjunction with irrigation. Seed tapes require special planting equipment and must be kept dry during storage, so the tape doesn't dissolve prematurely. A more recent development has been the use of plant tapes. Seeds are sown in soil substrate blocks which are connected by a paper tape at specific plant-to-plant spacing and arranged in a planting tray. When the plants reach their appropriate size, the plant tapes are transferred to the field and are placed in specialized equipment for semiautomated transplanting.

Seed Testing Procedures

All vegetable seeds are tested multiple times beginning at harvest through to delivery to the market. Test may include germination, vigor, viability (tetrazolium, respiration, etc.), physical purity, genetic quality, and seed health. Certain species are required to be sanitized before they can be shipped from the production country to the eventual country of sale. For example, tomato seed must be acid treated with hydrochloric acid

Fig. 5.7. Seed sanitizers. Small seed samples are placed in mesh bags and larger seeds are placed directly in revolving cylinders (A). The systems are programmed to load the sanitizing solutions at either hot or cold temperatures. After the set time (typically 20 to 30 min), the solutions are drained, and a rinse cycle(s) completes the treatment (B, C). After treatment, sanitizing drums with seeds are lifted and centrifuged before forced-air final drying. (Courtesy of Tim Loeffler.)

before shipment to the EU. It is important that seeds be tested for germination prior to sale. There are official germination tests established by state, federal, and private seed labs to test the germination of seeds under optimal conditions. The International Seed Testing Association (ISTA) establishes standardized germination testing procedures and standards for seeds sold internationally. Standardized tests determine whether seeds are able to germinate, and results are reported as a percentage. On the other hand, vigor tests are used to tell how well seeds germinate under stress conditions but are not mandated by ISTA or federal or state laws in most countries. In the USA, the established germination tests are developed by the Association of Official Seed Analysts (AOSA) that publishes procedures for seed testing. These rules have been adapted by the federal and state governments as the standards for determining whether seed lots comply with the US Federal and State Seed Laws. Most countries have seed laws that establish standards for labeling seeds, minimum germination percentages, inert matter, date of packaging, and contamination from weed seeds; however, most seed company standards are significantly higher than the minimum country standards. Seed laws are designed to protect growers from purchasing inferior seeds.

References

Bello, P. and Bradford, K.J. (2021) Relationships of *Brassica* seed physical characteristics with germination performance and plant blindness. *Agriculture* 11, 220.

Bruggink, G.T., Ooms, J.J.J. and van der Toorn, P. (1999) Induction of longevity in primed seeds. *Seed Science Research* 9, 49–53.

Bruggink, H. and van Duijn, B. (2017) X-ray based seed analysis. *Seed Testing International* 153, 45–50.

Bruggink, T. and van der Toorn, P. (1995) Induction of desiccation tolerance in germinated seeds. *Seed Science Research* 5, 1–4.

Fortune Business Insights (2019) Seed & Seed Technology/Seed Treatment Market Size, Share and COVID-19 Impact Analysis. https://www.fortunebusinessinsights.com/industry-reports/seed-treatment-market-100156 (accessed 4 September 2023).

Gregg, B. and Billups, G. (2010) *Seed Conditioning*, Volume 2: *Technology. Parts A and B*, 1st edn. CRC Press, Boca Raton, Florida. doi.org/10.1201/b10312 (accessed 4 September 2023).

Gurusinghe, S., Powell, A.L. and Bradford, K.J. (2002) Enhanced expression of BiP is associated with treatments that extend storage longevity of primed tomato seeds. *Journal of the American Society for Horticultural Science* 127, 528–534.

Halmer, P. (2000) Commercial seed treatment technology. In: Black, M. and Bewley, J.D. (eds) *Seed Technology and Its Biological Basis.* CRC Press, Boca Raton, Florida, pp. 257–286.

Jett, L.W., Welbaum, G.E. and Morse, R.D. (1996) Effects of matric and osmotic priming treatments on broccoli seed germination. *Journal of the American Society for Horticultural Science* 121, 423–429.

McDonald, M.B. (2000) Seed priming. In: Black, M. and Bewley, J.D. (eds) *Seed Technology and Its Biological Basis.* CRC Press, Boca Raton, Florida, pp. 287–325.

Nehoshtan, Y., Carmon, E., Yaniv, O., Ayal, S. and Rotem, O. (2021) Robust seed germination prediction using deep learning and RGB image data. *Scientific Reports* 11, 22030.

Pedrini, S., Merritt, D.J., Stevens, J. and Dixon, K. (2017) Seed coating: science or marketing spin? *Trends in Plant Science* 22, 106–116.

Prajapati, S., Jamkar, T., Singh, O. P., Raypuriya, N., Mandloi, R. and Jain, P.K. (2015) Plant growth regulators in vegetable production: an overview. *Plant Archives* 15, 619–626.

Taylor, A.G., Allen, P.S., Bennett, M.A., Bradford, K.J., Burris, J.S. and Misra, M.K. (1998) Seed enhancements. *Seed Science Research* 8, 245–256.

Welbaum, G.E., Shen, Z., Oluoch, M.O. and Jett, L.W. (1998) The evolution and effects of priming vegetable seeds. *Seed Technology* 20, 209–235.

Wilson, H.T., Khan, O. and Welbaum, G.E. (2014) Chlorophyll fluorescence in developing 'Top Mark' cantaloupe (*Cucumis melo* L.) seeds as an indicator of quality. *Seed Technology* 36, 103–113.

6

Organic Seed Production

History

Demand for organically produced vegetable crops increased in many North American, European, and southern hemisphere markets toward the end of the 20th century and has continued to increase (Aschemann-Witzel and Zielke, 2017). The popularity of organic produce is less prevalent in many lesser developed countries. Modern organic farming was developed as a response to several concerns about high-input, industrial-scale crop production that came to dominate after World War II. Several high-profile incidents of consumer illness from pesticide-contaminated produce occurred in the 1980s (Pimentel, 2005). A chemical leak at a pesticide factory owned by a large international company in Bopal, India killed several thousand people in 1984, increasing public awareness of the potential dangers from chemical use in crop production (Varma and Varma, 2005). Consumer dissatisfaction with vegetable quality was often blamed on vegetable genetic improvement strategies and caused renewed interest in heirloom vegetables which consumers reasoned should have superior quality. Water pollution and soil erosion caused by intensive agricultural practices contributed to interest in alternative cropping strategies, including organic. Large multinational agricultural giants, perceived by some as a cause of high-input industrial crop production and its ills, were faulted, creating renewed support for small organic farmers. Consumer demand for safe, sustainably grown vegetables created a market for organically produced seed, which is the starting point for organic crop production in most certification programs (Aschemann-Witzel and Zielke, 2017).

Definition

An organic farming agricultural system is defined as using ecologically based pest controls, biological fertilizers derived from composted wastes, naturally occurring fertilizer such as rock phosphate, and nitrogen-fixing cover crops. In other words, organic vegetable production is based on natural practices free of concentrated chemicals that are manufactured synthetically (Fig. 6.1). Compared to conventional agriculture, organic farming typically uses fewer pesticides, decreases nutrient pollution of ground and surface water, and recycles composted wastes back into the soil, as was the long-standing practice before concentrated fertilizers were created. However, there are exceptions where organic practices add more inputs and management practices, such as increased plowing to control weeds. There are instances where organic production uses more pesticides than conventional. For example, for sweet corn production in some southern areas of the USA, *Bacillus thuringiensis* (*Bt*) must be applied almost daily under rainy, humid conditions during bloom and seed set to

DOI: 10.1079/9781789243260.0006

Fig. 6.1. Organic tomatillo (*Physalis philadelphica*) seed field in northern California, USA. Although approved practices may vary from country to country, organic vegetable seed crops must be non-transgenic and grown using natural practices free of synthetic herbicides, fungicides, insecticides, and concentrated industrial fertilizer.

achieve any level of earworm control. By comparison, transgenic corn expressing the *Bt* gene reduces applications to nearly zero. Toxicity levels of some organic pesticides, such as those containing copper, can be even higher than conventional pesticides. The widely held belief that "natural" is always safer is not necessarily the case.

Benefits of organic systems are countered by higher food costs for consumers and reduced yields for many crops. Yields of organic vegetables may be lower overall than conventionally grown crops, although this varies considerably with management practices used. Higher production costs and lower yields sometimes make organic vegetable seed more expensive.

As with any government regulatory program, an oversight organization or agency must monitor the authenticity of organic production. Many organic certification programs have been developed by countries around the world with varying standards (Prakruthi *et al.*, 2021). Multinational organic programs enable marketing of fresh and processed vegetables and their seeds both domestically and internationally to assure consumers that certified products meet consistent, reproducible, and uniform standards. Standards for organic products include a list of substances approved for and prohibited from use in certified production and handling. The International Federation of Organic Agriculture Movements (IFOAM) with offices in Bonn, Germany is the worldwide umbrella linking organizations in over 100 countries and coordinating organic agriculture movements internationally among seed organizations, including the seed branch of the Food and Agriculture Organization of the United Nations (FAO) and the International Seed Federation (ISF) (Prakruthi *et al.*, 2021). In the United States, the US Department of Agriculture (USDA) provides nationwide standards for production and labeling of organically grown vegetable products including seeds. The National Organic Program (NOP) is conducted under the direction of the Agricultural Marketing Service (AMS), a division of the USDA, which suggests that the US government believes that "organic" is primarily a value-added marketing tool. In the USA, organic certification is conducted by third-party agents that have been accredited by the USDA rather than employees of the agency. The NOP also authorizes states to oversee organic production and handling operations within their borders. In California, the State Department of Agriculture oversees and enforces the USDA organic regulations within California but does not provide certification. Some national programs have stipulations for importation of organic agricultural products with few restrictions from foreign programs deemed to have equivalent standards to their own. Equivalency agreements have been developed to facilitate the movement of organic products from one country to another for programs that are similar. For example, USDA's NOP has an equivalency agreement with the European Union (EU) which means the EU organic labeled products are accepted in the USA as organically certified.

Development of organic vegetable markets has increased demand for organically certified vegetable seed internationally. In most cases, both organic production for consumption and organic seed production is a *process* rather than a *product* certification. A process certification system is based on production planning and record keeping rather than inspection and testing of commodities after harvest. Most process certifications begin well before production begins. For example, a field must be farmed using organically approved methods for two or three years before products produced on it can be certified as organic. In addition, inspectors visit operations, often for a fee, to review production plans for an upcoming season. Production sites are randomly visited during the production season when products are spot checked for adherence to standards and freedom from prohibited materials. Particularly for organic seed production, this includes assuring defined borders and isolation from any sexually compatible genetically engineered crops. With a true product certification, records and plant material may also be checked before the season but emphasis is on inspection of a commodity to ensure the final product meets specified standards. The large size and scope of production in some countries makes product certification unfeasible. Even though testing of inputs such as soil, fertilizers, seeds, and products from organic production and handling is not a requirement, it is encouraged and sometimes may be requested by the certifying agent.

Producing Organic Seed

Organic seed producers and suppliers should have a fundamental understanding of the market for their products. Organic growers may desire different cultivar characteristics than conventional growers. Vegetable growers will choose cultivars according to their market requirements, e.g. suitability for fresh market, pre-packaging, or processing uses. Some large organic commercial vegetable growers produce exclusively for wholesale markets or export. Conversely, small farmers, who produce primarily for local fresh markets and often grow cultivars of local or regional interest, have much different interests. Both groups have different needs and offer different production challenges for seed companies. The organic fresh market grower may favor older heirloom, open-pollinated cultivars over F1 hybrids. Larger organic growers may want cultivars that have better genetic pest and pathogen resistance or tolerance of low-input organic systems.

Some challenges for organic seed producers include: fully understanding the certification process and establishing a relationship with the certifying agent, creating a detailed Organic Systems Plan (OSP) for the first time, initiating the soil conversion process for organic production, following correct procedures using only approved materials, and keeping appropriate required records for certification throughout the season (Table 6.1). Organic production carries potential greater commercial risks, especially the first several seasons, to learn more intensive techniques like integrated pest management (IPM) and to develop best practices for organic production. Production costs are often higher than for conventionally produced seed of some vegetables, due to remaining production problems and pest challenges. However, differences in production costs between certified organic and conventionally produced crops that once existed have declined somewhat in recent years. Increased organic production has stimulated research, product development, and competition. This in turn has reduced costs, improved pest management strategies, and increased yields.

It is essential that organic vegetable seed production research continues to address production problems for organic growers. These include improving techniques to manage plant

Table 6.1. Some important steps in managing and producing an organic vegetable seed crop.

1. First step to organic certification: identify and become familiar with the processes required or prohibited in organic production.
2. Complete business plan and define target markets, species, production, handling, processing, storage, and distribution of organic seeds. Determine needed investments and costs for certification.
3. Select/contact the certification entity and apply for certification. Certification is typically annual with reapplication each year.
4. Develop an Organic Systems Plan (OSP) for each site that will produce, handle, process, and store seeds. List inputs and management practices to be documented. The OSP needs to be updated with changes in the program and submitted for approval.
5. Receive certification approval from certifying agent. Production land not previously certified will require a conversion period. Investments at each site need to be complete before certification can be granted.
6. Create an organic vendor and grower list and keep it up to date.
7. Differentiate conventional and organic processing, inventory, and storage areas for storage and tracking. All organic materials must be clearly labeled, contained, and stored in a marked area.
8. Developed approved organic seed labels for your products.
9. Establish an organic warehouse pest management plan and immediately treat any infestation in organic warehouses.
10. Create seed production contract(s) with grower price structure and a base production rate.
11. Negotiate and execute seed contracts that include your quantity, quality, and costs expectations with the seed producer.
12. Assist and support your grower(s). Inspect production fields and conduct soil tests before delivering or planting stock seed. Seed production fields must have proper borders and/or isolation distances to prevent contamination from non-organic production. Follow a proper crop rotation program to reduce contamination from previous seed crops.
13. Soil building and maintenance is paramount for "true" or "authentic" organic production. Make soil building with cover crops and compost part of your fertility plan.
14. Develop a fertility plan to optimize soil health through permissible inputs. This extra work will pay off in the long run. Proper field preparation and bed creation will improve efficiencies throughout the growing season and harvest.
15. When needed throughout the growing season, monitor pest levels weekly in all organic seed fields and apply only approved crop nutrition and protection products.
16. Utilize best practices and advanced integrated pest management (IPM) techniques during the production season. Act early to prevent pest infestations because organic production has fewer tools for pest control. Stay ahead of any pest problem. Keep updated on organically registered crop materials that enter the market or are removed each year. Many new biologicals are available for organic production each year. Test promising new ones in research plots to identify new tools for pest management and growth enhancement. Be knowledgeable with restricted inputs and materials as they may change annually with your certification program. Update the OSP and request approval from the certifying agent to use new or existing products not previously requested.
17. Immediately treat any infestations with approved products and document amounts and timing of use.
18. Be prepared for the proper harvest timing to obtain the best possible quality, including prepping harvest and handling equipment.
19. Consider use of beneficial insects on a pre-scheduled program throughout the growing season. Plant insectary habitat strips, insect guard or indicator rows near each organic seed field.
20. Consider an approved preventative fungicide program for disease control for all species, especially for biennials.
21. Watch for signs of pesticide drift onto your certified organic crops. Report any issues to the certifying agent to resolve and mitigate non-conformities.
22. Keep and maintain seed movement, harvesting, washing, and drying, as well as conditioning logs. These includes all seed milling, grading, and receiving processes. Follow established warehouse pest control procedures.
23. Always clearly label organic seed on pallets and freight documents.
24. Work with certifying agent to issue certificates for all seed movements between production, handling, processing, and storage sites regardless if they are domestic or international.
25. Phytosanitary inspections are required for exported crops. Use approved post-harvest seed-borne disease sanitation treatments for organic seed crops if necessary.
26. Consider organic seed treatment, priming, and pelleting to enhance quality organic seed.
27. Always label or highlight organic seed when shipping and include "organic seed" and all freight documents.
28. Remember many organic certification programs are process certifications rather than product certification. Understand which type of program you are participating in and keep close communication with the certifying agent.

nutrition and plant pests during the growing season and developing organic seed treatments to improve stand establishment and productivity of vegetable crops. Van Bueren *et al.* (2005) discussed the need to broaden crop genetics to meet the cultivar requirements for organic growers. Development of cultivars that fulfill needs of all organic vegetable growers is needed with superior pest resistance, the ability to mine soils for less concentrated nutrients, and that perform better in lower input systems such as no-till cropping systems as examples. At the same time, however, some organic advocates argue to severely limit the tools that breeders and seed companies can use in cultivar development, such as tissue culture (Lammerts van Bueren *et al.*, 2003).

Organic seed crops should be produced using agricultural practices that create a sustainable, microbially diverse, agro-ecosystem that relies on naturally derived or recycled nutrients, biological control measures for pests, crop rotations that integrate beneficial non-seed crops, and timely shallow cultivation to control problem weeds (Welbaum *et al.*, 2004). However, all too often, industrial organic production techniques relying on intensive inputs approved for organic programs are the basis of both vegetable food and seed crops instead of a more holistic "true or authentic organic" approach that benefits the environment. Industrial organic techniques include more intensive fertilizers with concentrated inputs like Chilean nitrate and rock phosphate, pyrethroid and/or *Bt* spray treatments for insect control, and moldboard plowing between crops rather than conservation tillage. To protect against this approach, the OSP is the tool used to ensure proper organic practices are being followed and continuous improvement processes are in place to ensure proper soil management practices utilizing soil amendments, green manure crops, and crop rotation. During the inspection or auditing process, the certifying agent and producer work together to identify and avoid repeated and overuse of questionable practices. High-input industrial organic systems are easier for conventional seed companies to adapt and implement compared to more natural authentic organic systems which were championed by the originators of the organic movement and require more intensive management.

Organic seed crops generally must be produced to the same seed regulatory and market standards as conventional vegetable seeds but are not a requirement for organic certification. The quality of organic seeds continues to be regulated by existing country, state, and local seed laws, the same as conventional seeds. Organic and conventional seed quality differences are usually small for most crops. There are many unique requirements for producing organic vegetable seeds: inputs used such as field cropping history and the genetic makeup of seeds produced. Organic seed producers must adhere to strict guidelines established by the country where seeds are to be used. Approved products that must be used for certified organic production may change frequently. Before production begins, a grower must do extensive planning, research, and be familiar with the standards of the market for which seeds are being produced (Table 6.1). Organic standards are often long and complex. In most cases, the certification of vegetable seed is conducted annually by the same accredited agents. Accredited organic certification specialists often certify other productions in addition to crops and seeds. Because of this, some certifiers may lack an intimate knowledge of seed production unless an agent with seed crop experience is retained. Fees charged for certification are passed along to product end users and are one reason for the higher cost of organic seeds.

Obtaining Organic Certification

Obtaining organic certification is an involved process. To be certified organic, growers must develop a detailed OSP that includes how each crop will be produced, handled, and processed. The OSP includes the practices and procedures, inputs, monitoring practices, record keeping, and additional information that may be required by the certifying agent. This document along with a fee is submitted to the certifying agent who will approve it and will follow up with an initial certification audit. The inputs include the use of all soil amendments, fertilizers, sanitation, and pest control materials and their sources. The OSP is owned and prepared by the producer and anytime a crop or practice is changed, it needs to be updated and approved by the certifying agent. The OSP is the document the certifying agent uses to complete the audit and reports any inconsistencies as findings. Approved and prohibited materials may vary widely among certification programs, so a producer must have a working knowledge of all program requirements. An important consideration is the cropping history of production fields where the certified crop will be grown. There is a transition period also known as the conversion period for switching from conventional to organic production fields. The conversion period starts with an approved OSP followed by a certification audit and annual inspection and audits. Land used for organic seed production must be free of prohibited chemicals for a specified period and detailed records must be maintained during the conversion period. This is often from two to three years after the last conventional harvest, depending on the program, before a field is eligible for organic production certification. The soil must be managed using approved crop rotations, cover crops, and addition of natural materials. Plant and animal materials must be managed to avoid contamination to crops, water, or soil. Raw uncomposted manure can be used in advance of crop establishment after a significant waiting period specified by the program. Thoroughly composted waste can be applied to fields after a minimum internal temperature is reached and other protocols are followed. Sewage sludge is not allowed as an amendment. Mined fertility inputs such as rock phosphate or certain K^+ sources are often allowed depending on how they are processed. Sometimes a grower may need to develop a separate source of water with approved use of a water treatment system for organic production. In many programs, natural mulches are acceptable if produced organically. One exception is the US NOP where plastic mulch films are acceptable if they are removed from a field at the end of the season. Many programs do not permit photodegradable or biodegradable films. Lumber treated with toxic preservatives may not be used in transplant houses, cold frames, or hotbeds (Fig. 6.2).

A critical concept of managing organic production and handling is to prevent the organic crop from being contaminated from unapproved substances. This requires a risk analysis of all processes and mitigating those risks. Best practices require all equipment to be properly cleaned, inspected, and records kept. All inputs such as soil amendments, fertilizers, and seed are clearly labeled and protected to avoid unapproved substances from contaminating the materials or the organic products. Conventional

Fig. 6.2. Organic pepper seed production in an enclosed structure in southern France. (Courtesy of Tim Loeffler.)

production must be separated from organic production by a physical barrier or with an unrelated buffer crop to prevent contamination cross-pollination from other crops or drift from restricted pesticides. Isolation distances between conventional and organic crops are important. This is especially important for wind-pollinated seed crops like sweet corn (*Zea mays*) and beets (*Beta vulgaris*) which require significant isolation distances to prevent crossing with non-organic crops. Most programs require separation either in time or space for cleaning, processing, and storage facilities for organic and conventional seeds. This requires extensive procedures for cleaning and recorded keeping. Thus organic seed processing, handling, packing, and storage usually occurs in separate buildings to comply with program regulations and to prevent confusion with conventional lots. There is mandatory verification of organic production fields, processing, packaging, and storage facilities through on-site inspections and review of plans and records starting before production begins.

Accurate records must be kept of all substances applied to crops, soils, or seeds (Table 6.1). Only substances approved by the certifying agency may be used. Guidelines may change frequently and one brand of a product may be acceptable but the same product of a different brand may not. Certifying agents may have differing opinions on which products are acceptable. For organic seed production, organic seeds must be used if available. Most programs have provisions that allow producers to use untreated non-organically produced seeds if organic seeds are unavailable in sufficient quantities. This provision is changing and is expected to become stricter among several organic programs. For example, the EU organic regulations have committed that by 2035, all organic production must be produced using organically produced seeds. Documentation is usually required to prove that organic seeds were unavailable and various programs have different degrees of flexibility for dealing with organic seed shortages. Under the US NOP, treated seed can only be used when no equivalent organic cultivar is available. Input price has no influence on sourcing organic seed and growers must first attempt to source organic seed from three suppliers before using treated seed. If transplants must be used for seed production, they must be produced organically.

Certification in multiple countries

A particular challenge within the vegetable seed industry is that seeds are produced, handled, processed, and distributed across multiple countries and require organic certification at each location. An important criterion in order to supply organic seeds to the market is to maintain integrity of organic certification through the entire supply chain, which must be clearly documented. For example, seeds produced in Thailand, processed in the Netherlands, and sold in the USA or Canada technically require certification from each country's organic program. The use of equivalency agreements between these countries and authorization of foreign certifiers helps reduce the need to maintain multiple country certifications. The transfer or movement of organic seeds from one country to another requires certificates or e-certificates to track and reconcile seed quantities and usually requires a fee by the certifying agents. In the example above, using the EU certification as the more conservative organic certification program, the EU certification and label would be accepted for the target countries stated.

Seed companies with separate locations for production, handling, seed processing, and storage require certification at each location with its own OSP. This is typically managed by authorized foreign certification agents or a global certifying agent service provider with support in each country where production and handling activities take place. Thus, it is critical to understand the target countries in which seed is to be sold before any production is placed and to confirm that the seeds can be produced, processed, and stored to meet the organic regulations and maintain the organic integrity of the seed until it is delivered to the customer.

Limitations

A fundamental restriction for organic seed is that genetically engineered cultivars may not be included. This applies to cultivars carrying a transgene incorporated by *Agrobacterium*-mediated transfer, ballistics, conventional breeding making crosses with transgenic lines, RNAi gene silencing technology, or CRISPR (clustered regularly interspaced short palindromic repeats) gene editing technology. In many programs, cultivars developed by selection after sexual crossing, F1 hybrids made by crossing inbred lines, or spontaneous mutations are acceptable. Interestingly, seedless watermelons, a triploid, are permitted in many organic programs. Therefore, it is not the degree of genetic modification that is important in determining organic suitability but rather the technology used. In the case of seedless watermelon, a tetraploid breeding line is created by treating normal diploid plants with colchicine, a natural chemical derived from corms of crocus (*Colchicum autumnale*) that increases ploidy. The colchicine-derived tetraploid line is crossed with a diploid to create a triploid seedless cultivar that is planted for seedless watermelon fruit production. Seedless cultivar seed can be certified as organic despite the change in ploidy because colchicine is a naturally occurring compound and the final cultivar is developed by traditional plant breeding.

Biological contamination is transmission of living organisms and their antigens that cause human disease. Biological contamination is a major concern for growers of edible vegetables and is the major cause of food-related human disease. Most certified organic programs do not inherently protect against biological contamination of vegetable commodities. Some would argue that the large quantities of animal manures sometimes used may put organic produce at greater risk of biological contamination, specifically if composting guidelines are not followed. Good agricultural practices (GAP) were specially created to reduce biological contamination in our food supply including vegetables (Oya *et al.*, 2018). During production, biological contaminants can be spread by poor worker hygiene, contaminated water, and improper composting that fails to meet the minimum temperatures required to kill pathogens. However, under certain circumstances biological contaminates can be transferred via seeds (Cui *et al.*, 2017; Liu *et al.*, 2018). Historically, contaminates like *E. coli*, *Listeria*, *Salmonella*, etc. have not been transmitted via seeds from one generation to another to cause human illness. A series of mutations could change human disease-causing organisms to be more resilient and able to survive the dry, nutrient-poor environment of orthodox vegetable seeds in the future. Therefore, seeds produced for propagation are not GAP-certified or routinely tested for biological contaminates. Seed transmission of plant rather than human pathogens is a greater concern for both organic and conventional seed producers. Enzyme-linked immunosorbent assays (ELISA), polymerase chain reaction (PCR), and grow-out tests are routinely conducted on both organic and conventional seeds to ensure that seeds are free of plant pathogens before sale (Kumar and Gupta, 2020).

Some consumers and agriculturists object to the cost and bureaucratic nature of government-run organic certification programs. These concerns have led to establishment of alternatives offered by individual growers or producer organizations that lack official certification status (von Sehlen, 2007). These informal programs are prohibited from using the official logos and terminology of government programs such as "certified organic". Instead, alternative language such as "sustainably grown", "biologically grown", or "naturally produced" along with a pledge of production standards or general philosophy describe seed production standards. These informal grower-based programs often reject genetically engineered crops and emphasize natural production free of synthetic materials that have minimal environmental impact. Such philosophy statements and pledges provide consumers with assurances that sustainable practices were followed during production without the added cost and paperwork of government certification. However, since practices may vary among farmers and organizations, standards are not always consistent among programs.

References

Aschemann-Witzel, J. and Zielke, S. (2017) Can't buy me green? A review of consumer perceptions of and behavior toward the price of organic food. *Journal of Consumer Affairs* 51, 211–251.

Cui, Y., Walcott, R. and Chen, J. (2017) Differential attachment of *Salmonella enterica* and enterohemorrhagic *Escherichia coli* to alfalfa, fenugreek, lettuce, and tomato seeds. *Applied and Environmental Microbiology* 83, e03170-16.

Kumar, R. and Gupta, A. (eds) (2020) *Seed-Borne Diseases of Agricultural Crops: Detection, Diagnosis & Management*. Springer, Singapore.

Lammerts van Bueren, E.T., Struik, P.C., Tiemens-Hulscher, M. and Jacobsen, E. (2003) Concepts of intrinsic value and integrity of plants in organic plant breeding and propagation. *Crop Science* 43, 1922–1929.

Liu, D., Cui, Y., Walcott, R. and Chen, J. (2018) Fate of *Salmonella enterica* and enterohemorrhagic *Escherichia coli* cells artificially internalized into vegetable seeds during germination. *Applied and Environmental Microbiology* 84, e01888-17.

Oya, C., Schaefer, F. and Skalidou, D. (2018) The effectiveness of agricultural certification in developing countries: a systematic review. *World Development* 112, 282–312.

Pimentel, D. (2005) Environmental and economic costs of the application of pesticides primarily in the United States. *Environment, Development and Sustainability* 7, 229–252.

Prakruthi, M., Mahesh, M.S. and Shekhara, N.R. (2021) Certification process and standards of organic farming. In: *Biopesticides in Organic Farming*. CRC Press, Boca Raton, Florida, pp. 357–362.

Van Bueren, E.L., Van Soest, L.J.M., De Groot, E.C., Boukema, I.W. and Osman, A.M. (2005) Broadening the genetic base of onion to develop better-adapted varieties for organic farming systems. *Euphytica* 146, 125–132.

Varma, R. and Varma, D. R. (2005) The Bhopal disaster of 1984. *Bulletin of Science, Technology and Society* 25, 37–45.

Von Sehlen, J.A. (2007) Beyond organic: defining alternatives to USDA certified organic. Graduate Student Theses, Dissertations, & Professional Papers 94. Department of Communication Studies, University of Montana. https://scholarworks.umt.edu/etd/94 (accessed 5 September 2023).

Welbaum, G.E., Sturz, A.V., Dong, Z. and Nowak, J. (2004) Managing soil microorganisms to improve productivity of agro-ecosystems. *Critical Reviews in Plant Sciences* 23, 175–193.

7

Family Amaryllidaceae Subfamily Allioideae

Overview

Onion is an ancient crop, recorded from as early as 600 BC (Zohary and Hopf, 2000). Garlic has a crop history of over 7000 years paralleling onion. Onion (*Allium cepa* L.) is the most widely grown of all Alliums, which includes other important vegetables described in this chapter. Production occurs worldwide with the greatest concentration in the northern hemisphere.

Botany

Alliums have unique plant architecture. The plant consists of a short subconical stem which produces a compact series of leaves with the sheaths of the outermost leaves enclosing the younger ones (Fig. 7.1). All leaves lie in a single plane (½ phyllotaxy).

The sheathing base of each leaf completely encircles the stem (Fig. 7.1). Thickening of the fleshy stem bases, together with the lack of internodal elongation, results in bulb formation. Alliums, except leeks and chives, are composed of these fleshy, enlarged leaf bases or scales. Onion is a biennial and has a long, hollow, leafless stem that arises from the terminal bud and bears the inflorescence after vernalization. The terminal inflorescence develops from the ring-like apical meristem. One or more flower stalks called scapes, peduncles from compressed stems, elongate above the last foliage leaf from 30 cm to more than 100 cm (12 to 39 in.) depending on genotype. At first, the scape is solid but, after differential growth, becomes thin-walled and hollow. The onion scape has a characteristic spherical hollow bulge along its length. The total number of developing scapes per plant depends on the number of sprouted lateral buds which are also able to develop into scapes (Rubatzky and Yamaguchi, 1997; Brewster, 2008).

A spherical umbel develops atop each scape and can range from 2 to 15 cm (0.8 to 5.9 in.) in diameter. Early in development, a spathe (sheathing bract) initially encloses the inflorescence (Fig. 7.2).

The umbel is an aggregate of generally 50–1000 small individual florets each about 5 mm (0.2 in.) or less long (Fig. 7.3). Flowering within each umbel may continue for four or more weeks but individual florets are fertile for only one week. Flowers are perfect, with six white petals, six stamens, a single style, a three-carpel pistil, and an ovary with two ovules. Bulbils are produced in the inflorescence of some cultivars instead of florets.

Leek (*A. ampeloprasum* var. *porrum*) forms a pseudo-stem that is used in cooking and does not form a bulb. They are popular and important in Western European cuisine. Leek is a biennial and in the second year the plant produces one or two scapes with a large umbel and perfect

DOI: 10.1079/9781789243260.0007

Fig. 7.1. Vegetative onion plants before vernalization. New leaves emerge from the center of the plant and sheaths of the outermost leaves enclose younger ones.

Fig. 7.2. Spathe encloses an Allium inflorescence atop a scape.

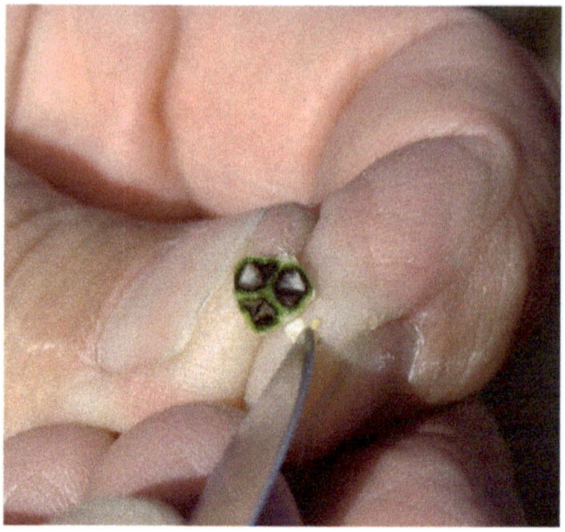

Fig. 7.3. Cross-section of an onion floret nearing seed maturity. There are three carpels with two ovules each that produce six black seeds per floret.

flowers like onions. Leeks are self-compatible and sometimes can form bulbils (Navazio, 2021). Although there are some differences in seed production for leek compared to onions and due to the space constraints, this text will focus on seed production for onions. Most leek seed production takes place in France, Belgium, and the Netherlands.

The true garlic (*A. sativum*) umbel inflorescence develops at the top of a straight and solid scape, is subspherical, and usually contains only bulbils or a combination of bulbils and flowers, which rarely set seed in commercial clones. The infrequently formed flowers are lavender and usually wither and abort before any seeds are set. Viable seeds have been obtained from certain garlic clones particularly those belonging to subspecies *ophioscorodon* of *A. sativum*, which are being used to develop novel garlic types (Pooler and Simon, 1994; Rabinowitch and Currah 2002; Volk *et al.*, 2004).

Types and Cultivars

Onion (*Allium cepa* L.) is the most common vegetable from the genus *Allium*. Other important vegetables include *Allium cepa* (shallot, top set onion, multiplier onion), *A. ampeloprasum* var.

porrum (leek), *A. sativum* (garlic), *A. ampeloprasum* (elephant garlic or great head leek), *A. schoenoprasum* (chive), *A. fistulosum* (Welsh onion, Japanese bunching onion), *A. chinense* (rakkyo), *A. tuberosum* (Chinese chive), and *A. cepa* × *A. fistulosum* (Beltsville bunching onion). These monocots were once classified in the family Alliaceae but have been reclassified into the subfamily Allioideae of Amaryllidaceae (Welbaum, 2015).

Onion is a biennial or a perennial by means of its bulbs in areas mild enough to overwinter. Chives are also perennials. Alliums are cool-season crops, tolerant of frost but not prolonged freezing temperatures below the range from −9.5° to −6.5°C. Optimum temperatures for growth and development are in the range from 13° to 25°C, somewhat narrow compared to other vegetables. Flowering occurs after vernalization in contrast to bulbing, which is a day-length response (Rubatzky and Yamaguchi, 1997; Brewster, 2008).

Growth and Development

Onion food crops are annuals consumed when immature (green onions) or mature (dried bulbs) (Welbaum, 2015). However, for seed production, onions require two seasons. Like other biennial crops, the first season is vegetative growth and builds plant size. After vernalization, reproductive growth is initiated to produce seed to complete the life cycle. The initial growth rate is slower than many other cool-season crops because of slow leaf area development and poor light interception that limits photosynthesis. As seedlings grow, new foliage and roots develop, along with a slight elongation and widening of the compressed stem (Rabinowitch and Currah, 2002; Brewster, 2008). Onion seed production requires low-humidity conditions during spring and summer when seeds mature. Disease management, pollination, and seed maturation are enhanced by warm temperatures and low relative humidity. Foliage diseases are more common under humid conditions and bee pollination is reduced during rainy weather. Pre- or post-harvest drying is enhanced in climates with low humidity. Climates that are cool in the winter with warm to hot spring and summer seasons with low humidity and limited rainfall are near ideal for onion seed production. Although onion seed production requires two growing seasons, both can occur within a single 10- to 12-month window if the crop is seeded in late summer or transplanted in early fall. The first season the plants grow vegetatively from seed and must reach a sufficient size past the juvenile stage for vernalization to occur. Over the winter, plants are vernalized, flower, and produce seed the following spring and summer. The optimum range of temperatures for best growth and development are 20° to 25°C.

Bulbing

Bulbing is a change in leaf morphology initiated when sufficient exposure to a critical day length is exceeded, although temperature also has an influence. Each cultivar has a critical day length for bulb induction. The duration of light exposure is most important, and the exposure process is cumulative. A brief exposure to the appropriate day-length stimulus is not sufficient to initiate the bulbing process. When cultivars reach their critical day length before adequate vegetative growth is achieved, resulting bulbs will be small. Cultivars that require long days to bulb will not bulb when grown during short days (Rubatzky and Yamaguchi, 1997; Brewster, 2008).

Onions are identified as short-, intermediate-, long-, or very-long- (also known as long, long) day cultivars. Short-day cultivars bulb when day length is equal to or greater than 11–13 h. Intermediate cultivars bulb in response to day lengths equal to or greater than 13–14 h, long-day cultivars bulb in response to day lengths of 14–16 h, and very-long-day cultivars in response to 16 h or longer. These designations have a positive correlation with latitude. For bulb production, short-day plants are usually grown at less than 30° latitude, intermediate between 30° and 38°, those grown at between latitudes of 38° and 48° are long-day types, and very long plants are at greater latitudes. All cultivars are long-day plants for the bulbing response, because they bulb in response to increasing rather than decreasing day length (Rabinowitch and Currah, 2002).

Photosynthate partitioning differs with various growth phases. During development prior to bulbing, leaf blade growth is greater than that of leaf sheaths. Induction of bulbing causes the mobilization of food reserves into the leaf bases, resulting in an enlargement that results in an increase in diameter at the base of the plant so that leaf sheath growth accelerates compared to leaf blade growth.

Immediately after induction, successive new leaves tend to be longer and have wider leaf bases (Fig. 7.4). Leaves and roots continue to be produced at a relatively uniform rate, although during bulbing, inner scale or bladeless leaf growth becomes dominant (Rabinowitch and Currah, 2002; Brewster, 2008).

Flowering (bolting)

Bolting is the formation of a seed stalk and associated inflorescence (Fig. 7.4). Cultivars differ greatly in their response to low temperature and the duration of exposure needed for vernalization, the induction of bolting. Exposure to 5° to 10°C (40° to 50°F) for four to six weeks is adequate to vernalize many cultivars (Rubatzky and Yamaguchi, 1997). Chilling degree days or hours can be calculated from weather or storage room growth data using 10°C (50°F) degrees as the threshold or base temperature for chilling hour accumulation. For some cultivars, temperatures between 10° and 13°C (50° and 55°F) stimulate bolting. Insufficient vernalization lowers seed yields. It is possible to devernalize a plant by exposure to high temperatures after vernalization.

Fig. 7.4. Onion bulb in cross-section. After exposure to an inducing day length, leaf sheaths expand to form a bulb. New leaves emerge from the apical meristem in the center. After vernalization, the meristem produces a scape that rapidly elongates and forms an umbel flowering structure on top.

Sufficient plant size is required for flower induction to occur. Once beyond the juvenile seedling, low temperatures can induce bolting. Large plants are more responsive to vernalization conditions than small ones. Plants with less than four or five leaves or "neck" diameters less than 10 mm, less than a pencil diameter, are usually considered to be juvenile and unresponsive. Plants grown for seed production should achieve ample leaf area before vernalization to support bolting. To get the vegetative growth needed for high seed yields, establishment should occur in mid- to late summer so that large plants are sufficiently developed before inductive vernalization temperatures occur in the fall and winter. Seed stalks normally elongate as temperatures increase in spring following vernalization (Rubatzky and Yamaguchi, 1997).

The pattern of foliage growth and development is also altered after seed stalk development is initiated because the developing seed stalk is a strong sink for photoassimilates. It is possible to have bulbs and seed stalks developing simultaneously although bulb size is usually reduced due to limiting photoassimilates. However, rapid vigorous bulbing can suppress seed stalk emergence and growth even after initiation due to photoassimilate competition.

In bulb-to-seed production (described below), bulbs require a cool dormant period for floral primordia initiation. The storage temperatures of bulbs influence sensitivity to bolting. Storage at either 0° or 25°C is less conducive to bolting than intermediate temperatures (Voss *et al.*, 2013). Subsequent seed stalk development is enhanced by large bulbs. The number of flower stalks per plant depends on the number of lateral bud shoot apices and large bulbs produce more. However, when plants are grown directly from seed, i.e. seed-to-seed production, one seed stock per plant is more commonly formed.

Soil

Site selection and field preparation

Alliums can be grown for seed production in a wide range of soil types but prefer fertile, well-drained, non-crusting mineral soils that are high in organic matter and have good moisture retention. Loam, sandy loam, or silty loam soils are preferred as they have better soil aeration and retain less water to reduce problematic diseases. Heavy clay soils and light sandy soils should be avoided unless frequent irrigation is available (George, 2009). Onions grow best in the pH range of 5.3 to 6.5. Above pH 6.5, deficiencies of copper, manganese, and zinc may occur (Smith *et al.*, 2011). Onions should only be grown in the same field once every four to five years to help prevent the buildup of soil-borne disease, nematodes, and weeds. In some areas, soil fumigation may be required to reduce disease pressure. It is important to know the previous cropping history of fields and to avoid fields where chemical carry-over could be an issue, especially with previous crops like corn. The soil must be finely tilled to a depth of at least 18 cm (7 in.) using rotovators or disks and free of clods, stones, and other impediments, especially for direct seeding, to ensure uniform emergence and good stand establishment. Pelleted seeds improve singulation and spacing with precision belt or air seeders. Planting Allium seed crops on raised beds improves drainage and encourages salt accumulation away from the root zone. Major seed production areas for onions include Argentina, Australia, Chile, France, India, Italy, South Africa, and California and Pacific Northwest of the USA.

Fertilization

Onions are a shallow-rooted, cool-season crop that responds well to fertilization. Generally, Alliums have a high requirement for micronutrients and moderate need for N-P-K. Alliums are moderately sensitive to salinity especially during germination and as small seedlings and soils with an electrical conductivity (EC) of 2.5 dS/m should be avoided. Salt tolerance increases with maturity. Growth is reduced when salinity ranges from 4 to 5 mmhos/cm (Voss *et al.*, 2013) and can result in severe yield loss. Alliums have a shallow fibrous root system restricted to the top 30 cm (12 in.) of the soil, so precision fertilizer placement is essential. Fertilization needs vary depending on many factors such as temperature, cultivar, spacing, soil type, and available water. Fertilization should be in accordance with soil test results to reduce waste.

Tissue and soil analysis combined with cropping history are important for determining nitrogen fertilization needs (George, 2009). Quick tests of leaf tissue provide rapid assessment of N availability at different times during the growing season. Nitrogen requirements vary depending on N availability from soil, irrigation efficiency, and the amount of N loss due leaching and the N cycle. Alliums are generally sensitive to ammonia. With efficient irrigation, total of 260 kg/ha (231 lbs/acre) N should maximize yield on most mineral soils, with less needed on soils with high residual N. However, needs may vary from 110 to 450 kg/ha (98 to 402 lb/acre) N depending on soil, cropping history, and water received. Higher amounts may be justified in fields receiving significant rainfall or lower nutrient-holding capacity. Excessive applications of N result in excessive vegetative growth, delayed flowering, and increased lodging. Excess nitrogen also promotes unwanted secondary growth (Rosen *et al.*, 2008). Nitrogen fertilizer should be delivered in multiple applications throughout the season, with no more than 25% of the seasonal total applied in a single application. Some growers row-band all P or P and K fertilizer in the bed before planting. N, and K if needed, is applied in various combinations such as a preplant band application at seeding, a top dressing approximately four weeks into the season as bolting begins, or weekly if required by fertigation in conjunction with petiole analysis (Smith *et al.*, 2011). Composted manure (11 to 22 t/ha) can supply nutrients and fulfill early-season N requirements; however, this is not a common practice for seed production.

Soils with bicarbonate-extractable P greater than 30 ppm require a preplant application of no more than 56 kg/ha (50 lbs/acre) of P_2O_5, while soils testing at 10 ppm P or less may require as much as 224 kg/ha (200 lbs/acre) of P_2O_5. With adequate preplant application, in-season P additions are unnecessary (Smith *et al.*, 2011).

Soils exceeding 150 ppm ammonium-acetate-exchangeable K are unlikely to respond to additional fertilization. However, if soils test <100 ppm, potassium additions of up to 170 kg/ha (152 lbs/acre) of K may be needed for seed crops (Smith *et al.*, 2011).

Pollination

Onion and certain other Alliums exhibit protandry, the shedding of pollen before the stigma is fully developed and receptive. Protandry promotes out-crossing by insect pollinators. It takes about two weeks or more for all the florets on an umbel to open completely. Most of the pollen is shed on the first and second day a floret is open and pollen viability declines quickly after opening. The stigma develops more slowly usually after pollen release typically on days 3 to 4 after anthesis. Stigmas remain receptive until day 6 or 7 after opening. The flowers are self-fertile so pollen from one floret can pollinate another on the same umbel. Where male-sterile lines are used for seed production, pollen must be moved between umbels for pollination by insects (Rubatzky and Yamaguchi, 1997; Brewster, 2008).

Allium flowers are visited by a range of insects, including bees, flies, and other insects, that collect pollen and nectar (Parker, 1982). The duration of anthesis of an individual umbel is roughly four weeks. Honeybees reluctantly visit onion flowers to collect both nectar and pollen. Onion flowers, and specifically their nectar, are one of the least favorites of honeybees who prefer flying longer distances to visit flowers of other plant species instead. Only nectar foragers will visit both male-sterile and male-fertile lines in hybrid onion seed production fields. The sugar concentration of the nectar may be increased by potassium fertilization. However, elevated potassium in nectar may be another reason why bees find onion nectar unattractive. Therefore, seed producers should not plant crops more attractive to bees near onion seed fields. If honeybees are used, colony stocking rates as high as 30 hives/hectare (12 hives/acre) or more have been suggested (Goodwin, 2012). The number of hives normally ranges from 10 to 25 hives/ha (4 to 10 per acre) and production areas with more natural pollinators require fewer hives. Placement of the hives should begin when the male parent is 5 to 10% in bloom and should be placed uniformly around and within the field when practical, with easy access to place or remove the hives. Bee activity and hive strength should be monitored during the pollination period. Sometimes beehives are rotated with new hives when pollination activity is less than desired. It is important the new beehives do not come from another onion field to avoid contamination. Beehives should be removed when the male parent is no longer producing viable pollen and there is visual seed set in the female parent.

Isolation distances are based on the type, shape, and color of onions. The isolation distances between cultivars typically range from 1600 m (1 mile) to 4800 m (3 miles). For example, the distance between differences in color can be 4800 m and when the same type, shape, and color exists, the distance may be 1600 m. In some countries, there are declared zones in which only cultivars of a specific bulb color can be grown for seed (George, 2009).

Honeybees are the primary insect used for pollinating seed crops. Alternatives to honeybees have been investigated for pollinating onion seed crops. Flies can pollinate onions particularly during plant breeding and producing small lots of seed (Goodwin, 2012) when contained in net cages or houses. Ground nesting sweat bees (*Halictus farinosus*) will pollinate onion seed crops. Farmscaping may increase sweat bee populations. Managed native bees, such as alkali bees (*Nomia melanderi*) and leaf cutter bees (*Megachile* spp.), are alternative pollinators of onion seed fields (Delaplane and Mayer, 2000; Voss *et al.*, 2013).

Insecticides negatively impact pollinators and pollen/stigma interactions. Some products dramatically reduce pollen germination and pollen tube growth (Gillespie *et al.*, 2014). For hybrid onion seed production, moderation of insecticide use is advised when controlling onion thrips (*Thrips tabaci*). Overuse of insecticides in onion seed fields may reduce yield by impacting pollinator behavior and post-pollination processes (Gillespie *et al.*, 2014).

Irrigation

Soil in onion seed fields should remain moist after seeding so germination and early seedling development are uniform. Soil crusting delays or inhibits seedling emergence, reduces uniformity, and potentially delays seed maturity. Moist soil prevents crusting. Since seeding for seed-to-seed production occurs in the summer, often at high temperatures, sprinkler irrigation is used to cool the soil and prevent crusting by evenly hydrating the seed bed. A recent trend is to transplant seedlings into the field in the early fall to avoid the risks of poor stands from direct seeding, especially for hybrid seed production. However, sprinkler irrigation is not as water efficient as drip or subsurface drip. Since onions have a shallow root system that is composed of straight non-branched roots, plants must receive a consistent supply of water throughout the growing season. Onions cannot access water at soil depths below 60 cm (24 in.). Most available water is extracted from the top 30 cm (12 in.) of soil (Voss *et al.*, 2013). The general recommendation for at least 2.5 cm (1 in.) of water per week, especially during bulbing, applies to onion seed crops on medium-textured soils. Photosynthesis and expansive growth are reduced by even mild water stress because, unlike many crops, onions cannot reduce their leaf water potential by osmotic adjustment to compensate for water stress. Stressed onion plants exhibit poor flower and pollen development, reduced seed yields, lower seed weights, reduced nectar production, and lower seed vigor when germinated (Voss *et al.*, 2013). The precise amount and frequency of irrigation depend on soil type, stage of development, and environmental conditions (humidity, wind, irrigation methods, rainfall, evapotranspiration, etc.). Evapotranspiration from seed fields can be monitored by pan evaporation to determine water loss. Tensiometer measurements of soil determine when water application is needed. Crop coefficients determined through research and published for various locations determine how much of the evaporative losses must be reapplied when the soil moisture tension threshold is reached. Irrigation is usually needed when 25% of available water in the top 60 cm (24 in.) has been depleted. In general, onion seed crops use 65 to 90 cm (26 to 35 in.) of water per growing season. With 70 to 80% efficiency, approximately 90 to 115 cm (35 to 45 in.) of water is needed to produce an onion seed crop (Voss *et al.*, 2013).

Rogueing

Rogueing for seed-to-seed is done in the field. It is important to use high-quality stock seed with genetic test data to confirm type, shape, and color. These characteristics are difficult to impossible to fully detect in the seed production field, especially considering the seeding rates are much higher than bulb-to-seed production. During fall of the first season, remove weeds, plants with off-type foliage, off-type plants or stem color, and plants that bolt prematurely the first year. During flowering the following spring, check inflorescences and flower characters for gross abnormalities. Flowering heads of plants infected with *Ditylenchus dipsaci* tend to be bent over, so infected plants should be removed and burned. Prior to overwintering, each field should be rogued to remove plants with off-type foliage or stem color, plants bolting prematurely in first year, and late-maturing plants (George, 2009).

An advantage of the bulb-to-seed production method is the ability to inspect and rogue bulbs for improper color and off-type shapes that occurs before planting for seed production. Rogueing in a bulb-to-seed production system occurs primarily when bulbs are inspected after lifting and before replanting. However, rogueing is important during plant growth and bulb development before they are lifted to remove diseased and off-type plants with unusual leaf growth and premature bolters.

Genetics

F1 hybrid cultivars are more uniform and higher yielding than open-pollinated onions. Therefore, F1 onion cultivars are widely grown and popular with bulb onion growers. Hybrid onion seed production requires special considerations. Inbred lines sometimes lack vigor and are poor seed producers and seed yields of F1 hybrids can be highly variable. To obtain more consistent and higher seed yields sometimes a male-sterile

F1 hybrid or sister line crosses are used for the female parent to create three-way or modified single-cross hybrids. Another factor is to develop inbred lines for seed production that have suitable vigor and flower simultaneously. Unfortunately, inbred lines often lack these important characteristics. The ideal situation is to have the male pollen parent flower shortly before, during, and after the female parent flowers. However, the time from seeding or bulb planting to flowering often differs among inbred lines crossed to make F1 hybrids, especially when onions from different parent lines from different day lengths are crossed. To achieve simultaneous flowering or "nicking" of male and female parental lines, planting in many cases must be staggered. An alternative approach is to direct-seed one of the parental lines and plant bulbs of the other. To achieve pollen production over an extended period, split plantings are sometimes employed. When planting bulbs for both parents, one practice is to plant half of the male parent first and then plant the remainder of the male and the female parents later. For the best results, seed companies must share information about flowering characteristics of hybrid parental lines with seed producers (Voss *et al.*, 2013).

Modern onion cultivars are F1 hybrids produced using a male-sterile inbred line as the female parent to ensure crossing by bees will result in the desired hybrid for the market. Bees tend to find male-fertile lines more attractive than male-sterile lines and tend to move up and down rows instead of crossing between male-fertile and male-sterile lines, which may reduce pollination. Because only bees foraging for nectar will visit both male-fertile and male-sterile lines, colonies introduced to onion fields should have large numbers of adult bees and should not be fitted with pollen traps or be fed with sugar syrup, as both methods promote pollen collection instead of nectar foraging (Goodwin, 2012).

Most leek cultivars are F1 hybrids and are based on male sterility. Bees are the predominant pollinators, but flies can be used in protected environments. Typically, the parent lines for hybrid seed production are grown from bulbils which are used to produce transplants. Nearly all hybrid leek seed production is produced in cages or tunnels.

Most true garlic (*A. sativum*) production is clonal via cloves. Great head or elephant garlic (*A. ampeloprasum*) has head anatomy similar to true garlic but with larger heads and cloves and milder flavor. Great head is more closely related to and has characteristics similar to leek and is often grown from true seed. Research on *A. sativum* from Central Asia has advanced our knowledge of garlic genetics but has not led to significant cultivars of true garlic grown from seed or F1 hybrid garlic (Shemesh *et al.*, 2008). Some interspecific hybrids among Alliums have been successful. Probably the best example is Beltsville bunching onion (*A. cepa* × *A. fistulosum*) which is developed for green onion use and is grown from seed.

Seed Production

The three basic methods of seed production of onions are seed-to-seed, bulb-to-seed, and seedling-to-seed. The factors involved in deciding which method to use vary by production area and include type of onion, day length, parent seed quality, environmental risks, historical success and failure rates, and costs. The planting ratio for hybrid seed production ranges from a male-to-female ratio of 1:2 to 1:5 with 1:3 being most common and depends on the parent lines, location, and method of production.

Seed-to-seed

Onion and related seed crops can be grown directly from seed or from bulbs. For the seed-to-seed method, plants are established in late summer by direct seeding, vernalized, bolt, and seed produced in one field (Fig. 7.5).

An advantage of seed-to-seed production is that it is less expensive because labor is reduced since bulbs are not lifted or replanted (Voss *et al.*, 2013). Optimal onion seed germination occurs at about 24°C (75°F). Summer planting allows plants to develop beyond the juvenile phase for fall or winter vernalization. Before planting, minimum isolation distances from surrounding *A. cepa* seed fields should be verified. Seed-to-seed onion seed crops are often planted on raised beds 100 to 105 cm (39 to 41 in.) wide although other widths are possible. For example, single rows may be planted on smaller, 75 cm (30 in.)

Fig. 7.5. Onion seed-to-seed field in California's Sacramento Valley, USA in early July. The crop was established from seed planted in early fall the previous year and vernalized in the field. Seed matures the following summer.

wide raised beds. Plate, belt, or vacuum precision seeders can establish seed fields at optimal spacing, so thinning isn't necessary. The soil should be finely tilled with small ped size and kept moist until emergence. Target seeding depths are 10 mm to 2.5 cm (0.04 to 1 in.) deep with deeper planting in lighter, drier soils. The target seedling rate for seed-to-seed production is 800,000 to 1,000,000 seeds per ha (325,000 to 400,000 seeds per acre). Seed-to-seed production causes many short-day cultivars to develop multiple scapes, while long-day cultivars tend to produce a single scape. Seed-to-seed production usually outyields bulb-to-seed. Seed quality from bulb-to-seed production is superior because of the additional selection and rogueing for off-color and off-types before planting bulbs for seed production. However, excellent quality can be obtained from seed-to-seed production, provided stock seed is grown from bulbs that are rigorously rogued for off-types (George, 2009; Voss *et al.*, 2013).

Bulb-to-seed

The bulb-to-seed method is common for production of F1 hybrid cultivars in some areas as well as stock or parental seed. Bulb characteristics are assessed and selected to maintain genetic purity, uniformity, and trueness-to-type in the bulb-to-seed method. Planting dates of parental lines can be varied using the bulb-to-seed method to synch flower times for hybrid production. Bulb-to-seed requires planting of mature non-dormant bulbs that have been vernalized to induce bolting (described below). The seeding for bulb production usually occurs in early to late spring depending on the day length of onion and area of production. The size of the bulb produced is controlled by the plant population and higher densities are preferred to produce smaller bulbs to permit for higher planting populations in the seed production field. Production of larger bulbs requires more space when transplanted and results in lower plant populations and tends to have a higher storage rot. At maturity, the bulbs are under-cut and allowed to cure in the field prior to harvesting. After the bulbs are harvested, they are stored until they are transplanted in the fall or the early spring depending on the production area. Use of forced air to dry the bulbs placed in storage helps prevent neck rots and speeds the curing process. Bulbs should be well-ventilated until they are removed from storage. Bulbs not transplanted in the fall should be stored at 0°C (32°F) and 70% relative humidity. During the transplanting process it is important all the bulbs are placed in the upward position to permit uniform emergence and growth and usually requires manual labor.

Because bulbs are established more quickly, planting can take place in fall in areas where the winters are mild. Vernalization of bulbs exposed to temperatures below 10°C (50°F) for extended periods can occur in a field. However, prolonged exposure to potentially lethal temperatures of −7°C to −9°C (19°F to 16°F) or lower must be avoided. The lethal temperature for onion seed crops depends on genotype, plant size, soil conditions, and temperature duration. Generally, bulbs of long-day cultivars are harvested in fall and stored at low temperatures for spring planting. The resulting seed crop is harvested in late summer or early fall the following year. Short-day cultivars are sown in the fall and grown over winter or seeded in late winter and bulbs are harvested in spring to early summer. Stored bulbs are replanted in fall with seed harvest occurring in late spring or early summer the following year. When bulbs are lifted for inspection, the shape, color, and size are examined for trueness-to-type. At replanting, these characteristics can be re-checked as well as signs of early sprouting bulbs. Bulbs are generally planted in single rows on beds at a spacing of 5.0 cm (2 in.) or closer depending on bulb size. The bulbs are placed in a shallow furrow and covered with 2.5 to 3.0 cm (1 to 1.2 in.) of soil. Target population for bulb-to-seed production is 120,000 to 200,000 bulbs per ha (48,500 to 80,000 per acre) depending on

Fig. 7.6. Bulbs dropped ready to be covered manually for bulb-to-seed production. (Courtesy of Tim Loeffler.)

the type and cultivar. Bulbs tend to produce multiple seed stalks compared to seed-to-seed production (George, 2009; Voss *et al.*, 2013).

Seedling-to-seed

This method is used for both bulb and bunching onion seed production in areas where the fall and early winters are mild, and the crop can continue to develop. The open-pollinated cultivars or parent lines are sown either in plug trays in a greenhouse or an open-field nursery to produce bare-root transplants. The plants are grown for 60 to 100 days after sowing depending on the cultivar and day-length requirements and transplanted in the seed production field in early fall. The transplants should be 5 to 8 mm in diameter and have four to five true leaves at the time of transplanting. Advantages of this method include avoidance of loss of bulbs during storage, especially for short storage cultivars, reduced logistics of managing bulbs and associated costs, and elimination of stand loss risk from direct-seeded crops.

Seed Harvest

After pollination and prior to seed harvest, the male rows in hybrid seed production fields are destroyed to prevent seed mixtures. Once the male rows have been removed, the remaining female parent should be inspected for noxious and problematic weeds and removed when identified. Determining when to harvest is one of the biggest challenges of onion seed production. This is largely due to conflicting objectives of maximizing seed maturation on the mother plant and yield and minimizing loss of seeds from mature umbels due to shattering. Crop lodging from diseased scapes and high winds in combination with low humidity rapidly dry umbels and increase seed loss. The optimum time for seed harvest is based on cultivar characteristics, weather, and the harvest method. The seeds are black when mature and are clearly visible on an umbel. Many growers harvest onion seed when about 10% of black seeds are visibly exposed in the umbel (Fig. 7.7).

At this stage, seed or whole umbel moisture content is approximately 65%. Shattering increases dramatically as umbel seed moisture content falls below 55%. Harvest should commence before 5% of seeds have been lost from the umbel (Voss *et al.*, 2013).

Harvesting may be direct or indirect. The indirect method of harvest is preferred for hybrid seed production and involves manually or mechanically cutting the umbels, and to thresh the seed later mechanically after the umbels have dried.

Fig. 7.7. Umbel with 10% of mature black seeds showing.

A single operation that collects seeds in one pass through a field is called direct harvesting and this method is used when manual labor or mechanical cutting of umbels is not available or for open-pollinated cultivars. One method of direct harvest is the fields may be swathed, sometimes on paper and dried for two to three weeks before the crop is threshed (Fig. 7.8).

A once-over destructive harvest with a combine when the majority of field is ready is more common for open-pollinated cultivars. Direct harvest results in significant seed losses in fields that lack uniform maturity and are usually lower quality. For direct harvesting, the beaters on the combine are removed to reduce shattering. The best time for mechanical harvesting with the least shattering is when the seed moisture content is 60–70%. Adhesive materials are applied to umbels in some production areas to delay mechanical harvest until optimum seed quality is achieved and reduce shattering. Direct harvesting increases seed losses but costs associated with labor are significantly reduced. Indirect onion seed harvest remains popular with many companies despite its high labor costs because it results in higher-quality seeds and less seed losses during harvesting (Voss *et al.*, 2013).

Indirect harvesting occurs when crops are harvested in two stages. Umbels are first hand-harvested, dried, and threshed in separate operations. For indirect harvesting, seed heads with approximately 10 to 20 cm (3.9 to 7.9 in.) of scape or seed stalk attached are cut with a sharp knife (Fig. 7.9).

During cutting, the umbel is grasped in the palm of the hand to avoid loss of seed. If the crop is uniform, a single once-over hand-harvest may be possible and is the primary method for large fields. Multiple successive hand-harvests may be necessary if a crop lacks uniformity to obtain maximum seed yield and are more common for smaller seed fields or protected production in net cages or houses (Fig. 7.10).

For indirect harvest, the seed heads are further dried on tarpaulins or cloth sheets either in the open or in structures (Fig. 7.11). Canvas is breathable and preferred to plastic for its superior air circulation. For drying, umbels are placed 15 to 25 cm (5.9 to 10 in.) deep on a large canvas tarp and dried under ambient conditions for

Fig. 7.8. Hybrid onion seed field swathed to field-dry before threshing. The male pollinated rows have already been removed from the field. (Courtesy of Tim Loeffler.)

two to three weeks in areas dry with low humidity and no rain. The piles of umbels are turned by hand to facilitate drying.

Onion seed generally has a relatively short storage life and viability decreases rapidly at high temperatures and/or higher seed moisture contents. To retain the highest seed quality, the seeds should dry quickly after harvest without an excessive heat buildup while on the tarps. Frequent turning and shading will maintain seed quality. Air circulation is key. When air is stagnant, low-speed fans speed drying. An alternative system that reduces labor is when mechanical umbel harvesters are used to cut seed stalks and umbels and load them into a truck, drying wagon, or bins used for drying. The cut umbels with seed stalks are mechanically collected. The biomass is then dried on tarps or drying bins with forced air (Fig. 7.12A, B).

Fig. 7.9. Hand cutting scape and individual umbel in an onion seed field in northern california, USA.

The attached stems add weight to the piles, increasing heating and requiring more frequent turning to facilitate drying prior to threshing. In less arid regions, indoor drying may be required. Some producers use stacked boxes or a crate system to encourage air circulation. In climates with high humidity, forced-air drying or drying beds may be necessary (Fig. 7.12A, B; Voss *et al.*, 2013; Bashir *et al.*, 2016). After drying, umbels or umbels with stalks are threshed using conventional combines and partially cleaned (scalped) before transport to a milling facility for post-harvest processing (George, 2009).

Fig. 7.10. Picking crew making a pass through an onion seed field hand-harvesting mature umbels and placing them into red plastic buckets. Less mature umbels will be harvested in a later pass.

Fig. 7.11. Buckets of harvested umbels are collected and dried in the field or transported by trailer to another location for additional drying. Dried umbels will be threshed and seeds cleaned.

Post-harvest Seed Conditioning

Cleaning

Dried umbels are ready for threshing when the seeds can be easily separated from their capsules by hand. To avoid damaging the brittle seeds, threshing should not be delayed beyond this stage. Several threshing methods have been adopted for onions depending on the scale of operation and include flailing, rolling, threshing machines, and combining. Onion seeds are very easily damaged during processing and frequent checks should he made to ensure that the seed coats are not accidentally cracked during processing. Examination of seeds with a hand lens can confirm physical damage. Effective processing should separate seeds from the dried flower tissues without excess debris. The light debris ("tailings") must also be examined to ensure it is free of good seed (George, 2009).

After threshing, the seed is transported to a cleaning and milling facility for further processing. When seed arrives at a processing plant the moisture content should be 8 or 9% to prevent heat buildup in bulk storage. If the moisture content is above this threshold, seeds should be dried using forced air. Raw seed is milled to remove debris, which may account for as much as 10 to 20% of a seed lot by weight. After threshing, the initial cleaning uses an air screen machine. Density gravity tables are used to further upgrade onion seed lots. If the seed is not sufficiently clean after these operations, or if it contains noxious weed seed, the lot may be run through the previous steps again or more specialized equipment used. Flower pedicels may be removed by either color sorting equipment, a magnetic separator (iron powder is added which adheres to the pedicels and not the seeds), or by flotation. For flotation, only the lightest fraction should be placed in water to avoid seed damage. During flotation, the good seeds sink while the poor-quality light seeds and pedicels float to the top. Flotation should not exceed three minutes per seed lot. After spin drying and further drying on racks, additional debris is removed by an air screen cleaner. After cleaning is complete samples are analyzed for moisture, germination, purity, and tested for genetic quality and seed health. If results are substandard, further milling may be required to upgrade the lot. Seed cleaning

Fig. 7.12 Drying umbels under cover (A) and in an umbel drier (B). (Courtesy of Tim Loeffler.)

removes some good seed as well as contamination, so only the minimum number of steps required to meet the desired purity standards should be used. Once minimum germination is achieved, 85% in many countries, moisture content is again checked and dried to 8% or less moisture content depending on storage method and packaging (George, 2009; Voss *et al.*, 2013).

After conditioning, onion seed may be treated with fungicides, insecticides, biologicals, and pelleted with coating materials to improve singulation during planting, and packaged in metal cans or plastic buckets for larger quantities or resealable foil or plastic packets for smaller quantities. Since onion has a relative short shelf life, seeds must be protected from heat and humidity during storage and shipment using temperature-controlled storage and resealable packaging that keeps seed moisture low (Voss *et al.*, 2013).

Seed Yield

The best yield from open-pollinated crops produced under ideal conditions is 2000 kg/ha but 1000 kg/ha is more common. In some areas yields as low as 500 kg/ha may be deemed satisfactory. The yield from F1 hybrids is usually lower than from open-pollinated crops and is often as low as 50–100 kg/ha. There are approximately 9000 onion and leek seeds in one ounce (George, 2009).

Pest Management

Managing pests in an important aspect of Allium seed production. Both chemical and non-chemical strategies can be effectively employed. Crop rotations to non-Alliums for multiple years are a simple way to control many soil-borne diseases, insect pests, certain weeds, and nematodes. Site history should be evaluated to make sure fields do not harbor noxious weed, insect, or disease pests of onion. Weeds growing in nearby fields must be eradicated to remove alternate hosts that could harboring pests that could migrate into seed crops. Solarization and sustainable fumigation techniques should be considered if less invasive techniques do not effectively control pernicious soil-borne pests (CABI, 2017; Ontario Ministry of Agriculture and Food, 2022).

Weeds

Onions compete poorly with weeds. The relatively slow-growing onion plants take longer to achieve and provide less foliage cover due to their round leaves and ½ phyllotaxy orientation. The long growing season required for onion seed crops enables successive flushes of weed growth (Voss *et al.*, 2013). Since many onion seed crops are planted in late summer and remain in the ground through the winter and spring in mild climates, they may encounter different types of weeds such as perennials, winter annuals, summer annuals, grasses, and broadleaves. The limited availability of pre- and post-emergence herbicides makes site selection and preplant management via cultivation important components of an effective weed management program for onion seed crops (Voss *et al.*, 2013). Control in areas adjacent to production fields will reduce weed seed introduction into seed production fields. For annual broadleaf weed control, the stale bed technique may work effectively. This technique is practiced by preparing the bed for planting and applying irrigation to encourage weed growth. After the first flush of weeds emerges, they may be killed by shallow application or herbicides before the onion seed crop is planted. During winter and spring months, wet soils may interfere with cultivation. Post-emergence herbicides, where available, may be applied at the three- to four-leaf stage to provide effective control. Morning glories and specifically perennial bindweed (*Convolvulus arvensis*) are particularly bothersome weeds because their seed is similar in size, shape, and color to onion seed and are difficult to remove during post-harvest processing. In some countries, perennial bindweed is a federally designated noxious weed with zero tolerance for contamination of onion seed lots (Voss *et al.*, 2013).

Pathogens

Diseases encountered when producing onion seed are similar to those encountered for edible bulb production. Downy mildew is caused by the oomycete organism *Peronospora destructor*, which infects first the leaves and later bulbs of onions and other Alliums in mild, humid weather in spring and early summer as seed stalks develop. Downy mildew is often worse on direct-seeded crops. White rot (*Stromatinia cepivora* or syn. *Sclerotium cepivorum*) results from a soil-borne fungus that persists in the soil for many years and may be controlled by crop rotation.

Some seed production areas are quarantined for white rot and bulbs or plants from other production areas are not allowed. Purple blotch (*Stemphylium vesicarium* and *Alternaria porri*) are potentially serious fungal diseases that cause purple discoloration of foliage. Pink root (*Phoma terrestris*) is a soil-borne fungus that discolors roots and stunts onion plants. Neck rot (*Botrytis aclada* and *B. allii*) is primarily a post-harvest storage disease. Symptoms include sunken areas and scales that turn gray to dark brown during storage for bulb-to-seed production. Proper curing can prevent this disease. Moist conditions favor botrytis leaf blight (*B. squamosa*). Botrytis causes leaf spots (lesions), leaf dieback, and blighting, reducing onion growth and seed yield. Onion smut (*Urocystis cepulae*), a soil-borne seedling disease common in northern areas, is typified by distinctive narrow elongated dark streaks, usually on the cotyledon or first true leaf. Infected seedlings fail to emerge or usually die shortly after emergence (Schwartz and Mohan, 2008; Chaput, 2011; CABI, 2017). Iris Yellow Spot Virus (IYSV) is a more recent disease problem and can cause significant damage in onion seed crops as infected seed stalks can lodge resulting in yield loss. IYSV can overwinter and can serve as a source of the disease from one year to the next. It is vectored by onion thrips (*Thrips tabaci*) and control is important to prevent its spread.

Insects

Field scouting is important to identify the first appearance of troublesome insects. Both onion and flower thrips (*Frankliniella occidentalis*) feed on leaf sap, causing distortion, reduced seed set, and viruses such as IYSV (*Tospovirus*: *Bunyaviridae*). Thrips cause economic losses when they become too numerous on umbels. Onion maggots (*Delia antiqua*) can be controlled by crop rotation, feed on onion roots, and overwinter in soil. Corn seed maggots, wireworms, leafminers, armyworms, mites, and cutworms may also attack onion seed fields. Broad-spectrum chemical insecticides can control many onion insect pests but must be used with extreme caution during bloom to protect insect pollinators (Gillespie *et al.*, 2014). Nematodes can also be a problem for onion seed production in some areas. The stem and bulb nematode (*Ditylenchus dipsaci*) and root knot nematodes (*Meoidogye* spp.) can stunt growth and reduce seed yields. Fields with a history of nematodes should be avoided.

References

Bashir, A.A., Sinha, J.P., Jha, G.K. and Chopra, S. (2016) Drying kinetics for vegetable seeds with Zeolite beads. *Indian Journal of Agricultural Sciences* 86, 1630–1634.

Brewster, J.L. (2008) *Onions and Other Alliums*, 2nd edn. CAB International, Wallingford, UK.

CABI (2017) *Allium cepa* (onion). In: *Crop Protection Compendium*. CAB International, Wallingford, UK. https://www.cabidigitallibrary.org/doi/10.1079/cabicompendium.4239 (accessed 22 September 2023).

Delaplane, K.S. and Mayer, D.F. (2000) Chapter 9. Alkali bees. In: *Crop Pollination by Bees*. CAB International, Wallingford, UK, p. 84.

George, R.A.T. (2009) Chapter 14. Alliaceae. In: *Vegetable Seed Production*. CAB International, Wallingford, UK, pp. 251–263.

Gillespie, S., Long, R., Seitz, N. and Williams, N. (2014) Insecticide use in hybrid onion seed production affects pre- and postpollination processes. *Journal of Economic Entomology* 107, 29–37. doi.org/10.1603/EC13044 (accessed 5 September 2023).

Goodwin, M. (2012) *Pollination of Crops in Australia and New Zealand*. RIRDC Publication No. 12/059. Wagga Wagga, NSW 2650, Australia. https://agrifutures.com.au/wp-content/uploads/publications/12-059.pdf

Navazio, J. (2021) *The Organic Seed Grower: A Farmer's Guide to Vegetable Seed Production*. Chelsea Green Publishing, White River Junction, Vermont.

Ontario Ministry of Agriculture and Food (2022) Identification of diseases and disorders of onions. Ontario Ministry of Agriculture and Food. omafra.gov.on.ca/english/crops/facts/95-063.htm (accessed 15 September 2023).

Parker, F.D. (1982) Efficiency of bees in pollinating onion flowers. *Journal of the Kansas Entomological Society* 55, 171–176.

Pooler, M.R. and Simon, P.W. (1994) True seed production in garlic. *Sexual Plant Reproduction* 7, 282–286.
Rabinowitch, H.D. and Currah, L. (2002) *Allium Crop Science: Recent Advances*. CAB International. Wallingford, UK.
Rosen, C., Becker, R., Fritz, V., Hutchison, B., Percich, J., Tong, C. and Wright, J. (2008) *Growing garlic in Minnesota*. Vegetable Management Series, University of Minnesota Extension. https://www.researchgate.net/publication/254442113_Growing_Garlic_in_Minnesota (accessed 6 September 2023).
Rubatzky, V.E. and Yamaguchi, M. (1997) *World Vegetables: Principles, Production and Nutritive Values*, 2nd edn. Chapman & Hall, New York.
Schwartz, H.F. and Mohan, S.K. (eds) (2008) *Compendium of Onion and Garlic Diseases*, 2nd edn. APS Press, St Paul, Minnesota.
Shemesh, E., Scholten, O., Rabinowitch, H.D. and Kamenetsky, R. (2008) Unlocking variability: inherent variation and developmental traits of garlic plants originated from sexual reproduction. *Planta* 227, 1013–1024.
Smith, R., Biscaro, A., Cahn, M., Daugovish O., Natwick, E., *et al.* (2011) *Fresh-market bulb onion production in California*. Vegetable Production Series, Publication 7242, University of California Vegetable Research and Information Center. https://escholarship.org/uc/item/5xr115cw (accessed 6 September 2023).
Volk, G.M., Henk, A.D., Richards, C.M. (2004) Genetic diversity among U.S. garlic clones as detected using AFLP methods. *Journal of the American Society for Horticultural Science* 129, 559–569.
Voss, R., Murray, M., Bradford, K., Mayberry, K. and Miller, I. (2013) *Onion seed production in California*. ANR Publication 8008, Division of Agriculture and Natural Resources, University of California.
Welbaum, G.E. (2015) Chapter 14. Family Amaryllidaceae, Subfamily Allioideae. *Vegetable Production and Practices*. CAB International, Wallingford, UK, pp. 267–288.

8

Family Apiaceae

Overview

Apiaceae is one of the largest families of flowering plants, consisting of approximately 434 genera and 3700 species (Stevens, 2012). The name Umbelliferae still persists in some older literature but is outdated and largely replaced (Rubatzky *et al.*, 1999). The family contains many culinary herbs, medicinal plants, and vegetables including: angelica (*Angelica* spp.), anise (*Pimpinella anisum*), caraway (*Carum carvi*), celeriac (*Apium graveolens*), chervil (*Anthriscus cerefolium*), coriander/cilantro (*Coriandrum sativum*), cumin (*Cuminum cyminum*), dill (*Anethum graveolens*; Fig. 8.1), fennel (*Foeniculum vulgare*), lovage (*Levisticum officinale*), parsley (*Petroselinum crispum*), parsnip (*Pastinaca sativa*), smallage (*A. graveolens*), and Queen Anne's lace (wild carrot, *Daucus carota*).

This chapter will focus on seed production of celery (*A. graveolens*) and carrot (*D. carota*) although much of the information will apply to other crops, particularly those of the same species.

Celery is believed to be native to temperate Europe and western Asia with a habitat and wild forms that extend over a large area from Sweden southward to Algeria, Egypt, and Abyssinia and to the mountains of India. The eastern Mediterranean region appears to be a center of domestication, but the broad distribution of wild types makes this questionable (Rubatzky and Yamaguchi, 1997). The earliest form was likely "smallage", a leaf type of primitive celery. Modern celery cultivars have been developed through plant breeding and selection.

CARROT

Botany

Carrot is a cool-season, biennial dicot, whose fleshy storage root is eaten. During the first year, carrots form a rosette of leaves, a storage root, and smaller fibrous roots for water and nutrient uptake. The stem is compressed like a plate, with foliage height ranging between 25 and 60 cm (10 and 24 in.). Leaves emanating from the flat stem have long petioles that flare at their basal attachment. Leaf blades are divided into many small, narrow, highly lobed sections. The ratio of foliage to root varies among cultivars. Cultivars developing large tops tend to produce larger roots but grow slowly, while cultivars with short tops often produce smaller roots that mature quickly (Rubatzky *et al.*, 1999).

The taproot comprises hypocotyl and primary root tissues, with the proportion of each varying among cultivars. For many cultivars, the enlarged, edible carrot "root" is largely hypocotyl tissue, although the upper taproot may be enlarged as well. Primary and lateral roots develop along the lower regions of the hypocotyl with highly branched fibrous roots, reaching

DOI: 10.1079/9781789243260.0008

Fig. 8.1. Open-pollinated dill seed field grown on raised beds in central California, USA.

depths of more than 75 cm (30 in.). It is often difficult to identify the transition from hypocotyl to root since much of the hypocotyl is root-like in appearance (Rubatzky *et al.*, 1999). Rapid root elongation begins 12–24 days after emergence, when the maximum length may be determined. Although yield potential is correlated with root length, those longer than 30 cm (12 in.) often break during harvest and handling.

Roots comprise an outer periderm layer, with a layer of pericycle beneath, and the interior is primary xylem and phloem tissues joined by a ring of cambium. The cambium produces secondary xylem toward the interior and secondary phloem toward the outside during root enlargement. Typically, xylem color is less intense than the phloem tissue, but both should be uniformly colored (Rubatzky *et al.*, 1999).

Hardened carrot plants can withstand short periods of temperatures a few degrees below freezing. The fleshy axis and short-stemmed rosette with silvery foliage develops into an elongated flower stalk after vernalization below 10°C (50°F) for two to ten weeks (Dickson and Peterson, 1958). The transition from juvenile to adult phase is not well defined but roughly equates to crown diameters of about 2 cm (0.79 in.) and eight true leaves (Atherton *et al.*, 1990). The amount of vernalization to inducc bolting varies with genotype. Development to flowering is more rapid with increasing thermal times of vernalization. Short photoperiods (12 h) or darkness during vernalization result in earlier and greater enhancement of flowering than long days (16 h). Plants induced to flower, but not yet undergoing stem elongation, can be devernalized by several days of high temperatures (28–35°C) (Hiller and Kelly, 1979). Flowering often occurs during the second year of growth. However, under the right vernalizing conditions, flowering may occur in the first year. To be vernalized, plants must have developed past the juvenile phase. Floral initiation requires a transformation of the apical meristem from relatively flat morphology that produces leaves, to a conical one capable of stem elongation and inflorescence development (Hiller and Kelly, 1979). During bolting, multiple floral stalks elongate, producing many rough branches covered with stiff hairs or bristles. Depending on the cultivar, carrot seed stalks elongate to roughly 1 to 2 m (3.3–6.6 ft) tall, forming a highly branched inflorescence with large nodes and hollow internodes (Rubatzky *et al.*, 1999; Spurr, 2003).

The carrot inflorescence is a series of compound umbels borne terminally on seed stalk branches (Fig. 8.2). The central main stem terminates with the primary or first order "king" umbel, which is typically 100 to 150 mm (3.9 to 5.9 in.) in diameter.

"Umbel" is a botanical name for the distinctive compound flower cluster that forms at the end of a seed stalk. The large primary umbel of a flower stalk may contain 50 umbellets each with up to 50 flowers. Secondary lateral branches form in successive order after the primary umbel. These smaller secondary umbels are followed by tertiary, and quaternary umbels develop later at the terminus of the lateral branched shoots (Oliva *et al.*, 1988; Fig. 8.3). The tertiary and quaternary umbels are smaller, less productive, and their inferior quality seed may be small or partially developed (Hawthorn *et al.*, 1962; Hiller and Kelly, 1979). Generally, more than 90% of the seeds are borne on the first three umbel orders (Hawthorn *et al.*, 1962). Flowers are usually bisexual but with protandrous and centripal floral development, which requires insect pollination (Peterson and Simon, 1986). At maturity, peduncles of the outer umbellets curve inward and the umbels are concave.

CELERY

Botany

Celery is a cool-season biennial. Hardened celery plants can briefly withstand temperatures a few degrees below freezing. The stem of the vegetative celery plant is short, highly compressed, and fleshy. The plant is easily vernalized by exposure to below 10°C (50°F) after developing past the juvenile phase, which roughly equates to plant diameters of roughly pencil size (1 to 2 cm, 0.39 to 0.79 in.). The amount of chilling required for vernalization varies among cultivars. Some cultivars are vernalized after exposure to as little as ten days to two weeks below 10°C (50°F). Long days during vernalization inhibit flower initiation and bolting. Long days after vernalization accelerate flower stalk development. Non-juvenile plants can be vernalized, flower, and form seeds within a year, although this usually results in lower seed quality and yields. When the plant becomes reproductive after vernalization, the stem rapidly elongates, or bolts, and becomes fibrous with little additional new petiole development.

Celery stalks comprise several enlarged, long, thick, fleshy, and solid edible petioles in a rosette pattern. Younger petioles develop in the apical meristem and are hidden by the larger outer ones. Outer leaf petioles are long and

Fig. 8.2. Primary carrot seed umbel as individual florets begin to open.

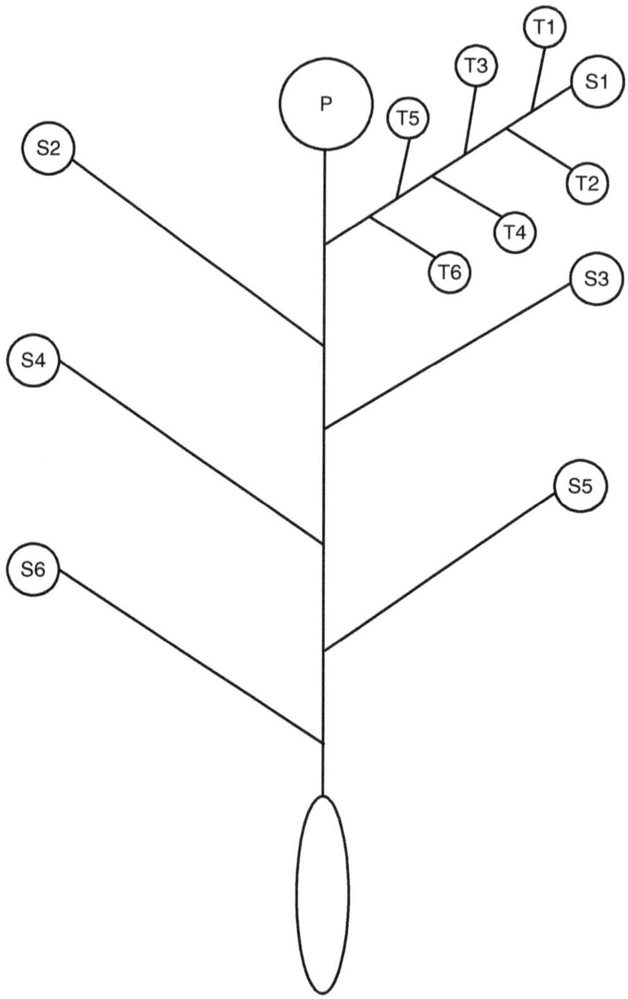

Fig. 8.3. Schematic diagram of the umbel arrangement on a flowering carrot plant: P = primary or king umbel, secondary umbels S1 to S6 =1st to 6th, and tertiary umbels T1 to T6 = 1st to 6th forming on the first secondary branch. Fourth and higher order umbels can also occur. Not drawn to scale.

straight with broad sheathing bases. Interior leaves become progressively smaller and are referred to as the celery heart. Heart petioles and leaves are pale green because of light exclusion. A well-developed mature celery plant has 7–15 clearly distinguishable petioles. The petiole is the most important edible tissue and marketable vegetative plants are trimmed and sold as celery stalks. Celery leaves have strong flavor and are not widely consumed. Celery petioles are crescent shaped in cross-section, with prominent ribbing due to a series of collenchyma bundles on the outside and a smooth inner surface. Vascular bundles inside the petiole and collenchyma bundles on the outside provide strength to the petiole and contribute to its unique fibrous, stringy texture (Esau, 1936).

The root systems of celery develop a large mass of fine, well-branched roots within a relatively short radius of the taproot. Except for the taproot, lateral roots are relatively shallow, developing within 30 cm (11.8 in.) of the soil surface (Rubatzky and Yamaguchi, 1997). During transplanting or insect damage, the taproot of celery can be damaged or destroyed, causing a proliferation of adventitious lateral roots. Some cultivars of celery and more commonly smallage tend to sucker, forming axillary stems around the base of the mother plant. Because suckers form late in development, they contribute little to the mother stalk and are usually trimmed off at harvest. The amount of tillering is greatly reduced in modern cultivars because of selection against this undesirable trait.

Umbels of celery, celeriac, and smallage are similar to carrot only smaller. Greenish white celery florets are borne in compound umbels as described above for carrot. Although individual florets are "perfect", containing both male and female parts, flower development limits natural selfing. Anthers develop and shed pollen before stigmas are fully developed and receptive to pollen, a type of flower development called protandry. Flowers are mostly insect pollinated. The fruit is a very small 1 mm (0.04 in.) schizocarp that splits when mature in two single-seeded mericarps. Celery plants are prolific seed producers.

Types and Cultivars

Carrots originated in present day Afghanistan about 3000 BC. By the year 1000, carrots had spread to Syria. The modern carrot probably evolved from a wild biennial plant. Mutation and selection are likely more important in carrot cultivar development than hybridization with wild plants (Banga, 1957, 1963). *Daucus carota* subsp. *carota* is a wild carrot known as Queen Anne's lace in North America and is a weed in the USA, Canada, Europe, and Asia that probably resembles progenitors of domesticated carrot (USDA, 2023). Queen Anne's lace foliage looks like commercial carrot cultivars while its white roots are narrow and irregularly shaped with a strong carrot odor. Other wild carrot species common around the world include *D. carota* L. subsp. *maritimus*, *D. pusillus* Michx., *D. carota* subsp. *hispanicus* (Gouan) Thell, *D. carota* subsp. *gummifer* (Syme) Hook, *D. carota* L. subsp. *drepanensis* (Arcang.), and *D. carota* subsp. *major* (Vis.) Arcang. (Haywood, 1983; USDA-ARS, 2023).

By the year 1100, yellow and purple carrots were known in Spain, and by the 1200s, they had spread to Italy and China. During the 1300s, white, yellow, and red types were grown in France, Germany, and the Netherlands. By the 1500s, orange-colored roots were first reported, probably from a chance mutation, and carrots were known throughout Europe (Banga, 1957, 1963).

Plant breeding during the 17th century in the Netherlands improved root smoothness and orange color leading to the Long Orange and Horn-type carrots (Banga, 1957, 1963). Louis de Vilmorin, who laid the foundations for modern plant breeding, developed Nantes and Chantenay carrots in France during the 19th century. These types are still popular today (Banga, 1957, 1963).

Over the centuries, orange cultivars have dominated in North America, Europe, and elsewhere. Specific shapes and root colors such as purple, yellow, and white are preferred in certain markets.

Carrot cultivars are sometimes grouped into two distinct classes: Eastern or Asian, and Western or European. European cultivars have strong biennial characteristics and require longer exposure to vernalizing temperatures. European cultivars are firm-textured, sweet, highly flavored, yellow-orange to bright orange in color, slow bolting, and grow best under cool conditions.

In contrast, Asian cultivars tend to have softer texture, are less sweet and flavored, prefer warm temperatures, bolt easily with minimal exposure to cold temperatures, and may be red or red-orange colored. Tropical Asian and South American cultivars exhibit an annual habit and appear to have a short-day preference for flowering (Rubatzky and Yamaguchi, 1997).

Western cultivars are often classified into three categories based on root shape and length. Older short-rooted or baby cultivars, which mature quickly (50 to 60 days) for early-season fresh market production, examples include heirlooms Oxheart, Parisian Market, Early Nantes, and Amsterdam Forcing. Medium-rooted cultivars are grown for main-season production for fresh market and processing. Medium cultivars mature in 60 to 75 days from seeding and examples include Berlicum, Flakkee, Autumn King, Danvers, and Royal Chantenay. Long-rooted cultivars such as Imperator are usually grown in thoroughly tilled deep sandy loam or muck soils and mature 60 to 75 days from seeding.

Cultivars derived from Imperator or which show similar characteristics are often referred to as Imperator types.

In addition to color, there is also a diversity of carrot root shapes and sizes across commercial markets (Rubatzky and Yamaguchi, 1997). Modern plant breeding efforts include the development of F1 hybrids that combine favorable traits from the various types to improve taste, texture, color, sweetness, diseases, and insect resistance. Specific market applications include developing cultivars for bunching, cello-pack, fresh market processing like cut-and-peel. Cultivars are also developed for traditional processing such as dicing, slicing, and more recent markets like juicing. Within each carrot type, modern cultivars include a range of maturities to meet the needs of the market.

Cultivars are often grouped into types that reflect similar morphology or horticultural characteristics. Many modern carrot cultivars are relatives of older successful commercial cultivars now referred to as carrot types or classes of carrot roots. Some of the more common commercial European types include the following:

Amsterdam carrots are slender cylindrical, fine rooted, and are for forcing, bunching, baby carrots, and processing in Europe. Root lengths range from 12 to 16 cm (4 to 6 in.). A popular older variety is Amsterdam Forcing. The cultivars are typically early maturity, and some are planted in the summer and can be overwintered. Sometimes the Amsterdam carrot is crossed with the Nantes cultivar to develop early varieties that contain favorable characteristics of both types.

Berlicum carrots are a heavier cylindrical root with a length of 20 to 25 cm (8 to 10 in.). They are generally main-season carrots with long storage capability and can be overwintered in areas with mild winters. They are used for the cello-pack market, processing, and juicing primarily in Europe.

Nantes is nearly cylindrical in shape, blunt and rounded at both the top and tip. Root lengths range from 18 to 23 cm (7 to 9 in.). Shorter cultivars Nantes Half Long and Early Nantes are special selections. Cultivars of Nantes types as are known for their flavor and eating quality. Nantes cultivars tend to crack when they are dropped on hard surfaces during mechanical harvesting or handling. Nantes is the most popular carrot type in Europe and has the largest number of cultivars with early-, medium-, and late-season applications. Most modern Nantes cultivars are used for a wide range of fresh market applications. There are also tropical cultivars available which are popular in countries like Brazil.

Chantenay have broad shoulders and taper towards a blunt, rounded tip. Chantenay roots are shorter than other cultivars with an average length of 15 to 20 cm (6 to 8 in.), but with diameter up to 8 cm (3 in.). Chantenay grow better on heavy soils compared to some other cultivars. Chantenay cultivars have been used primary in the traditional processing industry in North America. Modern hybrids are being used for European markets and are crossed with Kuroda carrots in Asia.

Danvers roots have a conical shape, well-defined shoulders, and taper to a point at the tip similar to Chantenay but less broad. The average Danvers root length is 15 to 20 cm (6 to 8 in.) and up to 5 cm (2 in.) in diameter. Danvers cultivars are somewhat tolerant of heavy soil and are used mostly for traditional processing. Danvers has become a lesser recognized carrot type and is often incorporated in crossing with other carrot types to enhance their characteristics depending on the market needs.

Flakkee carrots are 20 to 25 cm (8 to 10 in.) long and are conic, stump rooted, smoother skinned compared to Berlicum types, and have become longer and more slender than original cultivars. An early open-pollinated cultivar was Autumn King. They are used for fresh market, processing, and storage and there are several F1 hybrids available. Flakee has become an important main crop in Northwest Europe.

Imperator resulted from a cross between Nantes and Chantenay. It was introduced in the 1920s. Imperator roots are 25 to 30 cm (10 to 12 in.) longer than other cultivars and taper to a point. They are very durable and withstand mechanical damage well, so they are popular for long-distance shipping. The Imperator carrots have been the key focus of breeding programs in the USA stimulated by introduction of the cut-and-peel carrots as a convenient fresh product for the consumer. It represents most of the F1 cultivars in the commercial seed market in North America. Modern plant breeding has incorporated traits from other carrot types to improve texture, sweetness, and color as well as other factors. The roots of these cultivars vary from long tapered to semi-cylindrical. Imperator carrots are also used for the cello-market, bunching, juicing, and other fresh processing markets and can withstand mechanical harvesting.

Paris Market is an early-season, small round carrot used primarily in Europe by home gardeners and for forcing. Its roots are approximately 5 cm (2 in.) in diameter. Originally selected by home gardeners as Parisian Forcing, it first appeared in a seed catalog by Vilmorin-Andrieux in 1888–89 (Banga, 1963). They perform well in shallow soils but can be sensitive to splitting with uneven soil moisture. There are some modern open-pollinated cultivars available serving a small specialty market.

Colored carrots have been a recent interest and there are now red, yellow, white, and purple F1 hybrids of mostly Imperator or Nantes types available for commercial use and home gardeners.

For the Asian market the Kuroda carrot is a popular type and is the focus of several breeding programs in Japan, Korea, China, and India. The roots are smooth skinned, conical in shape, and are about 20 cm (8 in) long. They are used for fresh market, processing, and juicing. The Kuroda germplasm tends to bolt easier during the growing season and requires particular attention to cultural practices. There are both temperate and tropical F1 hybrids available. The Kuroda carrot is also crossed with Nantes and Chantenay parent lines to improve cultivar characteristics.

There has been considerable evolution of celery cultivars over the past 100 years. Although older cultivars like Golden Self Blanching and Pascal may be grown in some areas as heirloom cultivars, most modern stalk celery cultivars sometimes called Utah types have longer, straighter petioles, milder flavor, and more uniform dark green color. Traits that differentiate cultivars include growth rate, color, number of petioles produced per head, degree of suckering, plant diameter, petiole shape and depth of curvature, heart development, stalk compactness, susceptibility to bolting, keeping ability both in the field and during shipment, petiole length, and susceptibility to disease and physiological disorders.

F1 hybrid celery cultivars are a recent development and are gaining interest in some fresh market production areas. There are also cultivars with pink or red coloration that are used for flavoring and decoration but not eaten fresh (Rubatzky and Yamaguchi, 1997).

Growth and Development

Both root and foliage growth for carrots is optimum between 16° and 21°C (61° to 70°F). Growth slows at temperatures below 10°C (50°F) (Lorenz and Maynard, 1988). Temperatures above 21°C (70°F) tend to make roots short and thickened, while temperatures below 16°C (61°F) tend to favor long slender roots. Foliage growth is less affected by temperature than root growth. The foliage tolerates higher temperatures, but above 30°C (86°F) leaf development is inhibited, and roots have stronger flavor. Large diurnal temperature fluctuations are conducive to rapid growth. In regions with cool night temperatures, carrots may be grown in the tropics (Rubatzky and Yamaguchi, 1997). The length and the thickness of the roots decrease with increasing density.

Carrot is a half-hardy vegetable meaning acclimated plants resist light freezes of short duration (Lorenz and Maynard, 1988). Young seedlings tolerate low temperatures and frosts better than older plants. Carrot seedlings can survive frosts and temperatures as low as −6.5°C (20°F) if properly acclimated. Leaf growth is slow below 4°C (39.2°F). Crowns are killed by exposure to prolonged subfreezing temperatures (Rosenfeld *et al.*, 1997). Soil tends to buffer roots from fluctuating low air temperatures that damage leaves so the crown can continue to produce new growth. Once the soil is frozen, crown damage is visible as small hairline and horizontal cracks on the surface of the roots.

The optimum temperature range for carotene synthesis, the orange pigment in root tissue, is from 16°C to 25°C (61°F to 77°F) and poor below 16°C (61°F) or above 25°C (77°F). Pigment synthesis occurs after root growth, a reason why young roots are pale. With continuing root growth carotene accumulates, attaining a maximum after about 90–120 days and then usually remains constant or may slowly decline (Rubatzky and Yamaguchi, 1997).

Carrots are vernalized below 10°C (50°F) for two to 12 weeks depending on the cultivar, with bolt-resistant cultivars needing longer cold exposure. Some tropical Asian cultivars can be induced to bolt at temperatures less than 15°C (59°F). Plants must have developed to the mature stage that roughly correlates to eight or more leaves and a diameter greater than a pencil (4–8 mm, 0.16–0.31 in.) (Dickson and Peterson, 1958). The juvenile stage of development lasts longer for large-rooted cultivars. Seeds mature four to five months after plants have been vernalized (Rubatzky and Yamaguchi, 1997).

Celery is challenging to grow as a seed crop because of its slow growth and specific environmental requirements. Fertile soils and a temperature range between 16°C and 21°C (61°F to 70°F) are optimum for celery growth (Rubatzky *et al.*, 1999). Some modern cultivars show greater tolerance of high temperatures. Celery is sensitive to freezing temperatures, while celeriac and smallage are more tolerant. When acclimated, celery can tolerate a light frost of −2°C (28°F) for a short time with little or no damage (Rubatzky *et al.*, 1999).

When grown continuously at 15°C (59°F) or higher, celery plants remain vegetative. Seedlings with less than four or five true leaves are considered juvenile and insensitive to vernalization. Bolting is initiated when mature plants with at least six to eight or more leaves are vernalized by low temperatures, usually < 10°C (50°F). Thresholds for vernalization temperature and duration vary among cultivars. Some tolerate low temperatures and long exposure better, while others are sensitive to vernalization at temperatures as high as 13°C to 14°C (55°F to 57°F). Plant stress can also increase sensitivity to vernalization (Rubatzky *et al.*, 1999).

Celery seed fields should be planted so that mature plants are exposed to < 15°C (59°F) for several weeks to ensure all plants are vernalized. Acclimating transplants to warm temperatures of 25°C to 30°C (77°F to 86°F) for 10 to 20 days prior to transplanting reduces transplant sensitivity to bolting and may prevent premature flowering (Rubatzky and Yamaguchi, 1997).

Soil

Ideal soils for carrot seed crops are deep, friable, fertile, well-drained sandy loams. Soils with high clay content may crust when dry, impairing emergence. Shallow-rooted cultivars are more tolerant of heavier soils. Long-rooted cultivars are markedly shortened by shallow or compacted soils. Compacted soils change root shape making it difficult to identify off-types. Direct-seeded carrots for seed production should be planted in lighter soils to minimize heaving and cracking during the freezing temperatures in the winter. Freezing temperatures can cause the roots to be exposed to freeze injury and dehydration, resulting in stand loss. Fields selected should be sloping to provide good air drainage and the potential for standing water should be avoided. In areas with light soils and prone to windy conditions, fields may require planting with spring barley into blank rows to provide protection from wind damage. Root-to-seed production is preferred in areas where there is higher risk due to freezing and loss of stands from seed-to-seed carrots. After bolting, tall plants must be well anchored to prevent lodging under windy conditions particularly when soils are wet. Carrots are tolerant of a wide range of soil pHs but generally a range of 5.8 to 6.8 is preferred.

Celery requires relatively high levels of nutrients, especially nitrogen. Fertile fine-textured loam, or silt loam soils are preferred. Celery prefers a soil pH of around 6.5 on mineral soils. Celery needs a consistent supply of water. Poorly drained or waterlogged soils should be avoided (Rubatzky *et al.*, 1999).

Fertilization

Carrot seed crop fertilization requirements vary dramatically with soil type and cultivar. Applications should be based on test results and periodic foliar analysis throughout the season. Soil nitrate testing prior to planting assesses N fertility but values are transitory. On deficient soils, a split application of 56 to 78 kg/ha (50 to 70 lbs/acre) N is applied preplant to support field establishment and early growth (Lorenz and Maynard, 1988). Excessive nitrogen fertilization tends to promote foliage growth over root development, leading to lodging of flowering plants that are not properly anchored. To provide increased N needed for reproductive growth, the remaining N may be applied as side dressing when plants bolt (Lorenz and Maynard, 1988).

If fertigation is used, all phosphate and some N and K may be applied preplant. In this system, the bulk of N and K is applied during weekly fertigations often in conjunction with foliar analyses.

Celery seed crops established after other crops may benefit from residual N depending on soil and environment. Nitrate levels greater than 20 ppm in the top 30 cm (12 in.) of soil are adequate for celery establishment and early growth (Daugovish *et al.*, 2008). Little biomass is produced during the first month of establishment either by transplants or seeding. If soil tests are low, 22 to 34 kg/ha (20 to 30 lbs/acre) N is applied preplant to support establishment and early growth (Lorenz and Maynard, 1988). Excess N stimulates vegetative growth at the expense of reproductive growth. However, N requirements increase with maturity, especially at the time of bolting to support reproductive growth. During this growth phase, 17 to 22 kg/ha (15 to 20 lbs/acre) of N per week are required and may be added through drip irrigation or alternatively side dressing if fertigation is not available (Daugovish *et al.*, 2008). A celery seed crop, including seeds and biomass, removes approximately 280 kg/ha (250 lbs/acre) N from soil.

Phosphate fertilization should be based on soil testing levels of extractable phosphorus. Levels above 60 ppm are adequate for growth. For lower test results, 45 to 90 kg/ha (40 to 80 lbs/acre) of P_2O_5 are recommended, particularly prior to planting in cold soils (Daugovish *et al.*, 2008). Phosphate uptake by a celery seed crop, including both seeds and biomass, is in a range from 56 to 67 kg/ha (50 to 60 lbs/acre).

Potassium fertilization needs are determined by soil testing before any nutrients are applied. Carrots have a high requirement for K. An average carrot seed crop, including plant biomass, removes about 381 kg/ha (345 lbs/acre) K from the soil. Soils with greater than 150 mg/kg (150 ppm) of ammonium-acetate-exchangeable potassium are sufficient for carrot seed production. A split application of K at planting with a side dressing/fertigation of K at bolting is common practice (Daugovish *et al.*, 2008). Alternatively, if fertigation is available little or no K may be applied at planting with the bulk applied weekly during irrigation. Potassium uptake by a celery seed crop ranges from 390 to 505 kg/ha (350 to 450 lbs /acre).

Celery is subject to nutrient deficiencies besides N-P-K. Calcium deficiency, particularly in conjunction with drought stress, causes the physiological disorder "black heart", which kills the central apical meristem (Daugovish *et al.*, 2008). Boron deficiency results in lateral stem cracking across petioles (Rubatzky *et al.*, 1999). Fertilizer supplemented with boron is common in areas where soil levels are chronically low. Zinc fertilization is recommended if extractable soil levels are less than 1.5 mg/kg (1.5 ppm). Zinc fertilizer is routinely added to soils with high phosphorus, which reduces zinc uptake (Daugovish *et al.*, 2008). Insufficient magnesium is a cause of leaf chlorosis.

Pollination

Celery and carrot flowers have both male and female flower parts. However, they are protandrous, so flower parts don't develop simultaneously. Anthers mature and senesce before pistils are receptive to pollen. Protandry is a strategy to increase cross-pollination by insects, primarily bees. Pollen may come from different plants (cross-pollination) or from later maturing florets on umbels from the same plant (self-pollination). Since a high percentage of flowers naturally cross-pollinate, there is considerable genetic variation among seeds in open-pollinated cultivars.

Wild carrot is classified as a noxious weed in some areas (USDA, 2023). Wild carrot pollen readily crosses with the cultivated carrot and may contaminate seed fields. Contamination of seed crops by wild carrot pollen is a major reason for genetic deterioration of seed stocks in some areas of the world. Wild carrots should be eliminated from spaces surrounding seed fields to prevent unwanted crossing. Because of the high possibility of cross-pollination with other carrot seed fields and wild populations, isolation distances for commercial carrot and celery seed crops should be a minimum of 8 km (5 miles). Isolation distances between fields is determined by shape and type of carrots of the same color with distances ranging from 0.4 to 3.2 km (0.25 to 2.0 miles) to prevent cross-cultivar contamination. Carrot cultivars of a different root color like red, yellow, purple, or white may require up to 8 km (5 miles) from the common orange color.

In regions specializing in carrot seed production, cultivars of the same root type are grown in specified zones, while maintaining the above separation distances, to minimize pollen contamination.

Hives should be placed in carrot and celery seed fields as soon as the first flowers open and for hybrid seed production, placement should begin when the male pollinator is at 5 to 10% bloom. Placement should be near the middle of small fields of 1 ha (2.4 acre) or less. For larger fields, placement in groups of three hives should be uniformly distributed around the field and downwind, when possible, from another flower source. A minimum of 10 strong hives/ha (4 hives/acre) ensures adequate pollination; however, as many as 25 hives/ha (10 hives/acre) may be required depending on the pollen availability, natural insects, hive strength, and competing crops in the production area. Hives should remain until after the completion of flowering of the tertiary umbels for open-pollinated production or the male-sterile line for hybrid production. Honeybees tend to pollinate up and down the same row rather than randomly across rows. Such pollen foraging patterns can affect F1 hybrid pollination since bees are more likely to visit the male-sterile row rather than cross-pollinate both male and female parents (Cresswell *et al.*, 1995). Increasing bee populations or planting male-sterile and male-fertile lines closer can improve cross-pollination between parents. After hives are taken out of a field, male-fertile lines can be removed by flailing or mowing.

Irrigation

Carrot and celery seed production are grown in areas where there is low annual rainfall with no rain during the active growing season. Uniform soil moisture until maturation drying begins is essential to reduce water stress and support uniform growth of carrot and celery seed crops. Water stress early in development reduces photosynthesis leaf area, lowering seed yields. Severe drought stress at pollination may reduce seed set and increase abortions after fertilization. In areas with insufficient rainfall, overhead irrigation benefits crop establishment, activates dry fertilizer applications, and prevents soil crusting but increases foliar disease pressure and the potential for seed-borne pathogens. Furrow irrigation can be used where water is plentiful but can be problematic with certain pathogens causing root-related diseases and has become less preferred. A growing trend is to install drip irrigation to avoid foliar disease and to provide uniform soil moisture and save water. Carrot seed crops require 30 to 50 mm (1.2 to 2.0 in.) of water per week or 450 to 600 mm (17.7 to 23.6 in.) to produce a seed crop depending on soil type and evapotranspiration (Rubatzky and Yamaguchi, 1997). Irrigation is needed whenever about 40% of available moisture has been depleted from the root zone. Replacing 100% of evaporative losses results in the highest yields. Excessive soil moisture can be detrimental as well. Too much water increases disease and lodging of flowering plants. To ensure best drainage for root development, seed crops are often grown on raised beds.

Rogueing

For seed-to-seed carrot production, only rogueing of above ground plant tissue is possible since the crop is grown in the same field without root inspection. Thus, only parent seed with known genetic quality meeting high quality standards should be used for seed production. Seed-to-seed fields should be inspected to remove early bolting, atypical, or diseased plants.

The advantage of the root-to-seed method is that roots are dug and inspected. This allows off-type roots to be identified and discarded before seed is produced. During the vegetative growth stage, plants displaying atypical foliage should be removed. When producing carrots with strong biennial characteristics, plants that bolt prematurely without vernalization should be rogued. After the roots have been dug, they should be carefully inspected for trueness-to-type, according to root shape, color, and size. Discard roots showing poor color, incorrect color, colored shoulders (purple or green), split, twisted or forked roots, and roots with rough surfaces. To adequately inspect the harvested roots the roots should be grown in deep, well-drained, sandy loam soils. Heavy soils may cause off-type root phenotypes rather than plant genetics. Root color may also be affected by age and temperature.

Celery seed fields are usually rogued to evaluate vegetative growth and again after seed stalk formation following vernalization. During vegetative field growth, bolters and plants with abnormal leaves (unusual leaf color, size, and length) should be removed. Petiole characteristics should be scrutinized for shape, size, length, and color. Additional undesirable vegetative traits include basal sucker shoots, narrow hearts, and open stalks. Characteristics to examine during stalk maturity and reproductive growth include susceptibility to *Septoria apiicola* Speg. and other seed-borne pathogens, plant vigor, petiole characters, such as width of transverse sections, petiole ribbing, and general petiole characters including pithiness.

Genetics

Researchers have traditionally developed new cultivars of carrot and celery using traditional sexual-based crosses and selection methods. Although technically feasible, genetically engineered carrot and celery cultivars have yet to significantly impact global markets due to high developmental costs, lengthy and expensive government approval, and lack of consumer interest. There has been increased adaption of F1 hybrid carrot and to a lesser extent celery cultivars for commercial production because of their greater uniformity and heterosis. Use of cytoplasmic male sterility (CMS) in one inbred parent ensures cross-pollination for F1 hybrid seed production. Even though F1 hybrid carrots are now preferred in most major commercial production areas, open-pollinated carrot cultivars are important in much of the world. Open-pollinated cultivars remain popular when the extra cost of F1 seeds is not worth the investment in uses such as for traditional processing.

Seed Production

Although carrot production for consumption can be produced under a wide range of climatic conditions, the requirements for seed production are more specific. Seed production takes place in areas where annual rainfall is 250 to 350 mm (10 to 15 in.) with no rain during the active growing season with low relative humidity of 60 to 80%. Summer temperatures should not exceed 40°C (105°F) during pollination and seed development. Major seed production areas include the US Pacific Northwest in Oregon, Idaho, and Washington, the Mendoza and San Juan areas of Argentina, central Chile, southeastern Australia and Tasmania, South Africa, France, and Italy.

Two methods are used to produce carrot seed: root-to-seed and seed-to-seed. The seed-to-seed method is the most straightforward. Stock seed is direct-seeded in late summer. For seed-to-seed production, roots overwinter, are vernalized in the field, flower in spring, and produce seed that is harvested the following summer. Seed-to-seed production requires that plants are past the juvenility stage but before full maturity prior to vernalizing conditions. Larger roots at full maturity are subject to freeze injury potentially resulting in crop loss. The field environment must be sufficiently cold to satisfy vernalization requirement but not so cold as to kill plants and ideally between 0°C and 4°C (32°F to 39°F) and extreme temperatures below −18°C (0°F) will cause freeze damage. Sometimes row covers are used on temperature-sensitive parent lines for hybrid production. The duration of the cold period should be sufficient to vernalize the crop. In addition, temperatures should be consistently cold during vernalization without warm devernalizing periods.

The seed-to-seed method is less expensive and higher plant densities are used. The seed crop is planted in early fall with adequate growing season for the carrots to reach adequate size for vernalization. The maximum seeding rate in the fall is 15 to 20 live seeds/meter to achieve a final stand of 10 to 12 plants/meter for a stand of 110,150 plants/ha (45,000 plants/acre). Thinning is required when stands the following spring exceed the target population. However, since roots are not lifted for rogueing, the seed produced may be more genetically variable. High-quality stock seed is important because there are fewer opportunities to remove off-types during seed-to-seed production.

For root-to-seed production, stock seed is established in late summer or early fall in areas where growing conditions are ideal. Seed is sown to achieve plants of 3 to 5 cm (1 to 2 in.)

apart in the row with optimum population dependent on type.

Developed roots called stecklings are dug prior to full maturity (Fig. 8.4) and rogued for off-types and diseases in fall or late winter. Depending on winter conditions, the stecklings may be vernalized *in situ* or in cold storage (Spurr, 2003). For roots placed in cold storage for vernalization, temperatures of –1°C to –0.5°C (30°F to 33°F) at 95 to 100% relative humidity with adequate air circulation should be used to maintain high-quality roots. The time in storage depends on the type of carrot and typically varies from four to eight weeks. Vernalized stecklings are transplanted into carrot seed fields the following spring, produce foliage, and bolt.

The spacing of the transplanted roots is 10 to 30 cm (4 to 12 in.) in the row with a target population of 75,000 to 115,000 plants/ha (30,000 to 46,500 plants/acre) depending on type of carrot (Fig. 8.5). Although more expensive, an advantage of root-to-seed production is the reliability of consistent seed yields compared to seed-to-seed production which can completely freeze out in some winters. Seeds develop and mature for harvest at the end of the second summer.

Although F1 hybrid carrot seed can be produced using the seed-to-seed or root-to-seed methods, growth of both inbred parents must be synchronized, also called nicking, so they flower simultaneously for successful cross-pollination by insects. Inbred lines may develop at different rates, so they must be established separately, either by seeding or stecklings, or by trimming the early parent to optimize nicking. Growth regulators such as gibberellic acid (GA_3) may be applied to speed bolting to improve nicking, although this is more of an emergency rather than a standard practice (Nieuwhof, 1984; Bandara and Tanino, 1995). The ratio of male-to-female rows for carrot and celery F1 hybrid production varies from two male for every four to six female rows depending on the male parent vigor and pollen production. Removal of beehives is after pollination is complete and there is visible seed set. Typically, the

Fig. 8.4. Carrot root harvest for root-to-seed production. (Courtesy of Tim Loeffler.)

Fig. 8.5. Vernalized stecklings planted for seed production after inspection for off-types. (Courtesy of Tim Loeffler.)

beehives are removed before pollination of the quaternary umbels since the seed produced is very small and unmarketable. Male rows are removed after the flowers have senesced prior to seed harvest (Fig. 8.6). For open-pollinated cultivars, the whole field is harvested at maturity.

To ensure good drainage carrot seed crops are often produced on raised beds. Some common bed sizes vary in height and width ranging from 1.2 to 1.9 m (4.0 to 6.2 ft) wide. Seed/steckling spacing varies depending on cultivar size and vigor. Weak inbred lines of carrot and celery may be planted as close as 10 to 15 mm (4 to 6 in.) in-row. More vigorous plants can be spaced 31 to 46 cm (12 to 18 in.) in-row. Closer spacing increases the percentage of primary umbels and suppresses formation of lower order umbels in a seed field (Oliva *et al.*, 1988).

Poor seeding performance is one of the biggest problems in establishing carrot and celery seed crops. For seed-to-seed carrot production sometimes coated or pelleted stock seeds are precision-planted to a final stand using belt and vacuum seeding equipment. Carrots are direct-seeded to a depth of 1.3 to 1.9 cm (0.5 to 0.75 in.). Because of the difficulty in establishing celery, plug transplants are used to improve crop uniformity (Fig. 8.7). Carrot stecklings are hand-transplanted or by using mechanical transplanters so only the terminal shoot is exposed or slightly covered (Fig. 8.5). Carrot seeds are very small, slow to germinate, and field emergence percentages are often at best only 80 to 90%. In addition to inferior quality seed, poor stands may occur because seedlings cannot grow properly due to soil crusting, adverse temperatures, weed competition, and dry conditions. Carrots germinate at soil temperatures above 4.4°C (40°F) but optimum germination occurs at 26.7°C (80°F). Carrot seed thermodormancy is another factor that may reduce germination at high temperatures (Nascimento *et al.*, 2008).

Fig. 8.6. Peak bloom in hybrid carrot production field. The two male rows in the center will be removed before F1 seed is harvested from the remaining rows. (Courtesy of Tim Loeffler.)

In areas where soil temperatures are high at planting, the fields are usually irrigated using overhead systems until seed germination has started. After planting, adequate moisture for uniform emergence, to prevent crusting, and support early seedling growth is critical. Seeding growth even for high-quality carrot and celery seeds is slow because emergence can vary from seven to more than 20 days and the first true leaves may not develop for three to four weeks after planting. Late-emerging seedlings may not reach the critical stage for adequate vernalization resulting in slower bolting and later flowering plants and create nicking problems compared to the first plants that emerge.

Transplanting can improve establishment when stock seed quality is poor, environmental conditions are less than optimum, or inbred lines are weak. Transplants may also be used to improve nicking of celery (Fig. 8.7).

Carrot and celery seed quality is variable by nature because seeds form on separate umbels that mature at different times. The first umbel to form is called the primary and produces larger seed. This is because the primary umbel is less source limited since its seeds form first. Secondary, tertiary, and quaternary umbels form later in the season and are derived from axillary buds lower on the flower stalk. Seed size and quality decrease with increasing umbel order. Seeds from a quaternary umbel are smaller and develop later than those on the primary umbel and usually are not marketable. Studies show that embryos of seeds formed on lower order umbels are less developed than on the primary umbel. If a grower waits for seeds on lower order umbels to develop, the higher-quality seeds on the primary umbel become overly mature and may be lost due to shattering. So, determining the best time to harvest celery and carrot seed crops is problematic (Fig. 8.8).

Opting for the highest seed yield means more risk with lower-quality seed since it is composed of smaller seeds from the lower order umbels. Plant spacing, as detailed above, can contribute to increased seed quality and reduce the spread in umbel maturities. Higher plant populations reduce the number of lower order umbels produced per plant. Each plant may produce only two umbels, one primary and a

Fig. 8.7. Celery transplants being established using a carousel planter to improve uniformity for open-pollinated seed production.

Fig. 8.8. Carrot maturation with male rows removed. (Courtesy of Tim Loeffler.)

secondary, so the number of seeds produced by each plant is reduced. But across the entire field, more primary umbels are produced so seed quality is improved and overall yield is similar (Oliva *et al.*, 1988).

Seed Harvest

Different harvesting procedures are employed for carrot and celery seed crops with varying levels of mechanization. Celery seed is notorious for preharvest shattering even under optimal conditions. Valuable seed can be lost if bad weather occurs near harvest, or the cutting time is misjudged. In some cases, breeder or stock seed is produced in tunnels and umbels are hand-harvested particularly in areas where rain near harvest may damage the crop. For small-scale commercial outdoor production, carrot and celery umbels are cut by hand as they ripen. If the umbels are fully mature and dry, they can be threshed directly or final dried on tarpaulins prior to threshing.

For larger scale carrot and celery seed harvest, entire plants are cut with a sickle bar mower and windrowed to allow final drying in a field to reduce shattering before harvest. Using this method, a decision must be made about the best cutting time for maximum seed yield. There is a tendency for some seed loss due to shattering from primary umbels if cutting is delayed allowing maturity of lower order umbels. Generally, the crop is cut when the second order umbels are a straw brown color, and the third umbel order is changing from green to straw color (Fig. 8.9). Cutting plants early in the day when the dew is still present will also reduce shattering. The cut crop continues to dry in windrows. Fully matured and dry seed is brown or gray when ripe and the umbel brittle. Plants can be sprayed with polyvinyl acetate to prevent shattering losses from primary umbels, although this is not a standard practice and may cause problems during seed conditioning. Typically, the optimum timing for harvest is when the seed moisture content reaches 9 to 14% which is typically three to four days after the crop has been windrowed.

For larger fields, celery and carrot seed crops are harvested by direct combining (Fig. 8.10). As with the windrowing method, combining should be done early in the day or while there is still some toughness to the stems to take advantage of higher plant moisture to reduce losses from shattering. Celery and carrot plants must not be too dry or in addition to seed loss, small

Fig. 8.9. Maturing primary umbel during the later stages of development shortly before seed drop. With maturity, the umbel color changes from white to pale yellow-green. At maturity, the outer umbellets curve inward and the brown umbels develop a cup shape.

Fig. 8.10. Carrot seed windrow harvest by combining. (Courtesy of Tim Loeffler.)

plant debris will accumulate with the harvested seed complicating post-harvest seed cleaning. The reel and beaters on the combine intake are removed to reduce shattering as plants enter the combine. Certain cultivar types as well as the less mature seeds are more subject to mechanical damage, and it is important to use proper cylinder speeds during seed threshing. Appropriately sized screens must be installed in the combine to prevent excessive seed loss. When combine harvesting is not feasible, the dry material may be processed using a stationary thresher.

Post-harvest Seed Conditioning

Carrot and celery (as well as dill, caraway, and others in this family) have spines or "beards" on the sides and end of the seed. These spines must be removed using a debearder before further cleaning operations. Debearding improves seed flow and reduces the volume of a seed lot. Once the crop has been debearded and precleaned, any seed above 7–8% seed moisture content should be dried to preserve seed shelf life. Further cleaning is achieved by air screen machines, screens, air column tunnels, gravity tables, and indent cylinders.

When celery umbels are dried on tarpaulins, especially under cover, the seed separates naturally from the head without machine threshing. Seed is further cleaned by an aspirated screen separator, but because the seed is very small, care must be taken in selection of the bottom screen.

Seed Yield

Carrot seeds are flat, ribbed, with small spines, and vary greatly in size depending on where they developed on the mother plant (Oliva *et al.*, 1988). Carrot seed weight is in the range of 400–1000 seeds/g or 11,300–34,000 seeds/oz. Open-pollinated carrot cultivars grown in the temperate regions should produce 600 kg/ha

(535 lbs/acre) with highest yields reaching 1200 kg/ha (1070 lbs/acre). Seed yields in tropical regions are usually lower despite vernalization at higher altitudes. High altitude tropical production of about 300 kg seed/ha (268 lbs seed/acre) can be expected for European types. The Asiatic types may yield less at about 250 kg seed/ha (223 lbs seed/acre) when produced in the tropics. Hybrid seed production is lower yielding compared to open-pollinated cultivars of the same type with yields ranging approximately from 140 to 280 kg/ha (125 to 250 lbs/acre). Seed yields of hybrid seeds is influenced by parent line, the female-to-male planting ratio, and pollination.

The oval celery seeds are smaller than carrot with approximately 2500 seeds weighing 1 g (0.04 oz) and 1000 seeds weight has been reported to be 0.5 g (0.02 oz) (Oliva *et al.*, 1988; Rubatzky and Yamaguchi, 1997). Like carrot, celery seed size and weight vary depending on umbel development position. Dried celery seed is also extensively used as a flavoring agent in cooking because of its stronger aroma and flavor compared with celery petioles (Ag Marketing Resource Center, 2023). An average seed yield for open-pollinated celery is about 500 kg seed/ha (446 lbs/acre).

Pest Management

Weeds

Carrot and celery do not compete well with weeds because of their slow early growth and small canopies. Integrated weed management strategies should be implemented prior to planting celery and carrot. It is best to locate production fields where weeds have historically not been a problem. Eradicating or mowing weedy areas surrounding fields can prevent weed seed from spreading by wind. Fields with a history of Canada thistle (*Cirsium arvense*) or dodder (*Cuscuta* spp.) are noxious weeds in many areas, and barnyard grass (*Echinochloa crusgalli*) should be avoided as these weeds are difficult to remove during the seed-cleaning process. Seed fields with the presence of these weeds should be rogued prior to swathing the crop when possible. Seed lots containing these weeds will be unusable and fields contaminated with these weeds should be controlled or hand-weeded when necessary. A weed management program usually includes: crop rotation, removal of weeds before they produce seed, preplant irrigation to stimulate germination followed by cultivation to kill the first flush of weeds, timing seed crop establishment to reduce weed impact, careful preparation and spacing of beds, and precision transplanting so that cultivation/spray equipment can be accurately aligned to a row (Daugovish *et al.*, 2008).

For celery, weed control is most critical until seedlings or transplants are established and form a canopy. Preplant and post-plant selective herbicides are effective against many weeds that can overwhelm celery and carrot fields, but most cannot be used for organic seed crops. Some effective pre-emergence herbicides for carrot and celery seed production include: pendimethalin, haloxyfop, fluazifop-P, or sethoxydim. Please consult local recommendations for herbicide use. Mechanical or hand weeding can be effective but requires labor or expensive equipment that increases cost. Weed-free straw or plastic film provides effective weed control while preserving moisture and modifying soil temperature.

Pathogens

Fungal pathogens responsible for carrot root diseases include Phytophthora root rot (*Phytophtora medicaginis*), violet root rot (*Helicobasidium brebissonii*), white rot (*Sclerotinia sclerotiorum*), gray mold (*Botrytis cinerea*), yellows (*Fusarium* spp.), purple root rot (*Helicobasidium brebissonii*), and Sclerotinia rot (*Sclerotinia sclerotiorum*). *Pythium* spp. are associated with many diseases of carrot, including damping-off, dieback, forking, brown root, lateral root dieback, rubbery slate rot, rusty root, pathogenic cavity spot (*Pythium violae*) and Pythium brown rot (*P. sulcatum*). Root diseases are more severe in heavy soils with poor structure. Control measures for root rot include rotations for at least five years between carrot and related crops, liming to prevent acid conditions, and well-drained soil (Nunez and Davis, 2016; CABI, 2023b; US Nematode Collection, 2023).

Fungal diseases affecting primarily carrot foliage include powdery mildew (*Erysiphe betae* and *E. heraclei*), white rust (*Aecidium foeniculi or Uromyces graminis*), Alternaria leaf blight (*Alternaria dauci*), and Cercospora leaf spot (*Cercospora carotae*) (Takaichi and Oeda, 2000; Nunez and Davis, 2016; CABI, 2023b). Alternaria leaf blight is also seed-borne and seed lots contaminated with this pathogen may require additional post-harvest seed treatments and testing for certain markets and to meet export requirements.

Bacterial diseases include soft rot (*Erwinia chrysanthemi, E. carotovora* subsp. *carotovora*), bacterial leaf blight (*Xanthomonas campestris* pv. *carotae*), carrot bacteriosis (*Xanthomonas campestris*), crown gall (*Agrobacterium tumefaciens*), hairy root (*Agrobacterium rhizogenes*), milky disease (*Bacillus popilliae*), and scab (*Streptomyces scabiei*) (Nunez and Davis, 2016; CABI, 2023b; US Nematode Collection, 2023). Bacterial leaf blight is also seed-borne and seed lots contaminated with this pathogen require post-harvest seed sanitation treatments and testing. Contaminated seed lots are not suitable for most commercial markets or for export.

There are many virus and viroid diseases that affect carrot. The list includes: alfalfa mosaic genus *Alfamovirus* (AMV), carrot latent genus *Nucleorhabdovirus* (CtLtV), carrot mottle genus *Umbravirus* (CMoV), carrot red leaf genus *Luteovirus* (CaRLV), carrot thin leaf genus *Potyvirus* (CTLV), carrot yellow leaf (CYLV), celery mosaic genus *Potyvirus* (CeMV), cucumber mosaic genus *Cucumovirus* (CMV), beet curly top genus *Hybrigeminivirus* (BCTV). Other viruses are those associated with carrot motley dwarf (CMD) disease, i.e. carrot mottle umbravirus (CMoV) and carrot red leaf virus (CRLV) (Nunez and Davis, 2016).

The following non-pathogenic disorders (and their causes) also affect carrot: crown rot disorder (no known pathogen), heat canker (high soil surface temperature), hollow black heart (boron deficiency), ozone injury (ozone pollution), root scab (physiological disorder), and speckled carrot (genetic disorder) (Nunez and Davis, 2016).

Celery transplants, particularly when grown in a greenhouse, often contract late blight (*Septoria apiicola*) and early blight (*Cercospora apii*). Late blight can be successfully managed with disease-free seed, fungicide treatments, or irrigation systems that keep foliage dry, e.g. drip or subsurface irrigation for transplant production (Daugovish *et al.*, 2008).

Fusarium (*Fusarium oxysporum* f. sp. *apii*) is a devastating soil-borne disease of celery. Several races exist, and the pathogen can persist for years without a host. Short crop rotations are ineffective. Soils are sometimes fumigated to control Fusarium and other celery soil-borne pests, especially in seed production areas where crop rotation options are limited. Pink rot (*Sclerotinia sclerotiorum*) and crater rot (*Rhizoctonia solani*) are soil-borne fungal diseases that affect the lower petiole when conditions are wet (Daugovish *et al.*, 2008). Fungicides sometimes provide effective control against both diseases. Other fungal diseases that attack celery include downy mildew, powdery mildew, crown and root decay of celery, brown spot, crater spot, phoma crown and root rot, southern blight, and violet root rot (Davis and Raid, 2002).

The most important bacterial diseases are bacterial leaf spot of celery (*Pseudomonas syringae*), bacterial blight, and soft rot (*Erwinia carotovora* and *Pseudomonas marginalis*). Limiting foliage moisture, using disease-free seed, and crop rotation can control bacterial disease. Bacterial blight often doesn't persist in celery seed fields.

Celery is susceptible to aster yellows mycoplasma (Daugovish *et al.*, 2008). Celery is also affected by a host of viral diseases with widespread distribution including celery mosaic, cucumber mosaic, calico mosaic, alfalfa mosaic, arabis mosaic, celery latent, celery yellow mosaic, tomato black ring, curly top, beet curly top, parsnip yellow fleck, southern mosaic of celery, strawberry latent ring spot, celery yellow net, strap leaf, and tomato spotted wilt. Control of insect vectors, such as leafhoppers, is essential for controlling viruses (Davis and Raid, 2002).

Insects

Lygus bug (*Lygus hesperus* and *L. elisus*) can devastate carrot and celery seed production in some areas. While lygus may feed on all plant parts, they are particularly damaging to flowers and seeds. Lygus bugs are piercing-sucking insects that feed on umbels and young developing seed. Adults and late-stage nymphs are the most injurious. Lygus may cause 100% loss of seed set or embryo-less seed. One lygus per flower or seed head may cause total crop loss. To control this pest, a strict integrated pest management (IPM) program to monitor the presence of the pest and regular insecticide sprays are followed.

The carrot aphid *Semiaphis dauci* (Aphididae), the willow-carrot aphid (*Cavariella aegopodii*), the carrot sucker (*Trioza apicalis*), and other aphids feed on foliage and vector virus diseases. Thrips (*Frankliniella tritici* and *F. occidentalis*), and flea beetles (*Systena blanda*), also feed on the foliage. The leafhopper (*Macrosteles quadrilineatus*) vectors aster yellows. Lady beetles, syrphid flies, and lacewing are predators of aphids. Aphid parasites should be preserved by avoiding unnecessary insecticide applications and providing habitat for beneficials.

Two serious pests of carrot seed crops are the carrot weevil (*Listronotus oregonensis*) and the carrot root or rust fly (*Psila rosae*). Larvae of both insects create feeding tunnels in the root but in different areas. The carrot weevil larvae tunnels in the upper one-third of the root, while the rust fly maggots tunnel mainly in the lower two-thirds of each root. The turnip moth (*Agrotis segetum*) and cutworms (*Agrotis* spp.) attack carrots. Armyworm (*Spodoptera exigua*) larvae feed on leaves and roots of carrot and celery.

Another pest of carrot roots is nematodes. Nematodes that affect carrot include the lance nematode (*Hoplolaimus uniformis*), lesion nematode (*Pratylenchus penetrans* and *Pratylenchus* spp.), sting nematode (*Belonolaimus longicaudatus*), cyst nematode (*Heterodera carotae*), and the root-knot (*Meloidogyne hapla*). Root-knot nematodes can be controlled by crop rotation and manure applications (Nunez and Davis, 2016). A crop rotation program should be no less than three years with five years preferred in most areas.

Depending on the region and the time of year, the most damaging insects to celery are leafminers, caterpillars, and aphids (Daugovish *et al.*, 2008). The black bean (*Aphis fabae*) and foxglove (*Aulacorthum solani*) and other aphids can damage celery. Several parasitic wasps in the genera *Diaeretiella* and *Lysiphlebus* provide natural control of celery aphids. The aster leafhoppers (*Macrosteles uadrilineatus*), which are vectors for the disease aster yellows, can be problematic (CABI, 2023a). The vegetable leafminer (*Liriomyza sativae*) is also a pest of celery. The primary damage from leafminers is caused by their larvae, which feed on leaf mesophyll tissue, forming hollow spaces in the center of affected leaves. Leafminer larvae affect both celery and carrot seed crops. Leafminers are parasitized by several species of wasp, which naturally suppress populations. Control strategies targeting larvae are generally more successful than the mobile adults, which are sometimes insecticide resistant (Daugovish *et al.*, 2008). When insecticides are used, different chemicals should be rotated to slow insect resistance.

References

Ag Marketing Resource Center (2023) Celery. Agriculture Marketing Resource Center, Iowa State University. www.agmrc.org/commodities-products/vegetables/celery (accessed 6 September 2023).

Atherton, J.G., Craigon, J. and Basher, E.A. (1990) Flowering and bolting in carrot. I. Juvenility, cardinal temperatures and thermal times for vernalization. *Journal of Horticultural Science* 65, 423–429.

Bandara, P.M.S. and Tanino, K.K. (1995) Effect of gibberellic acid (GA_3) on seed stalk development and in carrot (*Dauca carota* L. var. *sativa* DC). *Acta Horticulturae* 394, 251–261.

Banga, O. (1957) Origin of the European cultivated carrot. *Euphytica* 6, 54–63.

Banga, O. (1963) *Main Types of Western Carotene Carrot and Their Origin*. Tjeenk Willink, Zwolle, the Netherlands.

CABI (2023a) *Apium graveolens* (celery). Associated diseases. In: *Crop Protection Compendium*. CAB International, Wallingford, UK. https://www.cabidigitallibrary.org/doi/10.1079/cabicompendium.6599 (accessed 15 September 2023).

CABI (2023b) *Daucus carota* (carrot). Notes on pests. In: *Crop Protection Compendium*. CAB International, Wallingford, UK. https://www.cabidigitallibrary.org/doi/full/10.1079/cabicompendium.18018 (accessed 15 September 2023).

Cresswell, J.E., Bassom, A.P., Bell, S.A., Collins, S.J. and Kelly, T.B. (1995) Predicted pollen dispersal by honeybees and three species of bumble bees foraging on oil seed rape: a comparison of three models. *Functional Ecology* 9, 829–841.

Daugovish O., Smith, R. Cahn, M., Koike, S., Smith, H., *et al.* (2008) *Celery production in California*. Vegetable Production Series, ANR Publication 7220, Division of Agriculture and Natural Resources, University of California. https://anrcatalog.ucanr.edu/pdf/7220.pdf (accessed 15 September 2023).

Davis, R.M. and Raid, R.N. (2002) *APS Compendium of Umbelliferous Crop Diseases.* APS Press, St Paul, Minnesota.

Dickson, M.H. and Peterson, C.E. (1958) Hastening greenhouse seed production for carrot breeding. *Proceedings of the American Society for Horticultural Science* 71, 412–415.

Esau, K. (1936) Ontogeny and structure of collenchyma and of vascular tissues in celery petioles. *Hilgardia* 10, 431–476.

Hawthorn, L.R., Toole, E.H. and Toole, V.K. (1962) Yield and variability of carrot seeds are affected by position of umbel and time of harvest. *Proceedings of the American Society for Horticultural Science* 80, pp. 1–407.

Heywood, V.H. (1983) Relationships and evolution in the *Daucus carota* complex. *Israel Journal of Botany* 32, 51–65.

Hiller, L.K. and Kelly, W.C. (1979) The effect of post-veneralization temperature on seedstalk elongation and flowering in carrots. *Journal of the American Society for Horticultural Science* 104, 253–257.

Lorenz, O. and Maynard, D. (1988) *Knott's Handbook for Vegetable Growers*, 3rd edn. Wiley, Hoboken, New Jersey, p. 69.

US Nematode Collection (2023) Fungal Nomenclature Database. Mycology and Nematology Genetic Diversity and Biology Laboratory, Beltsville, Maryland. http://nt.ars-grin.gov/fungaldatabases/nomen/nomenclature.cfm (accessed 15 September 2023 but awaiting migration).

Nascimento, W.M., Vieira, J.V., Silva, G.O., Reitsma, K.R. and Cantliffe, D.J. (2008) Carrot seed germination at high temperature: effect of genotype and association with ethylene production. *HortScience* 43, 1538–1543.

Nieuwhof, M. (1984) Effect of gibberellic acid on bolting and flowering of carrot (*Daucus carota* L.). *Scientia Horticulturae* 24, 211–219.

Nunez, J.J. and Davis, R.M. (2016) Diseases of carrot (*Daucus carota* L. subsp. *sativus* (Hoffm.) Arcang.). Common Names of Plant Diseases, American Phytopathological Society, St Paul, Minnesota. http://www.apsnet.org/publications/commonnames/Pages/Carrot.aspx (accessed 15 September 2023).

Oliva, R.N., Tissaoui, T. and Bradford, K. (1988) Relationship of plant density and harvest index to seed yield and quality in carrot. *Journal of the American Society for Horticultural Science* 113, 532–537.

Peterson, C.E. and Simon, P.W. (1986) Carrot breeding. In: Bassett, M.J. (ed.) *Breeding Vegetable Crops.* AVI Publishing, Westport, Connecticut, pp. 321–356.

Rosenfeld, H.J., Baardseth, P. and Skrede, G. (1997) Evaluation of carrot varieties for production of deep-fried carrot chips. IV. The influence of growing environment on carrot raw material. *Food Research International* 30, 611–618.

Rubatzky, V.E., Quiros, C.F. and Simon, P.W. (1999) *Carrots and Related Umbelliferae.* CAB International, Wallingford, UK.

Rubatzky, V.E. and Yamaguchi, M. (1997) *World Vegetables: Principles, Production and Nutritive Values*, 2nd edn. Chapman & Hall, New York.

Spurr, C. (2003) Identification and management of factors limiting hybrid carrot seed production in Australia. Doctoral dissertation, University of Tasmania, Australia, pp. 281.

Stevens, P.F. (2001 onwards) Angiosperm Phylogeny Website. Version 14, July 2017. Missouri Botanical Garden, Missouri. http://www.mobot.org/MOBOT/research/APweb/ (accessed 6 September 2023).

Takaichi, M. and Oeda, K. (2000) Transgenic carrots with enhanced resistance against two major pathogens, *Erysiphe heraclei and Alternaria dauci. Plant Science* 153, 135–144.

USDA (2023) *Daucus carota* L.: Queen Anne's Lace. Natural Resources Conservation Service. https://plants.usda.gov/home/plantProfile?symbol=daca6 (accessed 6 September 2023).

USDA-ARS (2023) Germplasm Resources Information Network (GRIN). US Dept of Agriculture, Agricultural Research Service. www.ars-grin.gov/ (accessed 6 September 2023).

9

Family Asteraceae

Overview

Asteraceae is a large family of over 23,000 species, spread across 1620 genera (Jeffrey, 2007). This important family contains many familiar herbs, weeds, wildflowers, and ornamental plants including aster (*Aster amellus* L.), marigold (*Tagetes erecta* L.), calendula (*Calendula officinalis* L.), daisy (*Bellis perennis* L.), chrysanthemum (*Chrysanthemum indicum* L.), dahlia (*Dahlia pinnata* Cav.), and zinnia (*Chrysogonum peruvianum* L.) (Duke, 2013). This chapter will focus on seed production of lettuce (*Lactuca sativa* L.), a major world salad vegetable, with references to seed production of the related crops chicory (*Cichorium intybus* L.) and endive (Cichorium *endivia* L.).

Botany

Lettuce

The Latin name Asteraceae is derived the Greek word for star. An older family name for Asteraceae is Compositae, derived from the family's distinctive flower inflorescence (Zohary *et al.*, 2012). Lettuce (*Lactuca sativa* L.) is an erect, smooth, herbaceous annual plant grown for its crisp edible leaves that form continuously during the vegetative stage of development. Leaves are arranged spirally in a dense rosette. Plants have considerable diversity of leaf shape, texture, and margin among the many different forms. For example, glabrous leaves can be smooth, savoy, or crumpled. Leaf margins may be lobed, smooth, or finely divided, and leaf colors can vary from light to dark green with some cultivars having red or purple coloration (Welbaum, 2015).

Interior leaves of leafy cultivars tend to be lighter colored, while those of heading types are variations of white. Except for stem lettuce, the cylindrical stems are short and compressed. Upon bolting, induced by the combination of long days with warm temperatures, stems rapidly elongate, becoming erect, tall, branched with 10–25 florets (Rubatzky and Yamaguchi, 1997).

The inflorescence is composed of many individual florets that share a common receptacle that collectively make up the composite flower. The flower head is a structure called a capitulum, which is a contracted raceme composed of numerous individual sessile flowers, called florets (McKenzie *et al.*, 2005). The florets each open in the morning and are self-pollinating, although insect pollination occasionally occurs. Flowering continues for one to two months after induction. Florets produce achene-like fruit called cypsela with fused carpels, one locule, and one seed per fruit. The cypsela has a feathery mast-like appendage called a pappus attached at one end that acts as a parachute to disperse seeds through the wind (Welbaum, 2015; Fig. 9.1).

The cypselas are oblong, broadest towards the apex, and range in color from white to shades of yellow, brown, gray, and black (Rubatzky

DOI: 10.1079/9781789243260.0009

Fig. 9.1. Lettuce flowering and seed maturing with pappus. (Courtesy of Tim Loeffler.)

and Yamaguchi, 1997). A characteristic of many Asteraceae species is a white glue-like liquid latex released when tissues are cut.

Endive

Endive (*Cichorium endivia* L.) is a cool-season, hardy annual or biennial leafy salad, forcing (Belgian endive), or root vegetable producing a dense rosette of leaves on a short-compressed stem (Welbaum, 2015). High temperatures and long days cause bolting of annual types. Vernalization is required for biennial types. Upon bolting, the stem elongates and branches before flowering. The inflorescence bears many capitula, usually with attractive pale blue or white flowers (Rubatzky and Yamaguchi, 1997). The foliage resembles loose-leaf lettuces but tends to be prostrate. Endives are more heat tolerant than lettuce. Favorable growing temperatures are between 18°C and 20°C (64°F and 68°F) with seed harvest maturity occurring about 90–130 days after planting under optimal conditions.

Chicory

Chicory (*Cichorium intybus* L.) is a deep-rooted frost-tolerant perennial, grown as an annual or biennial for edible food production. After vernalization at two to three weeks at temperatures below 10°C (50°F), multiple branched stems elongate to produce blue flowers (Rubatzky and Yamaguchi, 1997). Chicories are cool-season plants that grow best at temperatures between 15°C and 18°C (59°F and 64°F) with seed maturity occurring about 100–130 days from bolting (Rubatzky and Yamaguchi, 1997).

Types, Growth, and Development

Lettuce

Most lettuces switch from vegetative to reproductive growth after exposure to long days and high temperatures. After induction the vegetative bud becomes reproductive. New leaf formation is replaced by rapid seed stalk development in a process called bolting (Welbaum, 2015). However, breeding and selection for bolting resistance causes many modern commercial cultivars to behave as day-neutral plants. Although the taproot may grow 1 m (3.3 ft) deep in some situations, lateral roots near the surface absorb the most moisture and nutrients. Because most roots are in the top 30 cm (1 ft), lettuces are considered by to be shallow rooted (Rubatzky and Yamaguchi, 1997).

There are four widely recognized morphological types of commercial lettuce: crisphead, butterhead, romaine or cos, and loose-leaf (Table 9.1). However, recent breeding efforts have created cultivars with intermediate morphological characteristics across these types. Stem, oil seed, and Latin lettuce are additional forms (Rubatzky and Yamaguchi, 1997). The Batavian lettuce is another crisphead type that predates iceberg with a smaller and less dense head (Miles, 2013).

Crisphead types are also known by their alternative names: iceberg or head lettuce (Table 9.1). Following early rosette development, subsequent overlapping leaf growth eventually entraps newly formed young leaves inside a dense, nearly spherical head weighing 0.7 to 1.0 kg (1.5 to 2.2 lbs). Crisphead is slow growing and takes 70–80 days for mid-summer plantings and as long as 130 days for late-fall plantings to develop solid mature heads and 185 days to seed maturity. Heads of modern cultivars may become very firm to the point that seed stalks have difficulty emerging (Welbaum, 2015). For some cultivars it is necessary to cut open and remove the head in a process known as "deheading" to ensure seed stalk emergence (Fig. 9.2). Crisphead lettuce for consumption does not grow well under warm conditions but prefers cool conditions, ample moisture, and full sunlight (Turini *et al.*, 2011).

Butterhead is a heading type with smaller tender leaves (Table 9.1). Leaves overlap to form a smooth soft head with an almost oily texture and delicate flavor. After exposure to long days and warm temperatures, plants transition to reproductive growth as the apical meristem elongates into a floral structure similar to other lettuces (Welbaum, 2015). Heads mature in 65 to 80 days while seed crops mature in 115 to 130 days. Bibb lettuce is sometimes considered a subtype of butterhead lettuce and typically is more semi-heading with thicker, dark green leaves and more prominent midribs. Batavia-type cultivars have characteristics intermediate between crisphead and butterhead types (Miles, 2013).

Cos or romaine heads have elongated, coarse, crisp-textured leaves with prominent broad midveins (Table 9.1). The long relatively narrow leaves tend to grow upright and may loosely overlap but do not form solid heads. Compared to crisphead, romaine is preferred by vegetable growers because it matures earlier than crisphead, takes less water to produce, handles shipping well, is more tolerant of warm weather, and is higher in some vitamins (Welbaum, 2015). Under good growing conditions, most romaine types produce mature heads in 55 to 75 days and seed crops in 115 to 140 days.

Loose-leaf or bunching cultivars are arguably the easy to grow. Loose-leaf cultivars exhibit considerable variation in leaf size, margin shape, color, and texture. Plants develop leaves in a tight rosette cluster (Table 9.1; Fig. 9.3).

Leaf texture varies from crisp to oily depending on the cultivar. Under cool, short day lengths, the plant remains vegetative. Bolting is favored by long days and warm temperatures

Table 9.1. Major types of *Lactuca* lettuce propagated from seed. (From Doležalová *et al.*, 2004.)

Species and group name	Common names	Description
L. sativum Group asparagina	Stem or asparagus lettuce, celtuce	Erect, thickened, edible stem 30–40 cm (12–16 in.) long
L. sativum Group capitata	Crisphead, iceberg	Large bud, dense rosette, cabbage-like overlapping leaves and crisp texture
L. sativum Group capitata	Batavian, summer crisp, French crisp	Smaller bud, less firm open head, compared to crisphead
L. sativum Group capitata	Butterhead, Boston, bibb	Smaller bud, flattened, less compact head, broad tender leaves with oily texture
L. sativum Group crispa	Loose-leaf, bunching	Loose rosette, fringed or crisp leaves varying in color
L. sativum Group longifolia	Cos or romaine	Large bud, upright head, narrow columnar, leaves ovate to oblong, with broad midrib
L. sativum	Latin lettuce	Rosette of loose, elongated, and soft cos-like leaves that may form a partly closed head
L. serriola	Oil seed	Leaves are not eaten, seeds contain 35% oil

Fig. 9.2. Crisphead lettuce plant deheaded to encourage seed stalk development. (Courtesy of Tim Loeffler.)

Fig. 9.3. Poor stand of red oakleaf non-heading leaf lettuce in a seed field prior to bolting. (Courtesy of Tim Loeffler.)

(Miles, 2013). Edible leaves mature in as little as 50 days from seeding and 30 days for small leaf or "baby" cultivars. Seed crops mature in from 100 to 120 days.

Stem lettuce is an important crop in Egypt, Japan, China, and other countries in Southeast Asia but it is not widely grown in other areas (Table 9.1). Plants are non-heading and grown for their erect-thickened, edible stem. Stem lettuce is less sensitive to bolting than other types and usually flowers during the summer after reaching mature size (Rubatzky and Yamaguchi, 1997).

Endive

Endives are hardy annual or biennial plants, which produce a dense rosette of leaves on a short-compressed stem. The foliage resembles some loose-leaf lettuces (Welbaum, 2015). Endive is a distinct salad vegetable from the same genus as chicory and shares many characteristics. Optimum temperatures are between 15°C and 18°C (59°F and 64°F). High temperatures and long days are conducive to bolting for annual cultivars, and high light intensity accelerates bolting. Vernalization is required for biennial cultivars (Rubatzky and Yamaguchi, 1997). Upon bolting, the stem elongates and branches. The inflorescence produces many pale blue or sometimes white florets. Flowers are self-pollinating and each floret produces an individual seed similar to lettuce. Harvest maturity occurs in about 70 to 100 days while seed maturity occurs in 100 to 130 days after seeding depending on the cultivar and environment.

Chicory

Distinctive forms of chicory include leafy salad, heading, forcing heads, and root types. The green leafy vegetable that resembles dandelion is a perennial called foliage chicory and is a popular vegetable in parts of Europe. Establishment for seed production is by direct seeding 5.0 cm (2 in.) apart in rows 46 cm (18 in.) wide. Foliage types are less sensitive to high temperatures and can be grown at 24°C (75°F) with cool night temperatures (Welbaum, 2015).

Chicory cultivars that produce a small red- or green-colored head, or are intermediate between green and red heads, resembling cabbage, are called radicchio. The head resembles a small head of cabbage but with prominent white leaf veins and weighs from 0.25 to 1 kg (0.5 to 2.2 lbs). Radicchio is biennial so plants require vernalization to produce seed stocks that emerge through heads. Radicchio is established by direct seeding or transplanting with a final spacing of 15 cm (6 in) apart in rows 46–61 cm (18 to 24 in.) wide. Radicchio is very sensitive to temperatures > 21°C (> 70°F) which causes leaf tip burn and head rot. Optimum daytime growth occurs < 21°C (70°F) and at night temperatures < 15°C (60°F) but above 4°C (40°F) (Rubatzky and Yamaguchi, 1997).

Chicory types forced indoors to produce tight, narrow, leafy heads called chicons are biennials with alternative names of: witloof, witloof chicory, French endive, or Belgium endive (Welbaum, 2015). Whitloof chicory roots are produced in open-field production similar to sugarbeets, harvested, and placed in cold storage until they are used for indoor chicon production. The roots are replanted in a growing room environment under dark, moist conditions and are harvested after the chicons reach the proper size for the market. Some perennial chicories are grown for their fleshy-enlarged root used as a coffee adulterant or substitute. Agronomic practices for root chicory are also similar to sugarbeet with open-pollinated seed production occurring outdoors.

Soil

Site selection and field preparation

Lettuce seed production occurs primarily in arid areas with warm temperatures with low relative humidity such as California, Chile, Australia, and France. Seeds are produced in open fields, net houses, and greenhouses with establishment by either direct-seeded or transplanted. Another major factor in the selection of production areas is the ability to control and manage insect pressure from aphids that transmit lettuce mosaic virus (LMV). This is a seed-borne and devastating disease in fresh market production; seeds sold in most markets must be free from LMV based on a 30,000 seed test. Lettuce thrives in a wide range of well-drained soils with good moisture-holding characteristics. Lettuce is sensitive to soil compaction, crusting, and acidity. A pH range of 6 to 7 is preferred and soils with a lower pH should be limed. Small seedlings are sensitive to salinity, but tolerance increases with maturity.

Optimum electrical conductivity of soil extracts is 1.0 to 1.5 dS/m and exceeding 2.5 dS/m may reduce germination and growth. For direct-seeded establishment it is important to select fields and prepare the soil to promote rapid and uniform emergence and development. Crusted soils interfere with stand establishment and uniformity. Wetting and drying cycles cause heavier soils with a significant clay fraction to crust. Preventing standing water during irrigation followed by rapid drying can reduce crusting. Shallow cultivations with a rotary hoe or other implements break crusts, as does frequent light irrigation that keeps soils moist. Mulching also reduces crusting. Well-tilled smooth seed beds, either flat or raised, encourage uniform seedling establishment and growth (Fig. 9.4).

The optimal temperature range for lettuce seed germination is between 15° and 20°C. Thermodormancy is induced in many genotypes when soil temperatures exceed 28°C. Using over-head irrigation to cool the soil or planting primed seed when higher soil temperatures occur improves stand establishment at high temperature. To avoid the challenges of direct seeding, transplants are preferred especially for indoor establishment. Transplanting provides several advantages including less consumption of parent seed, planting to stand rather than thinning, and uniform crop development for effective rogueing and seed harvest. Lettuce should be rotated a minimum of two years with unrelated crops to reduce buildup of soil-borne diseases and pests. In fields or areas where there is higher disease pressure, a longer crop rotation is preferred (Turini *et al.*, 2011).

Fertilization

Lettuce is a heavy feeder with a shallow fibrous root system. Precision application methods, timing, and placement avoid fertilizer waste.

Fig. 9.4. Uniform stand of butterhead lettuce plants established from transplants on well-tilled raised beds for seed production. (Courtesy of Tim Loeffler.)

Nitrates and/or phosphorus tend to accumulate in soils that are intensively cropped year after year. New applications should be applied only after soil testing reveals a need. Lettuce is sensitive to ammonium that can cause root burn and cavities inside older roots. Thoroughly composted manures and organic waste are sometimes applied to improve soil structure and nutrients for organic seed producers. Quantities of compost applied are in the range of 9 t/ha (4 tons/acre) (Turini *et al.*, 2011). Preplant nitrogen applications risk leaching nitrate-N. Small amounts, such as 22 kg/ha (20 lbs/acre) nitrogen, are often applied as starter fertilizer just prior to or at planting. At thinning, if needed, 56 to 90 kg/ha (50 to 80 lbs/acre) of nitrogen is side-dressed near the row (Turini *et al.*, 2011). Additional side dressings may be applied just after seed set and as needed in accordance with foliar testing. Fertigation is a more efficient application method. Weekly fertiligation can optimize nutrient applications in conjunction with irrigation. Soil nitrate levels greater than 20 ppm in the top 30 cm (12 in.) of soil are adequate for growing lettuce. A seasonal total of roughly 169 kg/ha (150 lbs N/acre) is typical for seed crops on heavier soils, while 224 to 280 kg/ha (200 to 250 lbs/acre) is applied for cooler conditions or lighter textured soils (Turini *et al.*, 2011).

Levels above 60 ppm P are adequate for lettuce growth. If soils test < 60 ppm, preplant applications of 45 to 90 kg/ha (40 to 80 lbs/acre) of P_2O_5 or at-planting applications of 22 kg/ha (20 lbs/acre) of P_2O_5 are recommended. Zinc fertilization is recommended if extractable soil test is < 1.5 ppm. Zinc may be added to soils with high phosphates since phosphates inhibit Zn uptake (Turini *et al.*, 2011).

Soils containing > 150 ppm K are sufficient for growing lettuce. Potassium fertilizer presents less of an environmental risk than excessive N or P. A lettuce crop will remove approximately 134 kg/ha (120 lbs/acre), so fertilizing to replace more K than what the crop removes is wasteful (Turini *et al.*, 2011). If foliar analysis is used as a diagnostic tool to determine fertilization applications, 6000, 3000, and 30,000 ppm N, P, and K, respectively, are considered sufficient on a dry weight basis for mature lettuce (Rubatzky and Yamaguchi, 1997).

Pollination

Lettuce produces a flower structure called an involucrate capitulum. Each capitulum can produce from 15 to 30 florets. Each floret is self-pollinating, nectarless, and produces a solitary seed (achene) per flower. Because of this, no F1 hybrid cultivars have been developed. All lettuce cultivars are non-hybrid, self-pollinated pure lines. It is extremely rare but up to 5% cross-pollination has been observed in lettuce although it is a self-pollinating crop, especially when there is high insect populations (George, 2009). Cross-pollination most likely occurs by insects that vector pollen. Seed production of different cultivars is often separated by an open space or a physical barrier to prevent mixing of seed more than insect activity. A minimum of 2 m should be maintained between different cultivars. Like lettuce, endive flowers are self-pollinating, and each floret produces a single seed. Chicory is cross-pollinated by insects. Flowers and seeds are similar in appearance and size to endive.

Irrigation

Since lettuce has a shallow root system, plants must receive a consistent supply of water until seed maturation. Irrigation is always needed for lettuce seed crops in arid and semi-arid areas. Drought stress particularly in combination with high temperature causes leaf tip burn from calcium deficiency leading to secondary infection by *Erwinia* spp. Excessive water or poor drainage may cause rot diseases and drip irrigation is preferred (Fig. 9.5).

In areas with heavy soils, lettuce seed crops are grown on raised beds to improve drainage and air flow. The recommendation for at least 2.5 cm (1 in.) of water every seven to ten days as a minimum applies to lettuce seed crops. More water is needed on sandy soils. To establish a seed crop, sprinkler irrigation activates fertilizer, prevents crusting, and ensures uniform hydration for good seed germination or transplant establishment. After establishment irrigation/fertigation continues during vegetative and reproductive growth with drip or furrow irrigation in flat areas with plentiful water resources. Approximately 46 cm (18 in.) of water, well distributed during growth, is usually adequate for growing a lettuce

Fig. 9.5. Lettuce greenhouse seed production using "bag" culture. Plants are fertigated by a spaghetti tube routed to each bag. Yellow sticky traps between supports capture flying insects. (Courtesy of Tim Loeffler.)

seed crop, although the amount varies widely with the type of lettuce, season, climate, and soil type. The quantity of water required to mature a seed crop in a hot, dry-summer, Mediterranean climate is approximately 750 to 1000 m^3/ha (1.5 to 2.0 acre-feet/acre), 1000 to 1250 m^3/ha (2 to 2.5 acre-feet/acre) for furrow irrigation, and 500 to 750 m^3/ha (1 to 1.5 acre-feet/acre) for more efficient drip irrigation (Turini *et al.*, 2011). Irrigation water with salinity levels of less than 1 dS/m are best suited for lettuce (Turini *et al.*, 2011). Water is usually withheld about four weeks before intended harvest to facilitate drying of vegetation. However, excessive drying may increase shattering.

Rogueing

There are three main stages for rogueing and selection:

1. Young vegetative plants at the four- to six-leaf stage.
2. The mature vegetative growth prior to and during head formation.
3. Reproductive growth after seed stock formation.

Probably the most important stage is prior to and during heading. In some cases, heading is the only stage for rogueing commercial seed crops. Rogueing is based on an assessment of morphological characters defined by the breeder during the vegetative stages of growth especially head formation. Diseased plants are removed from seed fields as well.

Genetics

Lactuca spp. have basic chromosome numbers of 8, 9, and 17. *L. sativa* is one of four interfertile or partially interfertile species having nine pairs of chromosomes. *Lactuca sativa*, the only domesticated

Lactuca species, is native to the eastern Mediterranean basin. Other species are *L. serriola*, *L. virosa*, and *L. saligna*. *Lactuca sativa* is not found in the wild and was likely derived from *L. serriola*, either alone or in combination with another species. The *L. sativa-serriola* complex is large, polymorphic, and capable of cross-breeding, so *L. sativa* may have been derived directly from *L. serriola* through selection (Jeffrey, 2007). So, *L. sativa* and *L. serriola* are cross-fertile and crosses have incorporated favorable genes in commercial lettuce. Genetic barriers make crosses with other species of *Lactuca* difficult. When successful, resulting progeny are few and often sterile. Nevertheless, some cultivated lettuces have been derived from *L. sativa* × *L. virosa* and *L. sativa* × *L. saligna* crosses (Rubatzky and Yamaguchi, 1997).

Lettuce produces one seed (achene) per floret and there is no male sterility, which makes production of F1 hybrid lettuce, endive, and chicory economically impractical because of the labor involved. All commercial lettuce and endive cultivars are pure lines. Lettuce and endive cultivars grow true-to-type, so seed can be saved for replanting from one season to the next when the proper seed production protocols are followed. Pure-line cultivars are developed from an initial cross and taken through several generations of self-pollination to ensure genes are "fixed" before release as new cultivars. Seed of stable cultivars can be saved from season-to-season with minimal variability. Genetically engineered transgenic lettuce has not been an area of development and currently there are no cultivars in the commercial market. Recombinant DNA technologies and/or CRISPR (clustered regularly interspaced short palindromic repeats) can improve disease resistance and other traits in the future.

Production and Harvest

Lettuce is a cool-season, frost-tolerant crop. Acclimated seedlings can survive as low as 2.0°C (28°F). Minimum survival temperatures of immature plants vary widely with genotype and degree of acclimation. However, heads are more easily damaged by frost (Rubatzky and Yamaguchi, 1997). Subfreezing temperatures may damage immature developing seeds prior to harvest; however, this is not a factor in the major seed production areas where frost risk occurs well after seed harvest is complete. Mature, drying, < 20% moisture content seeds are less susceptible to freeze damage while on the mother plant.

Stand establishment is a major production concern. Lettuce is often grown on raised beds to improve drainage, aeration, and reduce disease. A common configuration is raised beds on 100 cm (39 in.) centers. The top of the bed is typically 61 cm (24 in.) wide and the rows are spaced at the edges of each bed. Other bed dimensions are possible. Variation in emergence leads to variation in flowering, reducing uniformity of maturation at harvest, which can result in lower seed quality. Some cultivars are produced in net houses or greenhouses using transplants to attain higher-quality seed (Fig. 9.6). In addition, the germination of lettuce seed can be inhibited by high temperatures.

Exposure of hydrated seeds to temperatures over 24°C (75°F) for as little as 24 h may induce thermodormancy, a secondary dormancy that blocks germination. This is a particular problem in desert production areas when lettuce is direct-seeded into warm soils. Generally, the higher the temperature, the greater the number of seeds in the population affected and the deeper the dormancy. Recently progress has been made in understanding the genetic basis of thermodormancy in lettuce seed. Seed priming, controlled hydration followed by redrying, can increase the ceiling temperatures for lettuce germination (Schwember and Bradford, 2010). Genes in wild lettuce that allow lettuce to germinate at higher temperatures have been identified and are being transferred to *L. sativa* (Yoong *et al.*, 2016). In addition to thermodormancy, the Grand Rapids cultivar has a well-characterized, phytochrome-mediated photodormancy and requires light to germinate (Borthwick *et al.*, 1954). There have been successful breeding efforts to remove this dormancy out of most modern cultivars but this is still a factor for some cultivars and may affect certain heirloom cultivars (Yoong *et al.*, 2016). Some lettuce exhibits primary dormancy at seed harvest. Such seeds should be stored until after-ripening is complete and they are fully germinable and vigorous before planting. After-ripening is a series of biochemical changes that occur in a dry seed after harvest that increase viability and/or vigor. Loss of primary dormancy is accelerated at higher storage temperatures and moisture contents. However,

Fig. 9.6. Lettuce seed production in a screen house to prevent aphid entry. (Courtesy of Tim Loeffler.)

if storage temperatures or moisture contents are too high, seeds can rapidly transition from after-ripening into accelerated aging with loss of first vigor and then viability.

Lettuce seed crops are most often established by direct seeding for open-field seed production, but transplants are used for indoor production. Raw, coated, or pelleted seeds are established by direct seeding using air or belt seeders to ensure precision placement, good soil contact, and singulation because lettuce seeds are small, thin, and elongated. Prior to planting soil is finely tilled using rotovators or disks to eliminate clods and other impediments to encourage rapid germination and uniform emergence. In many cases seed crops are precision-planted to a final stand, but thinning is usually necessary to achieve optimum plant populations.

Transplants are used to ensure uniform stands of seed crops, reduce the in-field times, or establish stands under difficult conditions. Lettuce seedlings are sensitive to stress and not suited to bare-root transplanting. Lettuce transplants are grown in plug-transplant trays started in greenhouses or other protective enclosures and transplanted into seed production fields after several weeks' growth.

Growth is extremely slow at temperatures below 7.2°C (45°F). Optimum daily temperatures for growth are 18.3° to 21°C (65° to 70°F) with night temperatures in the range of 7.2° to 10.0°C (45° to 50°F). Temperatures greater than 30°C (86°F) usually stunt growth. Romaine and leaf lettuce are slightly more tolerant of high temperatures than iceberg and butterhead types (Turini *et al.*, 2011).

Generally, high light intensity and long days increase growth rate, hasten leaf development, produce broader leaves, and promote rapid head formation. Long days induce lettuce to bolt, and high temperatures accelerate the process. Some cultivars flower prematurely when planted in late spring or early summer because of the combination of high temperatures and long days. For indoor production, the crop may be de-leafed starting at the fresh market stage of growth. De-leafing involves the removal of the lower leaves to improve air movement to reduce the risk of fungal diseases prior to bolting. Some cultivars require the opening and removal of the heads in a process called deheading to promote uniform bolting. After bolting is initiated, flower stalks rapidly grow 0.6 to 1.2 m (2 to 4 ft) tall with little vegetative growth. The plants are usually trellised for indoor seed production as they are more prone to lodging and need support (Fig. 9.7).

Many modern lettuce cultivars are extremely resistant to bolting. Some cultivars are so resistant to bolting that production is difficult, resulting in lower seed yields (Rubatzky and Yamaguchi, 1997). Sometimes the plant hormone gibberellic acid (GA_3) is applied to promote earlier and more uniform bolting.

Florets open successively on inflorescences typically during a seven- to ten-day period (Fig. 9.8). It is best to avoid temperatures above 30°C to attain the highest seed quality and relative humidity should be less than 75% for good pollen viability for indoor production. This results in a steady sequence of ripening seeds at physiological maturity. The time from flowering to ripe seed on an individual capitulum is a relatively fast, 15 to 21 days, depending on the environment and genotype. High temperatures speed seed development and maturation and temperatures above 35°C should be avoided as much as possible.

One of the biggest challenges of lettuce seed production is timing of harvest to attain the highest physiological quality and maximum seed vigor. Lettuce is very sensitive to seed aging and harvesting at high temperatures and/or high relative humidity can significantly reduce seed storage shelf life. The optimal harvest stage is best when 85 to 90% of the crop is in the "cotton stage". Waiting until the crop is 95 to 100% cotton stage results in a higher percentage of older aged seeds. Seed production with uneven maturation may require a second harvest. The second harvest should be considered a separate

Fig. 9.7. Trellis system supports bolting plants to prevent lodging in a screen house. (Courtesy of Tim Loeffler.)

Fig. 9.8. Lettuce seed fields at various stages of flowering in the San Joaquin Valley, California, USA. (Courtesy of Tim Loeffler.)

seed lot from the first. If harvest is delayed past optimum maturity, rain or high winds will also result in significant seed losses. If harvest is premature, seed quality is adversely affected, and primary dormancy will often be deeper. If the seed producer waits for the development of seeds from the later inflorescences, the earlier developing seeds will probably be lost to shattering. The stage of ripeness at which the pappus is fully developed and dry is referred to as "feathering" (Fig. 9.9).

Since lettuce seed is highly susceptible to mechanical damage, most seed companies manually harvest the crop to meet the demand for high physiological quality. For smaller seed lots, one method of harvest is "direct shaking" the plants into a bag or container to collect the seeds. The method used for larger production volumes involves the cutting the plants and laying them on a cloth or plastic material to dry prior to threshing. A mower that causes minimal shattering should be used for cutting. Shattering is reduced if plants are cut when dew is present. The threshing process involves taking small bunches of plants and tapping them against the side of a container to remove the ripe seeds. The cut plants are ready to harvest typically in the afternoon once the relative humidity drops below 40%. Threshing can be the same day as cutting in warm arid areas or one to three days later under cooler conditions. After threshing, seeds are precleaned and dried to 6 to 7% moisture content within a few hours after they are collected. Lettuce seed can also be mechanically harvested by combining or stationary threshers placed in the field. However, direct harvest by combining usually results in lower-quality seed that is unacceptable for high-value markets. If standing lettuce seed fields are combined directly, the feed beaters are removed to reduce shattering and mechanical damage caused by beater impact. If the seed is combined directly from a standing lettuce field, some plant debris will be mixed in the seed lot. Green debris may undesirably increase seed moisture and must be pre-cleaned immediately after harvest to avoid a decline in seed quality.

Post-harvest Seed Conditioning

Lettuce seeds may continue physiological maturity after harvest. Many types of lettuce and

Fig. 9.9. Lettuce seed at harvest feathering stage when seeds can be easily lost due to shattering. (Courtesy of Tim Loeffler.)

endive exhibit varying degrees of primary dormancy at harvest. Storing seeds for at least two months allows after-ripening and development of full germinability and vigor.

The pappus, a feathery mast-like dispersal appendage, is a modified calyx that is milled off the cypsela during conditioning to yield a solitary achene that is the "seed" of commerce. Lettuce seed lots are notorious for containing large quantities of small debris after passing through a combine or thresher and usually the seed is precleaned immediately after harvest. Proper seed cleaning of lettuce, endive, chicory, and globe artichoke seeds is essential because the weight of chaff is as much as one and a half times the seed weight. The basic cleaning can be done with an air screen machine. Remaining small debris such as pieces of flower stems or pappus can be removed with a disk separator or an indent cylinder.

Seed Yield

On average, 1000 lettuce seeds weigh 1 g (28,571 seeds/oz). Typical yields for open-field production are between 112 to 224 kg/ha (100 to 200 lbs/acre). For endive, 900 seeds weigh 1 g (25,500 seeds/oz). For chicory, 800 seeds weigh 1 g (22,800 seeds/oz) (Rubatzky and Yamaguchi, 1997).

Pest Management

Weeds

Lettuce seedlings compete poorly with weeds so effective management is essential for successful seed production. The availability of effective herbicides is limited as lettuce is sensitive to herbicides. Pre-emergence herbicides are typically applied in a band 12.7 to 15.2 cm (5 to 6 in.) wide over rows after planting. The stale bed technique, which is irrigation to encourage weed emergence followed by shallow cultivation, is an effective non-chemical, preplant, control strategy. Burn-down herbicides or propane flaming may provide effective weed control in some situations. Additional strategies include mechanical cultivation, mowing, hand weeding, and eliminating weeds surrounding production fields. It is important to control weeds around seed fields. Weeds may harbor insects and diseases that could spread into production fields.

Pathogens

Lettuce infectious yellows, lettuce necrotic yellows virus, beet western yellows, turnip mosaic, and lettuce mosaic are some of the most important virus diseases of lettuce (CABI, 2013). The spread of LMV, as well as other viral diseases, may be stopped by targeting aphid vectors, the immediate rogueing of diseased plants from the field, and the use of healthy virus-free seed that has an LMV-free test result. Production and certification of seed that is free of LMV is a requirement for many fresh market production areas and requires strict testing and approval to meet requirements. Lettuce production areas such as in California and Arizona, USA require approval for each seed lot to be "indexed" or tested by a local or state governing authority before the seed can be planted. LMV-infected seed can be detected by enzyme-linked immunosorbent assay (ELISA) or polymerase chain reaction (PCR) testing.

Big vein (Mirafiori) lettuce virus makes leaf veins deformed, enlarged, and clear. The enlarged veins cause ruffled and deformed leaves or headless plants (UC IPM, 2023). The virus is soil-borne and introduced by the fungus (*Olpidium brassicae*) that attaches to lettuce roots. The severity of the disease varies greatly from season-to-season and is most prevalent during cool weather (UC IPM, 2023). Crop rotations are not effective in controlling big vein. Some cultivars are resistant.

Bottom rot (*Rhizoctonia solani*) usually occurs when field conditions are wet. The symptoms are a slimy rotting of the underside of the plant progressing upward into the head. Growing lettuce on well-drained soils and/or using raised beds, crop rotation, good sanitation, and fungicidal sprays are effective controls (UC IPM, 2023).

Lettuce drop (*Sclerotinia sclerotiorum*) is a soil-borne fungal disease that begins at the rosette stage. The disease begins at the stem base causing a wet rot of the entire plant. Lettuce drop is most severe during cold wet weather. The best control is proper sanitation, crop rotation, and good drainage (UC IPM, 2023).

Downy mildew (*Bremia lactucae*) is a serious disease in temperate areas causing light green to yellow angular spots on the upper surfaces of older leaves. White fluffy growth of the pathogen develops on the lower sides of these spots. Over time, lesions turn brown and dry killing severely infected leaves. Downy mildew is controlled by planting resistant cultivars or applying fungicides (UC IPM, 2023). Using effective irrigation practices such as drip irrigation to help reduce pathogen pressure. *Bremia lactucae* is complex and has many races or pathotypes and is a focus of most breeding programs.

Bacterial leaf spot (*Xanthomonas campestris* pv. *vitians*), varnish spot (*Pseudomonas cichorii*), and anthracnose (*Microdochium panattonianum*) are foliar diseases that affect developing plants. Favored by cool wet conditions, bacterial leaf spot can be controlled by fungicides. The varnish spot bacterium survives in reservoir water and is spread by sprinkler irrigation. Anthracnose is primarily a problem only when resting fungus resides in wet soil. Corky root (*Rhizomonas suberifaciens*) is a soil-borne bacterium controlled by crop rotations, resistant cultivars, and lower soil nitrogen (UC IPM, 2023).

Verticillium wilt (*Verticillium dahliae*) affects all types of lettuce. Severe infestations may destroy an entire crop. Verticillium microsclerotia, long-term resting structures, often lie dormant in soil for ten years or more. Fumigation, rotations with unrelated crops, and disease-resistant cultivars are control options (UC IPM, 2023).

Fusarium wilt (*Fusarium oxysporum*) of lettuce is controlled by avoiding the introduction of infected soil into clean fields. The disease is favored by higher soil temperatures. Damping-off (*Pythium* sp.), gray mold (*Botrytis* sp.), and leafspot (*Cercospora* sp.) also affect lettuce. Aster yellows is a phytoplasm (mycoplasma) transmitted by the six spotted leafhopper (*Macrosteles fascifrons*) (CABI, 2013).

Insects

Lettuce aphid (*Nasonovia ribisnigri*) and foxglove aphid (*Aulacorthum solani*) both feed on inner lettuce leaves. Since lettuce and foxglove aphids are protected within developing heads, timely detection and treatment are keys for control. The lettuce and foxglove aphids may not be adequately controlled with neonicotinoid insecticide applied by soil injection or foliar sprays. Green peach aphid (*Myzus persicae*) and potato aphid (*Macrosiphum euphorbiae*) accumulate on outer leaves making insecticide sprays effective. Natural enemies such as parasitic wasps help suppress aphid colonies. Hover fly larvae (Syrphidae, *Baccha* spp., *Criorhina* spp.) and other aphid predators such as lady bugs, also called lady beetles

(Coccinellidae), may suppress aphids on outer leaves (Turini *et al.*, 2011).

Leafminers (*Liriomyza* spp.) form visible tunnels inside leaves. Leafminer female flies feed on fluids obtained by puncturing leaf surfaces with their ovipositor. Parasitic wasps can suppress leafminer populations particularly from the genus *Diglyphus*. Insecticides can effectively manage larvae but not adults (Turini *et al.*, 2011).

Beet armyworms (*Spodoptera exigua*), cabbage loopers (*Trichoplusia ni*), and other adult caterpillars may sometimes damage lettuce. The beet armyworm and cabbage looper larvae are susceptible to natural enemies including diseases, predators, and parasitoids. Caterpillars can be treated with selective insecticides. *Bacillus thuringiensis* (or *Bt*), a gram-positive, soil-dwelling bacterium, is often used for biological control of caterpillars (UC IPM, 2023).

Silverleaf whitefly (*Bemisia argentifolii*), greenhouse whitefly (*Trialeurodes vaporariorum*), and banded whitefly (*Trialeurodes abutilonia*), among others, infest lettuce in warmer climates. Whitefly adults are small, around 1.5 mm (0.06 in.) long, with white wings and variation in body color (UC IPM, 2023). Tiny, oval whitefly eggs hatch into a mobile first larval stage with legs and antennae. The last nymphal stage, often called the pupa or the red-eye nymph, is easily identified. Whiteflies feed on lettuce tissue producing sticky honeydew on the leaves. A black, sooty mold often grows on the excreted honeydew. Silverleaf whitefly feeding can stunt and yellow head lettuce (UC IPM, 2023). Several wasps, including species in the *Encarsia* and *Eretmocerus* genera, parasitize whiteflies. Bigeyed bugs, lacewing larvae, and lady beetles (bugs) also prey upon whitefly nymphs. The lady beetle (*Delphastus pusillus*) can provide biological control. Insecticidal sprays may control whiteflies as well (UC IPM, 2023).

The adult western flower thrip (*Frankliniella occidentalis*) is a minute, slender-bodied insect with two pairs of long, narrow wings. Adult thrip bodies can be yellow, orange, brown, or black. The larvae are white, yellow, or orange. In spring, populations increase on weeds and other vegetation and migrate into lettuce seed fields as weeds decline. On lettuce, colonies rapidly reproduce into large populations (UC IPM, 2023). Thrips puncture leaves and suck sap, causing a silvery appearance that eventually turns to brown. Small, black fecal specks on damaged leaves confirm thrip damage (UC IPM, 2023). The western flower thrip is the most important vector virus disease. Managing vegetation in and around lettuce seed fields, biological control, insecticide sprays, and cultural practices are important controls for western flower thrips (Turini *et al.*, 2011).

Parasitic nematodes affecting lettuce include: root-knot (*Meloidogyne incognita*, *M. javanica*, *M. arenaria*, and *M. hapla*); needle nematode (*Longidorus africanus*); stunt nematode (*Merlinius* sp.); and spiral nematode (*Rotylenchus* sp.). Nematodes are microscopic roundworms that feed on lettuce roots and living in soil. Multiple species can coexist in the same field. Symptoms vary with species and crop type and are often non-specific such as yellowing or stunting. Root-knot nematodes, however, cause distinctive root galls. The geographical distribution of the different nematode species is highly dependent on temperature, soil type, and cropping history. Fumigation, sanitation, cultural controls, and crop rotation are possible nematode controls (UC IPM, 2023).

References

Borthwick, H.A., Hendricks, S.B., Toole, E.H. and Toole, V.K. (1954) Action of light on lettuce-seed germination. *Botanical Gazette* 115, 205–225.

CABI (2023) *Lactuca sativa* (lettuce). Production and trade. In: *Crop Protection Compendium*. CAB International, Wallingford, UK. https://www.cabidigitallibrary.org/doi/10.1079/cabicompendium.29609 (accessed 7 September 2023).

Doležalová, I., Lebeda, A., Tiefenbachová, I. and Křístková, E. (2004) Taxonomic reconsideration of some *Lactuca* spp. germplasm maintained in world genebank collections. *Acta Horticulturae* 634, 193–201.

George, R.A.T. (2009) *Vegetable Seed Production,* 3rd edn. CAB International, Wallingford, UK.

Jeffrey, C. (2007) Compositae: introduction with key to tribes. In: Kadereit, J.W. and Jeffrey, C. (eds) *Families and Genera of Vascular Plants*, Vol. VIII: *Flowering Plants, Eudicots, Asterales.* Springer-Verlag, Berlin, pp. 61–87.

McKenzie, R.J., Samuel, J., Muller, E.M., Skinner, A.K.W. and Barker, N.P. (2005) Morphology of cypselae in subtribe Arctotidinae (Compositae–Arctotideae) and its taxonomic implications. *Annals of the Missouri Botanical Garden* 92, 569–594.

Miles, C. (2013) Winter Lettuce. Washington State University Mount Vernon NWREC. agsyst.wsu.edu/WinterLettuce.html (accessed 7 September 2023).

Rubatzky, V.E. and Yamaguchi, M. (1997) *World Vegetables: Principles, Production and Nutritive Values*, 2nd edn. Chapman & Hall, New York.

Schwember, A.R. and Bradford, K.J. (2010) A genetic locus and gene expression patterns associated with the priming effect on lettuce seed germination at elevated temperatures. *Plant Molecular Biology* 73, 105–118.

Turini, T., Cahn, M., Cantwell, M., Jackson, L., Koike, S., *et al.* (2011) *Iceberg lettuce production in California*. Vegetable Production Series Publication 7215. UC Vegetable Research & Information Center. https://escholarship.org/uc/item/7w47j6zv (accessed 7 September 2023).

UC IPM (2023) Lettuce. Pest Management Guidelines, University of California. https://ipm.ucanr.edu/agriculture/lettuce/ (accessed 7 September 2023).

USDA (2023) Dr. Duke's Phytochemical and Ethnobotanical Databases. US Department of Agriculture. https://phytochem.nal.usda.gov/ (accessed 21 September 2023).

Welbaum, G.E. (2015) Chapter 12. Family Asteraceae. *Vegetable Production and Practices*. CAB International, Wallingford, UK, pp. 222–239.

Yoong, F.Y., O'Brien, L.K., Truco, M.J., Huo, H., Sideman, R., *et al.* (2016) Genetic variation for thermotolerance in lettuce seed germination is associated with temperature-sensitive regulation of ETHYLENE RESPONSE FACTOR1 (ERF1). *Plant Physiology* 170, 472–488.

Zohary, D., Hopf, M. and Weiss, E. (2012) *Domestication of Plants in the Old World: The Origin and Spread of Domesticated Plants in Southwest Asia, Europe, and the Mediterranean Basin*, 4th edn. Oxford University Press, Oxford, UK.

10

Family Brassicaceae

Overview

The family Brassicaceae is very large and includes many vegetable, field, and oil crops. Although *Brassica oleracea* L. crops look different during mature vegetative growth, reproductive characteristics are quite similar. Endemic and naturalized Brassicaceae can be found on all continents except Antarctica. The centre of origin for the family is likely the Middle East and Turanian region, where approximately 900 species grow in 150 different genera. Approximately 530 of the 900 are native. A secondary center of genetic diversity is the Mediterranean region, with around 630 species, 290 of which are native, in 113 genera. Brassicaceae is present but less prominent in the Saharo-Arabian region of Northern Africa, North America, South America, Southern Africa, and Australia and New Zealand (Schmidt and Bancroft, 2011).

Space does not permit a detailed explanation of seed production of all vegetables in this family. Common and specific representative crop characteristics and challenges of seed production will be given as examples for some of the more important Brassicaceae vegetables. Seed production for *B. oleracea* (cabbage, kale, collards, kohlrabi, Brussels sprouts, broccoli, and cauliflower) are similar and will be considered collectively in many cases. Asian Brassica vegetables are one of most diverse groups. Managing Asian Brassica seed crops through various environments, often outside their optimum temperature and moisture range, is particularly challenging.

Botany

Flower characteristics

The family Brassicaceae is sometimes called by its archaic name Cruciferae. Crops are collectively referred to as Crucifers for short in some literature (Nieuwhof, 1969; Rubatzky and Yamaguchi, 1997). Cruciferae means "cross" in Latin, taken from the characteristic cross-shaped flowers of this family (Fig. 10.1). Today many refer to crops in the family Brassicaceae collectively as "Brassicas", although this can be confusing since *Brassica* spp. are but one important genus within the family (Table 10.1). The inflorescence is a raceme that elongates rapidly with many small flowers forming at the terminus. The branched inflorescence may be 0.5 m (1.6 ft) across in some cases. Each floret has four opposed petals that form a square cross. Flower petals vary in color from white, cream, pink, or purple among species (Fig. 10.1; Nieuwhof, 1969). Flowers are usually cross-pollinated by insects, most often bees. Although flowers are perfect, they are not self-fertile because of pollen incompatibility (described earlier). Pollen can

DOI: 10.1079/9781789243260.0010

Fig. 10.1. Distinct cross-shaped flowers are a defining character of this family. (Courtesy of Tim Loeffler.)

only successfully germinate and grow on a stigma of a flower from a different plant. This is because pollen is genetically incompatible with flowers on the same plant where it was produced (Myers, 2006).

Flowers are bisexual, each with one pistil, six stamens, four long and two short. A superior ovary develops into a long slender fruit pod called a silique, 4.5 to 10 cm (2 to 4 in.) in length, with a thin, translucent inner membrane, the replum, separating the two chambers or carpels where seeds develop (Fig. 10.2; Nieuwhof, 1969). As siliques senesce, they dry and dehisce into two halves releasing from 10 to 30 dark purple seeds (Musil, 1948).

Some Brassicas have a distinct pungency caused by allyl isothiocyanate, an organosulfur compound. Allyl isothiocyanate is believed to be a natural chemical adaptation to protect plants against herbivory. Plants store allyl isothiocyanates as glucosinolates separate from myrosinase enzyme in different plant tissues including siliques and seeds (Fenwick, 1982). When animals or insects feed on Brassica tissues, the glucosinolate substrate and myrosinase enzyme mix releasing bitter tasting and goitrogenic substances such as isothiocyanates, thiocyanates, nitriles, and goitrin that repel predators and provide unique tastes for humans (Fenwick, 1982).

Fig. 10.2. Developing green silique opened to reveal seeds inside. (Courtesy of Tim Loeffler.)

GENUS BRASSICA

Brassica is a major genus, containing over 40 spp. of mostly annual, biennial, or sometimes perennial herbaceous plants or small shrubs of Old World origin (Nieuwhof, 1969). Different *B. oleracea* vegetables are difficult to visually differentiate in the seedling stage because they are closely related (Table 10.1). Unless noted otherwise, Brassicas generally lack deep taproots and

Table 10.1. Latin and common names of some of the more economically important crops in the family Brassicaceae. (From Griffiths, 1994.)

Genus, species	Common name
Armoracia rusticana	Horseradish
Sinapis alba	White mustard
Brassica juncea	Leaf mustard
B. napus (Napobrassica Gp)	Rutabaga
B. napus (Pabularia Gp)	Siberian kale
B. nigra	Black mustard
B. oleracea (Acephala Gp)	Kale, collard
B. oleracea (Albogiabra Gp)	Chinese kale
B. oleracea (Botrytis Gp)	Cauliflower
B. oleracea (Capitata Gp)	Green, white, or European cabbage
B. oleracea (Gemmifera Gp)	Brussels sprouts
B. oleracea (Gongylodes Gp)	Kohlrabi
B. oleracea (Italica Gp)	Broccoli
B. oleracea (Costata Gp)	Tronchuda cabbage
B. rapa (Chinensis Gp)	Chinese cabbage (non-heading), pak choi
B. rapa (Pekinensis Gp)	Chinese cabbage (heading), pe-tsai
B. rapa (Perviridis Gp)	Spinach mustard
B. rapa (Rapifera Gp)	Turnip
B. rapa (Ruvo Gp)	Broccoli raab, rapini
Lepidium sativum	Garden cress
Crambe maritime	Sea kale
Nasturtium officinale	Watercress
Raphanus sativus	Radish
Wasabia japonica	Japanese horseradish

have mainly fibrous roots confined to the upper layers of the soil profile. There is confusion about the correct botanical name for crops in this genus. This is due to molecular genetic studies that have shown that major morphological differences between different *B. oleracea* crops may be caused by single or few genes. Therefore, crops with much different phenotypes exist in the same species (MBGP, 2023). Over the past 25 years, there has been a growing consensus among plant scientists that cabbage, broccoli, and cauliflower should be sorted into distinctive "Groups". The term Group (Gp) is used for convenience by horticulturists to show different plants of horticultural significance within a species when major genic differences are lacking. Such differences were previously classified as botanical varieties or subspecies, terms which suggest more significant genetic differences than exist (Griffiths, 1994). For example, today cabbage is often classified as *B. oleracea* Group Capitata rather than botanical variety or subspecies capitata, which is common in older literature. This seemingly academic change in classification has real world significance for seed production. Because plants in *B. oleracea* can readily sexually cross, vegetables in different Groups must be isolated from each other as well as from escapees and wild *B. oleracea* during seed production. The following paragraphs describe basic botanical characteristics of some major Groups in *B. oleracea*.

European or green/white cabbage is a cool-season, frost-tolerant crop that is widely adapted and commonly grown throughout temperate and subtropical regions of the northern and southern hemispheres. A cabbage head is a single large terminal bud enclosed by overlapping leaves on the end of a short unbranched stem. As with other biennials, vegetative growth continues while plants pass through an immature juvenile phase when they cannot be vernalized. Only plants that have reached a mature stage of development corresponding to stem diameters roughly greater than 5 mm (0.2 in.) can be vernalized by prolonged exposure to temperatures between 1.7° and 10°C (35° to 50°F). Vernalization changes gene expression, causing the bud to become reproductive and the stem to rapidly elongate.

Annual broccoli leaves are thick, somewhat leathery, smooth oblong, simple, alternate, pinnate, and petiolate. Leaf colors range from grayish blue to green. Many fibrous roots are concentrated within the top 30 to 40 cm (12 to 16 in.) of soil (Nieuwhof, 1969). Sprouting broccoli (*B. oleracea* Group Italica) is an herbaceous cool-season, frost-tolerant mostly annual dicot, which is widely adapted and grown throughout temperate and subtropical regions. The primary broccoli inflorescence or head is highly branched and forms at the terminus of the elongated unbranched stem. The inflorescence forms with a compact, slightly dome-shaped head reaching up to 40 cm (16 in.) across at edible harvest maturity. After edible maturity, compactness is lost as the flower stalk elongates and the plant transitions to reproductive growth without environmental cues. Sprouting broccoli stems are taller than cabbage or cauliflower, ranging from about 40 to 100 cm (16 to 39 in.), and also have longer internodes. The green or purple bud clusters may be surrounded but are usually not covered by subtended leaves. Following growth of the primary inflorescence, smaller inflorescences form in the axils of the lower leaves. Development of secondary inflorescences is influenced by apical dominance of the terminal inflorescence and the degree of suppression is cultivar dependent (Rubatzky and Yamaguchi, 1997).

Cauliflower is an herbaceous dicot with summer, tropical, or winter types. Summer and tropical types have an annual growth cycle while winter cultivars are biennial. Leaves are oblong, pinnate, thick, somewhat leathery, glaucous, smooth, and usually alternate (Nieuwhof, 1969). The cauliflower curd consists of tightly clustered, undifferentiated, usually white, shoot apices (Fig. 10.3). The curd forms upon thick, hypertrophied, repeatedly branched fleshy terminal portions of the short thick stem, reminiscent of callus tissue (Rubatzky and Yamaguchi, 1997). Slow elongation associated with rapid thickening of lateral branches results in a short, thick, and compressed dome-shaped curd in most cultivars (Fig. 10.3). Color of the curd can be white, purple, orange, and green and some cultivars heads are pointed or pyramidal-shaped. The curd itself is not flower primordia and this is one characteristic that differentiates broccoli from cauliflower. A basal vegetative bud will change to reproductive after vernalization of biennial (winter) cultivars or through development control in annuals. As the plant bolts, a flower stalk will rapidly elongate through the spreading curd giving rise to a raceme composed of many tiny florets with four yellow petals. Flowers are attractive to bees which collect pollen and nectar. At full flower, the bushy plant will be roughly 1 m (47 in.) high and roughly 0.5 m (24 in.) across (Nieuwhof, 1969). Cauliflower has a moderately shallow root system with a branched taproot and many fibrous roots in the top 40 cm (16 in.) (Rubatzky and Yamaguchi, 1997).

Fig. 10.3. Late-maturing, summer cauliflower head in the vegetative stage showing domed head consisting of undifferentiated flower primordia. After vernalization and satisfactory growing conditions, a flower stalk will emerge as this biennial plant transitions from vegetative to reproductive growth.

Brussels sprouts (*B. oleracea* Group Gemmifera) are a cool-season, frost-tolerant, herbaceous dicot biennial (Griffiths, 1994). Leaves are thick, smooth oblong, simple, alternate, pinnate, petiolate, with a subtle undulating leaf blade texture. Leaf colors range from purple to green. Root systems are moderately shallow with a prominent taproot that branches with many fibrous roots concentrated in the top 30–40 cm (12–16 in.) of soil (Nieuwhof, 1969). The distinctive feature of Brussels sprouts plants are the numerous vegetative buds (sprouts) that develop in leaf axils along an elongated, erect, 5 to 8 cm (2 to 3 in.) diameter stem (Nieuwhof, 1969). After vernalization these axillary buds turn reproductive and rapidly elongate as the plant becomes bushlike when reproductive.

Kohlrabi is a cool-season, low-growing biennial. It develops a globular stem enlargement just above the soil line that is eaten like a turnip root during first year. The stem increases to 5 to 8 cm (2 to 3 in.) in diameter 55 to 65 days after seeding (Swaider *et al.*, 1992). Pinnate elongated leaves with slightly serrated margins form directly on the enlarged stem in a compressed spiral pattern with the youngest leaves forming at the apex. Smooth short petioles have a broad attachment to the bulbus stem with no abscission zone. The taproot is relatively slender and very fibrous with many small branching lateral roots (Rubatzky and Yamaguchi, 1997). Plants bolt and flower after vernalization.

Kale, collard, and mustard are cool-season, leafy greens that do not form a head. Kale and collard are biennials that require vernalization for several weeks at their mature stage to initiate flowering. Both can tolerate periods of subfreezing temperatures. Kale and collards form a concentric whorl of foliage toward the tip of erect unbranched stems. For many cultivars, the stem is relatively short, although stem heights of some may exceed 1 m (3.3 ft). Kale cultivars may have large upright heavily curled leaves. Savoy types have densely wrinkled leaf lamellae. Collard cultivars are similar to kale but with larger smooth leaves and margins. Leaves of some collards may overlap to resemble a loose head. Important kale and collard characteristics that differentiate cultivars are leaf numbers, shape, size, leaf texture, color, and stem length.

Mustards in general define a diverse group of plants from multiple species that may be annuals or biennials. Generically mustards may refer to vegetable as well as agronomic crops with common names like Indian, white, brown, black, or yellow mustard. Classification is confusing because of the many different forms, some of which are referred to as Chinese or ornamental mustards (Rubatzky and Yamaguchi, 1997). Mustard greens (*B. juncea*) are a cold-hardy annual upright vegetable producing medium green pungent leaves. Mustard greens are generally self-pollinated, while yellow mustard (*B. hirta*), used to make culinary mustard, is cross-pollinated. Examples of diverse types of *B. juncea* vegetable mustards include: heading, large, small, curled leaf, large petiole, green petiole, root, big stem, multi-shoot, and many other names (Rubatzky and Yamaguchi, 1997).

Asian Brassicas

The genus Brassica includes many vegetables that are widely grown in Southeast Asia and to a lesser extent elsewhere. A discussion of these vegetables is complicated by the many common names derived from different languages. However, Asian vegetables may be grouped as Chinese cabbage (*Brassica rapa*), Asian mustard greens (*B. juncea*), Chinese broccoli, or Chinese kale (*B. oleracea*) (Myers, 1991).

Chinese cabbage

Several scientific names describe Chinese cabbage. Genetic studies show morphological differences among *B. rapa* may result from fewer genes than the many diverse phenotypes suggest. Two distinct types of Chinese cabbage are often classified as Groups Pekinensis and Chinesis to show horticultural differences where major genetic differences do not exist. A few of the common names used to describe heading types of Chinese cabbage (Pekinensis Group) include: nappa or napa, hakusai, pai-tsai, won bok, pechay, and tsina. Common names for non-heading types (Chinesis Group) include: bok choy, pak choi, celery cabbage, celery mustard, chongee, petsay, and pei tsai. Heading types may be further classified according to head shape. Heading types are classified as open, erect or cylindrical, and ovoid or barrel (CABI, 2023; USDA, 2023).

Chinese cabbage is the most widely grown Asian Brassica and the most important in *B. rapa*. Chinese cabbage is a cool-season, relatively short-maturing annual, but some types are biennial (Wittwer, 1987). The vegetative head height of most cultivars ranges from 20 to 60 cm (8 to 24 in.). Chinese cabbage has a finely branched extensive root system that develops in the top 30 cm (12 in.) of soil. Moderate day and cool night temperatures result in rapid growth and development. Chinese cabbage does best when grown under daily average temperatures between 13°C and 21°C (55 and 70°F) (Rubatzky and Yamaguchi, 1997). Leaf blades of sessile, mostly ovate, and slightly wrinkled leaves extend to the bottom of the broad flat and lighter colored midrib. Many cultivars of heading Chinese cabbage resemble romaine lettuce. The head is a single large terminal bud comprised of tightly overlapping leaves attached to and enclosing the short unbranched stem. After vernalization of biennial types, the bud turns reproductive and develops a seed stalk and inflorescence.

Asian mustard greens

Asian mustard greens (*B. juncea*) are a large and diverse group that include Chinese green mustard or gai choi, mizuna, and various potherbs (USDA, 2023). Asian mustard greens are annuals and seed crops are grown as described for other cool-season greens. Climate, soil fertility, and irrigation requirements are similar to non-heading types of Chinese cabbage (Rubatzky and Yamaguchi, 1997).

Chinese broccoli

Chinese broccoli is also known as gai-lon (or gai-lohn), kailan, or Chinese kale. The Chinese broccoli plant resembles sprouting broccoli, except that the leaves are a bit broader, the stems longer, and the head is much smaller. Flowers form first in diminutive heads and then elongate rapidly into stalks bearing yellow or white flowers similar to broccoli raab. Chinese broccoli is grown as an annual like sprouting broccoli.

Brassicaceae Root Crops

Root crops include radish (*Raphanus sativus*), turnip (*Brassica rapa* Gp Rapifera), and rutabaga (*B. napus* Gp Napobrassica). Horseradish (*Armoracia rusticana*) is an herbaceous perennial. Turnip and rutabaga are biennials. All prefer cool temperatures and long days accelerate flowering.

Radish

Radishes are gown for their enlarged hypocotyl and root tissues and types are grouped by root maturity. The early spring or summer types develop small, 3–4 cm (1.2–1.6 in.) diameter round or long roots that are usually white, red, or bicolored. Spring and summer cultivars are direct-seeded, maturing into edible roots in 30 to 40 days after planting and seed crops in 100 days depending on environmental conditions. All are annuals that bolt in response to long days and high temperatures (Swaider *et al.*, 1992).

Spanish, daikon, and Chinese are called "winter" types. These take much longer to develop (60 to 90 days), have larger leaves, roots weighing up to 7 kg (15 lbs), a wide range of root colors, and much longer storage life, thus the name "winter". Long-rooted "winter" cultivars are more resistant to high temperatures and long days (Rubtazky and Yamaguchi, 1997). Many winter types exhibit biennial characteristics and bolt faster after exposure to cold.

Turnip and rutabaga

Turnip is a cool-season biennial. The leaves and petioles are pubescent and yellowish green (Rubatzky and Yamaguchi, 1997). Usually there is no hypocotyl "neck" transition between root and leaf attachment. Turnips have a distinctive

Fig. 10.4. China Rose winter radish. Three vegetative plants on the left and the longer-rooted plant on the far right has bolted and produced a seed stalk.

taproot with secondary roots. Turnip roots have several different shapes including round, top shaped, flattened, and cylindrical. Turnips roots occur in a wide range of colors (purple, green, white, pale yellow, or bronze) and may be solid or bicolor. Bicolor cultivars generally have different top and bottom root colors like the cultivar Purple Top White Globe.

Rutabaga is a cool-season, frost-tolerant biennial that grows best at temperatures below 22°C (72°F). Rutabaga produces large edible elongated storage roots, 2.3 kg (5 lbs) or more, with a pronounced "neck" (Rubatzky and Yamaguchi, 1997). Rutabaga is used for both human and animal consumption. The enlarged root is smooth with final laterals mainly associated with the taproot. Secondary roots arise from the underside of the enlarged root as well as the taproot. Leaves have thick petioles, smooth blades, and bluish green color that is often covered with bloom.

Horseradish

Horseradish (*Armoracia rusticana*) is an herbaceous perennial grown for its pungent enlarged roots. Horseradish is heterozygous and does not reproduce true-to-type from seed, so it is most often vegetatively propagated from root cuttings rather than seed.

Wasabi

Wasabia japonica is Japanese horseradish, also frequently called simply wasabi (Collins, 2003). Wasabi is an aquatic plant grown for its rhizomes and is a much different crop than horseradish although they have similar culinary uses. Wasabi does not breed true-to-type, so propagation is vegetative via offshoots rather than true seed. (Bratsch, 2009).

Types and Cultivars

Space does not permit detailed descriptions of all the diverse types and cultivars within the family Brassicaceae. Extensive descriptions of vegetables in this extensive family are provided in Rubatzky and Yamaguchi (1997) and Welbaum (2015). F1 hybrid cultivars are popular for *B. oleracea* vegetables, radish, turnip, Chinese cabbage, mustard, and others because of their improved uniformity and heterosis.

There is a great diversity of cabbage cultivars that vary in color, size, leaf texture, disease resistance, and cuticular wax often called "bloom". Leaf colors among cultivars include white, green, blue, red, or purple. Cultivars with savoy (crinkled- or blistered-leaf lamella) leaves are popular in some markets primarily in Europe. Red or purple cultivars are popular to add color to fresh salads and are also used for cooking and processing. Cabbage cultivars are often grouped based on head size, firmness, maturity, and shape (Rubatzky and Yamaguchi, 1997).

Although open-pollinated cabbages remain popular with home gardeners and in some markets, hybrids dominate commercial vegetable production in much of the world. The following are examples of some major cultivar classes of green or white cabbage and their defining characteristics:

Pointed or Wakefield: Small, pyramid-shaped head. Early maturing. Cold tolerant and resistant to bolting. Small framed plants are suited for close spacing. Popular in Europe as an early-season fresh market cabbage. Example cultivars include Early Jersey Wakefield

Copenhagen and Golden Acre: Round head with few wrapper leaves and small core. Early- to mid-season maturing. Can be susceptible to bolting. Golden Acre types are slightly earlier and suited for close spacing. Example cultivars include Copenhagen Market, Golden Acre, Badger Market, and others.

Flat Dutch and Brunswick: Large, flat, and solid head with numerous wrapper leaves. Maturity varies from early to late season. Usage ranges from fresh market, storage, and processing for sauerkraut. Example cultivars include Late Flat Dutch, Stein's Early Flat Dutch, Brunswick, Wisconsin All Season.

Danish: Moderate plant and head size, solid heads with few wrapper leaves, covered with moderate bloom. Late maturing for fall production. Good keeping qualities for fresh market, processing, and storage. Example cultivars include Danish Ballhead, Penn State Ballhead, Wisconsin Hollander No. 8.

Alpha: Small head. Early maturing, mostly for home garden market.
Langedijker: Large head, late maturing, mostly for storage and processing. Popular in Northern Europe.

Broccoli cultivars can be divided into either annual or biennial types. Most commercial types grown in the Americas, parts of Asia, and Southern Europe tend to be annuals and are sometimes called sprouting broccoli. Biennial types require vernalization for head formation and flowering, and are more common in Northern Europe (Rubatzky and Yamaguchi, 1997). Broccoli may also be classified as early-, medium-, or late-maturing. Late-maturing and overwintering cultivars are biennial and require vernalization. There are several distinctive types of broccoli cultivars that deserve mention. Early green sprouting is an annual with upright growth habit, no wrapper leaves, and rapid maturity. Purple cape is an overwintering single-headed biennial, while purple Sicilian is a pale purple' single-heading annual sometimes incorrectly called "purple cauliflower", with heads that form close to the ground surrounded by upright wrapper leaves. White sprouting is an overwintering branching biennial. Calabrese is an upright green sprouting annual (Rubatzky and Yamaguch, 1997).

Calabrese types are widely grown sprouting broccoli. Important characteristics that differentiate broccoli cultivars include head compactness and shape, extent of branching, size and color of individual flower buds, stem length, number and length of internodes, and auxiliary floral development. F1 calabrese type hybrids have largely replaced open-pollinated cultivars for large-scale commercial production (Swaider *et al.*, 1992).

In some newer literature, cultivars that produce multiple smaller heads for bunching are called sprouting broccoli. By contrast, cultivars forming a single large head on a thick stalk like calabrese are often called heading broccoli. However, using traditional notation both types are sprouting broccoli, separate and distinct from the overwintering true heading broccoli types that only develop heads after vernalization (Rubatzky and Yamaguchi, 1997).

Many sprouting broccoli cultivars produce heads within the canopy that emerge above the foliage at edible maturity. Broccoli buds eventually open as yellow flowers with advanced maturity. Some cultivars are lower growing, producing heads having cauliflower plant architecture with wrapper leaves extending above and around the heads. If the head comprises immature flower buds, such cultivars are considered broccoli. In contrast, a cauliflower head is not composed of flower buds but dense clusters of usually white or colored callus-like, undiffentiated, flower primordia. When cauliflower plants bolt, the flower stalks emerge from beneath and grow above the curd. Cauliflower heads are denser than broccoli (Rubatzky and Yamaguchi, 1997).

Other unique vegetables called broccoli have an appearance reminiscent of sprouting broccoli. Chinese broccoli (kan lan or jie lan; *B. oleracea* Group Alboglabra), also known as Chinese kale and white-flowered broccoli, resembles small sprouting broccoli growing between 40 and 50 cm (16 to 20 in.) tall. The edible portions of Chinese broccoli include the stem, accompanying leaves, and an early developing inflorescence, flowering without vernalization (CABI, 2023).

Broccoli raab (*Brassica rapa* L. Gp Ruvo) is a popular leafy vegetable in parts of Southern Europe. Broccoli raab is an annual resembling sprouting broccoli but with a much smaller plant and head with less compact inflorescence. Broccoli raab is also called Italian turnip, cima de rapa, or rapini (CABI, 2023).

Cauliflower types have been developed for production in a wide range of environments. Temperate cultivars are grouped as: early (summer or fall harvest depending on the growing season), intermediate (late fall and early winter), and late (winter and spring). Some biennial "late" cultivars of cauliflower will not form a head until vernalized. "Tropical" cultivars form tight compact heads even at high temperatures. Also called Asian cultivars, these are grown as early annuals in temperate and tropical regions (e.g. cv. Panta). The cultivar Cavolofiore di Jesi (cauliflower from Jesi) has distinctive spiral curds clustered in a domed head. Related cultivars have spiraled curd colored green (e.g. cv. Romanesco). Cultivars having green curds have been incorrectly marketed as "brocco-flower" or broccoli-flower. Brocco-flower is not a hybrid between broccoli and cauliflower, but simply a Cavolofiore di Jesi derived cultivar with chlorophyll in its curd. Northern European and North American early-maturing cultivars that form small edible heads the first summer include cvs. Alpha, Snowball, and others. Flowering and seed formation occurs after vernalization the second year. European biennial cultivars for late winter and spring production include cvs. Roscoff and St Malo. Important characteristics that define cultivars are plant size, wrapper leaf size and number, head size, shape, compactness, surface texture, and color (Rubatzky and Yamaguchi, 1997).

Early-maturing Brussels sprout cultivars tend to be shorter and produce fewer sprouts than tall-growing late ones. Tall cultivars are more prone to lodging particularly during flowering and seed production. Bud compactness, size, shape, color intensity, harvest period, and productivity are important attributes that differentiate cultivars.

Common widely grown kohlrabi cultivars include White Vienna (pale green) and Red Vienna (purple). Kohlrabi cultivars differ in the rate of fiber development especially as they grow larger. Characteristics that differentiate cultivars include bulb size, shape, and stem/leaf color leaf number, ruminant leaf scars, and tendency for stem cracking. Kohlrabi cultivars exist for both outdoors and greenhouses production.

Popular summer radish cultivars include Crimson Giant, Early Scarlet Globe, Scarlet Knight, Cherry Bell, Champion, Cavalrondo, and Comet (Swaider *et al.*, 1992). Common root colors of summer radish cultivars include red, white, or bicolor. Important traits that differentiate cultivars include top size, leaf pubescence, petiole strength, pungency, taproot transition, and crack resistance. The late-maturing winter types include Spanish, Daikon, and Chinese and these take much longer mature and produce larger leaves and roots. Characteristics that define radish cultivars include storage life, pungency, top size, leaf pubescence, neck size and strength, exterior and interior root color, pithiness, taproot length, and root hypocotyl transition zone.

Rutabaga roots are tan to yellow and larger than the white-fleshed turnips. Cultivars of rutabaga and turnip are also available with white- or yellow-colored roots (Rubatzky and Yamaguchi, 1997). The color on the root shoulder of both crops may be bronze, green, or purple. Some turnip cultivars are developed primarily for their edible leaves (turnip greens). Some have attractive edible roots, while others are classified as dual purpose because they have attractive edible roots and tops (Tokyo Cross).

Leafy greens

Greens are a diverse group of non-heading Brassica species, including mustard, turnip, kale, and collards. F1 hybrid cultivars are available. Greens are grown for their leaves that are either used fresh or cooked. Certain cultivars of turnips have been developed for use only as greens ('Seven Top') and do not form enlarged roots. Mustard greens differ in their leaf size, color, and pungency. The Siberian smooth-leaf and the Scotch, curly-leaf types of kale are both widely grown. Curly-leaf cultivars are showier and can have either blue or green leaves. Dwarf and full size kale cultivars are available (Rubatzky and Yamaguchi, 1997).

Chinese cabbage and greens

In some areas, cultivars producing erect cylindrical heads are called chichili, while round and compact heading types are called chefoo (Rubatzky and Yamaguchi, 1997). Having been grown in China since the fifth century, natural crossing has occurred resulting in distinctive and intermediate types in addition to chichili and chefoo. Chinese breeding has created a vast diversity of cultivars over its centuries. Some cultivars have greater heat tolerance (Fu *et al.*, 1991). Heat tolerant cultivars tend to have narrower leaves with a high ratio of midrib to leaf blade. F1 hybrid Chinese cabbage cultivars are available and widely grown throughout the world.

Growth and Development

Most Brassicas grow best at high light intensities. There are both annual and biennials in this family and in some cases, both occur within the same species. Some annuals, like sprouting broccoli, flower at a set time in development without environmental cues. For summer radish and Chinese cabbage, high temperatures and long days hasten the transition from vegetative to reproductive development. Preventing premature flowering in biennials can be a challenge, particularly in climates where temperatures fluctuate, and a few weeks of vernalizing temperatures can cause bolting before plants are sufficiently large to support reproductive growth. The amount of exposure needed to induce bolting and flower development depends on the crop and cultivar. For some species, as little as three or four weeks or less of low temperatures (between 1.7°C and 10°C or 35°F to 50°F) after the juvenile growth phase is completed can cause rapid flower stalk development (bolting) and flowering, particularly in early-maturing cultivars (Nieuwhof, 1969). The shoot apical meristem of biennial Brassicas undergoes drastic developmental changes as it transitions from vegetative to an inflorescence meristem. Genes responsible for meristematic changes associated with flowering are increasingly known and are described in the Genetics section of this chapter (Ridge *et al.*, 2015). Breeding for longer vernalization requirements is important to preserve biennial character. A strong biennial is less likely to bolt prematurely during both vegetable crop and seed production, which is especially important in regions with fluctuating temperatures. However, in the case of tropical cauliflower, the loss of vernalization helped adapt production to warmer climates.

The optimum mean temperature for cabbage growth and quality development is 15° to 18°C (60° to 65°F) with a minimum temperature of 4°C (40°F) and a maximum of 24°C (75°F). Generally, young plants are more tolerant of heat and cold than plants approaching maturity. Well-hardened transplants will withstand −7°C (20°F) for short periods. Hardening is an acclimation process when water is temporarily withheld, or plants are subjected to brief periods of non-lethal cold to improve stress tolerance prior to transplanting.

As with other biennials, cabbage does not respond to vernalization until its juvenile phase of development is completed. Maturing plants, with stem diameters roughly greater than 5 mm (0.2 in.), can be vernalized after prolonged exposure to temperatures between 1.7° and 10°C (35° to 50°F). Flowering begins with bolting, or stem elongation, that occurs after the heads split open allowing rapid elongation of the flower stalk (Fig. 10.5). After bolting has been induced, warmer temperatures speed the rate of seed stalk elongation. The flower stalk may elongate to up to 1 m (3.3 ft) or more in length before flowers open. Sensitivity to bolting is largely dependent on the cultivar with some requiring longer chilling exposure than others.

Many cultivars of Chinese cabbage are annuals. High temperatures > 15°C (59°F) and long days of 15 to 16 h or more for four to five weeks will induce flowering in mature plants. Plants large enough to support seed production can be induced to flower in late spring or early summer when day lengths and temperatures are increasing. It is more difficult to produce Chinese cabbage seed in fall or winter when temperatures are low and days are short (Fu *et al.*, 1991). Generally, young plants are more tolerant of heat and cold than plants nearing maturity. However, well-hardened transplants will withstand −7°C

Fig. 10.5. Bolting cabbage plant. Flowers will open after the seed stalk elongates or bolts. Since bolting makes fruiting plants top heavy, stakes are used to prevent lodging. (Courtesy of Tim Loeffler.)

(20°F) for short periods (Rubatzky and Yamaguchi, 1997). Maturing heads subjected to temperatures above optimum tend to deteriorate and rot from prolonged exposure especially with excessive nitrogen fertilization.

Annual cultivars form an inflorescence of many closely spaced immature flower buds clustered in a head 50 to 70 days after emergence without vernalization. Yellow flowers will open after 85 days if the heads are not harvested as a vegetable, and mature seeds are produced from 130 to 150 days depending on the climate and cultivar. Vegetative growth can occur over a wide range of temperatures. Annual broccoli is more heat tolerant than some other Brassicas, so seed can be produced in subtropical climates. However, best temperatures to assess head development are over a temperature range from about 12°C to 20°C (54°F to 68°F). Above 25°C (77°F), plants do not form compact heads because of excessive stem elongation and rapid floral development. Low temperature stress during early plant development may cause premature heading sometimes called "buttoning" in broccoli and cauliflower. Growth is slow at temperatures below 5°C (41°F). Properly acclimated plants can tolerate temperatures as low as −7°C (20°F) for short periods without damage. Mature vegetative tissues are more tolerate of freezing temperatures than young seedlings (Nieuwhof, 1969). Seed production at higher temperatures can cause genetic drift if practiced for many consecutive generations.

Biennial types, often called heading broccoli, require vernalization for head formation and flowering. These types are more commonly grown in Northern Europe (Rubatzky and Yamaguchi, 1997). For biennial cultivars, the stem elongates after plants have advanced beyond the juvenile phase and received sufficient chilling for three to five weeks below 13°C (50°F) to initiate floral development.

For cauliflower, small inner leaves initially envelop and protect the curd from discoloration due to sunlight. Cauliflower plant heights in the vegetative stage are variable, but most are 40 to 80 cm (16 to 32 in.) tall. Leaves are usually upright and oblong and longer and narrower than those of cabbage (Nieuwhof, 1969). As curd size increases and heads mature, the innermost leaves are unable to overlap and protect the head. The extent of inner and outer leaf cover varies greatly among cultivars. Late-maturing cultivars produce more and larger leaves and therefore provide better covering of the head. "Flora Blanca" types develop white curd resistant to yellowing. Large leaf and curd sizes are positively correlated. In late-maturing biennial cultivars, vernalization is required to produce reproductive shoot apices that develop into flowers. The required chilling period varies among cultivars (Rubatzky and Yamaguchi, 1997). Late-maturing cultivars require longer exposure and lower temperatures than early cultivars. Flower buds are formed in the axils of the elongated branches of the curd (Rubatzky and Yamaguchi, 1997).

Brussels sprout plants are freeze tolerant and survive mild winters as long as the ground doesn't freeze solid for extended periods. Brussels sprouts exhibit strong apical dominance. The tall 50–90 cm (20 to 35 in.) stem terminates with a large apical bud that strongly inhibits axillary bud development near the top of the plant (Nieuwhof, 1969). Axillary buds continue to develop from the bottom of the plant toward the top as long as the plant is actively growing. The production of more than 100 buds per plant is common (Rubatzky and Yamaguchi, 1997). Brussels sprout reproductive bud development only begins after prolonged vernalization. The inflorescence is a raceme that elongates rapidly from the terminal bud but also the axillary shoots give raise to additional racemes creating large bushy plants that flower over an extended period. Some seed producers will remove the apical bud to break dominance and encourage more uniform flower development by the axillary buds.

Excess nitrogen and water can cause stem cracking in some cultivars of kohlrabi. The optimum temperature for kohlrabi growth and development is 16° to 21°C (61° to 70°F). Temperatures below 7°C (45°F) result in vernalization and bolting, while high temperatures retard growth and promote tough, fibrous tissue.

Soil Fertility and Nutrient Management

Brassicas prefer deep, fertile, friable, sandy, or silt loam soils (Nieuwhof, 1969). Because of a need for uniform establishment, deep tillage and cultivation are often used to grow Brassica seed crops. Heavy clay soils should be avoided. Brassicas are generally shallow rooted, so soil should be well drained to avoid root diseases. Excessive moisture or poor drainage may contribute to the development of black rot (*Xanthomonas campestris*) and Phytophthora stem rot and other diseases (Rimmer *et al.*, 2007). Raised beds can improve drainage when soils contain 30–50% clay. When wet saturated soil conditions are possible, bed height can be increased and/or in-soil drainage systems can help remove excess water. Very sandy soils should also be avoided because water and fertilizer are more difficult to manage and can be easily lost to leaching.

Broccoli, cauliflower, and cabbage are only slightly tolerant to acid soils and do best with a pH range of 6.8 to 7.2. Acid soils limit availability of micronutrients such as molybdenum, but also increase the incidence of club root disease. Radish grows best at pH 6.2 to 6.8 and is more tolerant of low pHs (5.5–6.0) than broccoli, cabbage, cauliflower, or Chinese cabbage. The optimum pH is 6.2 to 6.8 for most other species in Brassicaceae. All Brassicas are moderately sensitive to salty soils with radish being the most sensitive and broccoli the least.

Soil testing prior to establishing Brassica seed crops is recommended. Fertility requirements of a given crop are site specific and dependent upon many environmental factors, including soil type and structure, native soil fertility, previous crops in the rotation, and length of the growing season (Swaider *et al.*, 1992). Excessive nitrogen application results in succulent less stress tolerant plants. Nitrogen applications should coincide with foliar and soil tests to meet documented nutritional needs. Spring applications are often made to replace nitrogen leached during the winter months for overwinter seed-to-seed crops. Manganese, boron, or molybdenum should be applied to encourage normal growth and development if soil tests warrant, or deficiency symptoms occur. Cauliflowers are especially susceptible to molybdenum deficiency that causes the whiptail disorder of leaf petioles.

A strong root system is essential to support good nutrient and water uptake throughout development and seed maturation. Structurally strong plants reduce lodging. Managing the fertilizer program from a focus on N during vegetative growth to a greater focus on K (flowering), Ca (fruit or pod development), and specific micros during generative growth facilitates balanced plants. The Brassica crops respond well to micronutrients in general, with B (pollen grain germination and pollen tube growth), Zn (phosphorus availability), and Mn (P and Ca availability) essential for seed production (George, 2009). Boron deficiency is associated with hollow stem or corky stem and curd browning. Most Brassica species have a significant boron requirement, higher than other vegetables, ranging from 1 to 4 kg/ha (0.9 to 3.6 lbs/acre). Sulfur, additional boron, and other micronutrients may be needed in sandy soils or soils that test low for these nutrients. Application of excess micronutrients such as boron may injure plants and reduce seed yields. For molybdenum deficiency treatment, a solution of sodium molybdate can be sprayed on young plants prior to transplanting in areas where Mo deficiency is common and soil testing show a need.

For both transplanted crops such as cabbage, broccoli, cauliflower, Brussels sprouts, and precision-seeded crops such as radish, turnips, rutabaga, and some Asian vegetables, banding part of the fertilizer recommendation 5–10 cm (2 to 4 in.) from young roots aids in establishment and early growth. For longer season cultivars with later maturity, from 1/3 to 1/2 of the total nitrogen requirement is supplemented post-plant during the rapid growth. For example, supplemental nitrogen for crops that form a head is often applied at the onset of head development or bolting. Excess nitrogen in Brassicas stimulates vegetative growth over reproductive growth, delaying head or curd formation and flowering. Kohlrabi fertilization and cultivation are like that needed for cauliflower. Excess nitrogen and water can cause rapid stem cracking in some cultivars.

Brassica root crops, such as radish, rutabaga, and turnip seed crops can be grown successfully on a wide range of soil types as long as they are well drained, deep, and fertile. Root crops are direct-seeded in a seed-to-seed system. A fine seed bed, free of stones and clods, is needed to facilitate precision seeding and uniform root development. Stones may increase the incidence of branched and crooked roots especially in radishes. Stones in the soil also increase bruising of root crops during harvest. Lack of deep rooting can cause lodging of seed crops before harvest. Establishment in cool, moist soil is preferred for winter radish seed production. An adequate supply of nitrogen, phosphorus, and potassium is required to promote rapid, continuous radish root growth. In deficient soils, boron may be required to prevent brown heart and stunted abnormal root development.

Pollination

The production of F1 hybrid Brassicas has changed in recent years from using predominately sporophytic self-incompatibility (SI) systems to greater utilization of cytoplasmic male sterility. Recent reviews of SI mechanisms provide detailed descriptions of molecular and genetic mechanisms (Iwano and Takayama, 2012; Muñoz-Sanz *et al.*, 2020). SI prevents self-fertilization based on discrimination between self-pollen and cross-pollination and its genetics are further described in the Genetics section below.

For F1 hybrid production in Brassicas, sporophytic SI genes were incorporated into the female parent inbred line to prevent selfing. This ensures that seeds were formed from a cross of the male and female inbred lines if the SI pollen discrimination system is working properly. However, environmental stressors like high temperature can cause a breakdown of SI resulting in an unacceptable amount of selfing in the female parent. Self-incompatibility is still utilized for Brassica seed production by some companies but due to its failure at high temperatures, SI is most often used in production areas with consistent mild temperatures during flowering. In some cases, both the female and the male parents have strong SI and both parents can be harvested as a reciprocal cross. One major advantage of this scenario is the cost of seed production is lower as seed from both parents can be sold as hybrid seed.

Cytoplasmic male sterility (CMS) is also available in most Brassica species and is more reliable under stressful conditions. Because of its reliability, CMS is replacing sporophytic SI as the preferred system to for hybrid Brassica seed production. While CMS is now widely used, this system has drawbacks as well. CMS is associated with flower deformities and particularly pale petal color, making flowers less attractive to bees. CMS has also been implicated in germination depression and reduced seed vigor, putting more emphasis on agronomic practices for successful stand establishment. Because of flower deformities associated with CMS, research has been conducted on the use of other pollinators, but honeybees remain the pollinator of choice for most Brassica seed production with CMS parental lines.

Since crossing is possible among the many crops in *B. oleracea*, care must be taken when locating fields for seed production. Common isolation distances between cross-compatible seed fields are 1.6 km (1 mile) and up to 8.3 km (5 miles) from another *B. oleracea* when the crop is produced outdoors. Small volume hybrid seed production and foundation seed increases are often conducted in insect-proof greenhouses, tunnels, or net houses. Radish has minimum isolation distance of 0.83 km (0.5 miles) between round red radish types and up to 3.2 km (2 miles) isolation from other seed fields with distinct characteristics as well as wild radish stands surrounding production fields. Since many commercial seed companies produce Brassica seeds in relatively few regions around the world with ideal climates, "pinning" location systems help growers identify and separate related seed crops (George, 2009).

Nicking of broccoli, cabbage, Brussels sprouts, and summer cauliflower inbred lines can be a challenging. The most common means of managing the synchrony of flowering between parental lines is planting the male and female at different times and sometimes planting the male at two different times, which is most common for broccoli seed production. The issues with nicking of tropical cauliflower cultivars are similar to summer cauliflower. For tropical cultivars, vegetative development up through curd formation has been selected to occur at high temperatures without vernalization. However, once flowering is initiated, tropical cultivars need milder temperatures similar to other Brassicas to develop highest quality seed. Finding locations to balance these two growth phases is challenging.

Seed production of long-season biennial cauliflower presents special challenges since these cultivars have vernalization and photoperiod requirements for both curd formation and flowering. For long-season cauliflower, nicking can be an issue if each line reacts differently to vernalization or photoperiod so that both male and female lines do not flower simultaneously.

Hybrid cabbage production is complicated by the frequent need to dehead to encourage seed stalk emergence (Fig. 10.6). Deheading is accomplished by scoring the head with an X through the top to allow flower stalks to emerge without hindrance. This is necessary because modern cultivars produce solid heads that are burst resistant, which negatively impacts bolting. Proper timing of deheading is critical to ensure successful nicking. When cutting open the heads, seed stalks should not be damaged. A flower stalk should emerge about two weeks after deheading. Additional considerations related to managing the nick in a seed crop will be discussed in the Seed Production section in this chapter.

Nicking is less of a problem with summer radishes because of their rapid growth and short life cycle. Some winter radishes have a semi-biennial character and flower faster after chilling. Seed production research and differential planting dates can ensure proper nicking of winter radish parental lines.

Irrigation

Ensuring developing plants have sufficient water during vegetative growth increases leaf area and the photosynthetic capacity to maximize seed yield. Irrigation is often needed to ensure uninterrupted plant growth and to promote uniformity. A rule of thumb is to ensure seed fields receive at least 2.5 cm (1 in.) of water per week. More is needed for sandy soils with low organic matter, particularly late in crop maturity (Davis *et al.*, 2007). Moisture requirements increase with Brassica development as evapotranspiration increases. Water stress during flowering can reduce seed set. Water stress in the *Brassica* spp. increases cuticle thickness and blue-green coloration in some cultivars (George, 2009). Phenotypic changes are most common under desert/high temperatures, increasing the difficulty of confirming some foliage characters during rogueing. Damage by nematodes, root diseases, insects, and weed competition is also more apparent on moisture stressed crops. Managing irrigation is critical to successful vegetable seed production and quite site specific. Irrigation should be reduced during the later stages of seed maturation and well before seed harvest. Depending on the climate, irrigation is often terminated two to three weeks before seed harvest to encourage drying.

Fig. 10.6. Deheading of cabbage to allow seed stalk emergence. This is necessary because modern cultivars have been selected to resist busting, making it difficult for the seed stalk to emerge. (Courtesy of Tim Loeffler.)

Furrow, subsurface trickle, or surface drip are used for Brassica seed crops in most production areas depending on the location. During stand establishment, after transplanting, or direct seeding, sprinkle irrigation is common but should be avoided the rest of the growing season to reduce foliar disease pressure. In humid areas, irrigation systems that wet foliage contribute to foliar disease and may need a more intensive spray program. Surface drip lines may be removed at maturity so they don't interfere with harvest equipment. Fertigation can be easily used with drip irrigation systems to accurately deliver fertilizer (Peirce, 1987; Swaider *et al.*, 1992).

Rogueing

Inspection and removal of off-types starts when transplants are in the nursery or greenhouse. Transplants with differences in leaf morphology and anthocyanin should not be used. The growth stages during seed production include fresh market, head breaking, elongation, and flowering. These stages are used to time rogueing, pruning, hive placement, and other management

Fig. 10.7. Furrow irrigation of hybrid cauliflower production in Chile. Furrow irrigation is being phased out but is still used on level ground where water is plentiful. (Courtesy of Tim Loeffler.)

activities. Out-crosses among broccoli, cauliflower, cabbage, and other *B. oleracea* crops are quite vigorous and should be identified and removed from transplant beds or fields early during vegetative growth. Another off-type is degenerate plants which occur due to aneuploidy or other chromosome abnormalities in some cultivars and parent lines. These plants tend to be semi-glossy, smaller in stature, and have shorter, narrow, and broad-veined leaves. CMS female parents are checked at the beginning bloom for fertile plants and usually flower before the sterile plants and must be removed. Different colored brassicas including wild radish and oriental radish/red radish out-crosses can be identified and removed early in development. Rogueing based on leaf differences is more difficult but may be used to eliminate off-types prior to flowering. As with any crop, rogueing of early bolters is necessary. The following are rogueing recommendations for individual crops (George, 2009).

Cabbage

The first rogueing stage for cabbage after field establishment and initial growth is to remove any obvious out-crosses with a different leaf and morphology before heading. Plants should be checked for general foliage, size, and cultivar characteristics. The second rogueing stage is at head formation, although this can be difficult since not all cultivars or parent lines form heads during the winter. When possible, heads should be checked for: shape, relative size, and firmness. Plants that are either too early or too late for the cultivar description should be removed.

Broccoli and cauliflower

The first rogueing stage is before normal head or curd development. Reject precocious or premature head or button curd development occurring before the normal maturity. Check foliage and plant growth habit for trueness-to-type. The second rogueing stage is when the curd or head has reached fresh market maturity. Check head or curd color, absence of bracts, solidity, shape, and form of head or curd, and wrapper leaf protection to ensure they are in accordance with the cultivar description.

Brussels sprouts

Brussel sprouts should be rogued at transplanting or before significant sprout formation occurs. Check for general plant vigor, relative height, foliage color, shape and size. The second rogueing should be conducted when lower sprouts or buds reach market maturity. Examine the general plant characteristics, productivity, and quality of the sprouts. Discard plants that are not within the proper maturity range or lack other characteristics consistent with the cultivar description.

Hypocotyl and root types

For bulb, root, and hypocotyl forming Brassicas, the first rogueing stage occurs when plants are thinned or established and have five to seven true leaves. At this stage remove any plants that are bolting prematurely and plants that are not true-to-type for foliage, bulb or hypocotyl characteristics, vigor, or pigmentation. The second rogueing occurs at normal market maturity of the bulb or root. Remove early bolting plants, check the bulb or root and foliage characteristics as well as pigmentation for trueness-to-type. If a seed-to-seed system is used, less attention is paid to root characteristics than in a root-to-seed system; however, only stock seed with high genetic seed quality should be used.

Genetics

Although many Brassica hybrids are now made using cytoplasmic male sterility, historically the SI system is very important for the development of hybrid Brassicas and is still widely used by some. Progress has been made in understanding the molecular biology of SI. Sporophytic self-incompatibility (SSI) occurs in Brassicaceae and is controlled by a single multiallelic S-locus, though systems controlled by two (or more) loci have also been described in other species (Muñoz-Sanz *et al.*, 2020). The S-locus encodes both male and female specificity determinants (S-determinants) whose products are predicted to interact and trigger the non-self and self-discrimination process (Iwano and Takayama, 2012). Most types of SI can be classified as sporophytic or gametophytic based on gene action in the stamen. The pollen phenotype is determined by the S-genotype of the diploid pollen-parent in sporophytic SI.

In Brassicaceae, S-locus genes encode serine/threonine receptor kinase (SRK) (Takasaki *et al.*, 2000) and cysteine-rich (SP11/SCR) proteins that, respectively, form the female and male S-determinants. Both S-locus genes, and a third gene encoding a glycoprotein (SLG) that may enhance SI expression (Takasaki *et al.*, 2000), are tightly linked and inherited as an S-haplotype (Sehgal and Singh, 2018). SRK protein is localized to the plasma membrane of the papilla stigmatic cells, and the small SP11 polypeptide is secreted from the anther tapetum, deposited onto the pollen coat, and finally acts as an SRK ligand upon pollination (Iwano and Takayama, 2012). Many Brassicaceae S-haplotypes have been identified. They exhibit complex hierarchical dominance relationships that are controlled by polymorphic small RNAs and their targets (Yasuda *et al.*, 2016). It should be noted that these dominance relationships complicate the observed compatibility patterns and are, thus, of considerable practical importance. However, in uncomplicated examples, specific interaction between SP11 and SRK from the same S-haplotype triggers a self-pollen rejection mechanism in stigma papillar cells (Muñoz-Sanz *et al.*, 2020). Downstream events are controlled by additional factors (Iwano *et al.*, 2015).

Progress has been made in understanding the genetics of vernalization and flowering in Brassicaceae based on studies of *B. napus*, *B. rapa*, Arabidopsis, and other species. MADS-box genes are involved in controlling male and female gametophyte development, embryo and seed development, as well as root, flower, and fruit development. MADS-box proteins are key players that regulate floral sex expression and flowering by binding directly to DNA in the first intron and promoter of Flowering Time (*FT*) and Suppressor of Overexpression of *CO1* (SOC1), respectively. This prevents transcription of the *FT* and *SOC1* integrator genes, thereby inhibiting flowering. At vernalizing temperatures, genes including Vernalization Insensitive 3 (*VIN3*), Vernalization1 (*VRN1*), and *VRN2* act together to physically repress Flower Locus C (*FLC*) mRNA transcription and translation, thereby allowing expression of *FT* and *SOC1*. *FLC* genes that play an important role in floral initiation based on studies of Arabidopsis and other species (Ridge *et al.*, 2015). *FLC* genes control flowering by blocking floral transition by binding directly to genes encoding activators of flowering and repressing their transcription. A study of biennial, annual, and rapid-cycling forms of *B. oleracea* found that a frameshift mutation in exon 4 causing an in-frame stop codon of *BoFLC2* was associated with early flowering behavior. In a study of cabbage, *BoFLC2* transcription was regulated by vernalization in a conserved manner. Also isolated from cabbage was a homologue of the important flowering time integrator gene *FT* whose expression in apices was inversely related to *BoFLC2* expression. These findings suggest that *BoFLC2* may underlie important differences between annual and biennial Brassica cultivars, particularly with respect to flowering time (Ridge *et al.*, 2015).

In annual Brassicas, functional alleles for late flowering (*BoFLC2*) and early flowering (*boflc2*) have been identified. Vernalization reduced transcription of *BoFLC* genes but increased transcription of floral integrator *BoFT*. Overall expression of *BoFT* was significantly higher in early-flowering *boflc2* lines, supporting the idea that *BoFLC2* plays a key role in maintaining the vegetative state in *B. oleracea* (Ridge et al., 2015). Further research will undoubtedly reveal more details about the initiation and control of flowering in Brassicaceae.

Agrobacterium-mediated genetic transformation protocols are available for Chinese cabbage, *B. oleracea*, and other crops in the family. To date, transgenic Brassica vegetables are not available in the Americas and are not allowed in the European Union because of consumer concern about their safety. Transgenic rapeseed or canola (*B. napus*) grown for cooking oil and meal containing engineered herbicide and insect resistance genes is grown in many areas. Precaution must be taken to avoid any potential out-crossing or seed mixes of transgenic canola and *B. napus* (e.g. rutabaga) vegetable seed crops.

Seed Production

Overview

Most Brassicas are cool-season crops both horticulturally and for seed production. These crops grow well in warmer weather during the early stages of vegetative growth. Once the plants transition to reproductive growth, mild, dry, and stable temperatures are key to successful seed production. The optimum diurnal temperature range is approximately 18°C to 26°C (64°F to 79°F) during flowering and seed development.

Grower acceptance and development of hybrid cultivars by seed companies has impacted commercial production. Hybrid cultivars of *B. oleracea* (cabbage, cauliflower, kale, broccoli, and kohlrabi), other *Brassica* spp., and radish are in demand throughout the world. Both sporophytic SI and CMS systems are employed to produce hybrid seeds as explained in the Pollination and Genetics sections in this chapter. The ratio of female to male parents varies depending on vigor of the male parent and its pollen production. Common ratios are: 1:1, 2:1, 2:2, or 3:1 respectively (Fig. 10.8). The percentage of sibs in F1 hybrid seed lots must not exceed allowances in seed laws where seeds will be sold, and many seed companies have set higher standards for important markets. Polymerase chain reaction (PCR) and electrophoresis can identify markers to confirm hybrid identity.

Although F1 hybrid seed is replacing open-pollinated cultivars, production of *basic* or foundation seed of non-hybrids and inbred lines is still important. For open-pollinated production, developing foundation or basic seed is the first step for later mass production of a cultivar for retail sale.

Basic seed

The following schedule describes basic or foundation seed production of open-pollinated cultivars. The first year, breeder seeds of biennial

Fig. 10.8. Hybrid broccoli seed field in Mexico configured with 2F:2M rows. (Courtesy of Tim Loeffler.)

Brassicas are sown for transplant production in early to mid-summer and transplants established in production fields in early fall. For crops that can be reliably established from seed, transplants are not used, and establishment is by direct seeding. During spring of the second year after overwintering and vernalization, plants are rogued for cultivar characteristics and quality. Off-types and disease plants are removed. True-to-type plants remain in the field to continue development or are pruned and transplanted to a greenhouse or high tunnels to grow in an optimized controlled environment that will exclude foreign pollen and diseases. Bees are supplied to pollinate the crop in a field and either bees or flies may be used in a greenhouse, tunnel, or net house (Fig. 10.9). Selected plants continue developing and produce a seed crop. For crops overwintering outdoors that produce a head the next spring or summer such as overwintering cultivars of cauliflower or broccoli, true-to-type plants will overwinter again at the end of year two to induce flowering which will occur in spring of year three. Mature seed is harvested during fall of year two or three depending on the location and whether a two- or three-year production cycle is used. After threshing, cleaning, and short-term storage, the seed is either used as stock for further seed production or stored for later additional field assessment to ensure best uniformity. Foundation or basic seed increase for annual Brassicas and radishes can occur in one year with earlier establishment and fall seed harvest.

Establishment

For commercial seed production for sale, cabbage, cauliflower, Brussel sprouts, and other biennial Brassicas are usually established in seed production fields using plug transplants. Transplants are grown in plug trays in artificial media often in a greenhouse or other protected structure to achieve control over timing of planting, uniform growth, fertilization, reduce disease, and control insects (Fig. 10.9). Germination is poor at temperatures exceeding 32°C (90°F), the ceiling temperature for many Brassica species (Jett

Fig. 10.9. Net house production of broccoli seed in Mexico. (Courtesy of Tim Loeffler.)

et al., 1996), which is another reason seeds are started in a greenhouse under controlled conditions and transplanted. Transplants are also a more effective use of valuable stock seed. Any used trays should be cleaned and sanitized prior to reuse for transplant production. Some greenhouse locations are designed with restricted access to control seed-borne diseases, like black rot. Stock seeds are tested to ensure freedom from seed-borne diseases or are hot-water treated before planting. Growing transplants in a greenhouse or in tunnels usually takes 35 to 45 days after sowing depending on the cultivar or parent lines. For hybrid seed production, split plantings are often required to synchronize the crop during pollination. There can be one to three planting dates for a male parent to be planted with a given female parent in the same hybrid seed production field. Timing of planting is important as too small a plant going into the winter months may not vernalize and too large a plant can be more susceptible to freezing injury. Bolting and flowering trials are often conducted in many production research programs to determine the timing of transplanting for successful nicking and pollination for each production area. In some areas, plants may be produced in specially prepared seed beds in the field and transplanted bare-rooted, but this results in less uniform stands, higher disease risk, increased labor, and is normally only considered for open-pollinated cultivars. Twenty-five grams (0.9 oz) of seed should produce about 3000 transplants. Some annual Brassicas like Chinese cabbage, and in some cases sprouting broccoli, are slow growing in the seedling stage and established by transplants as well. Transplant trays suitable for these crops have 2.5–3.8 cm (1–1½ in.) cells with from 200 to 338 cells per tray.

Final spacing depends on both the juvenile and mature size of the plants when flowering. There should be sufficient space to allow for plant inspection and remove of male plants after pollination. However, spacing of female plants should be optimized for maximum seed yield. Standard spacing recommendations cannot be given for Brassica seed production because many factors including environment, cultural practices, and genotype are involved. Optimal spacing for each crop should be determined experimentally through seed production research. Generic spacing suggestions would be rows 60 cm (24 in.) apart and 40 cm (16 in.) apart for small cultivars of cabbage or cauliflower. In-row spacing varies over a range of 30–60 cm (12–24 in.). Raised beds are preferable to improve drainage and to facilitate furrow irrigation when used. Bed widths vary with location but 122 to 152 cm (48 to 60 in.) bed widths are common (George, 2009). In general, plant density for broccoli and cabbage is 26,000 to 33,000 plants/ha, and cauliflower is 20,000 to 24,000 plants/ha depending on the cultivar, type of beds, and the growing environment.

Field establishment with transplants decreases the time plants are in the field and improves the synchrony of crop development. Transplants are hardened before field planting to reduce shock. Hardening is generally accomplished by withholding water, reducing nitrogen, and/or by exposing plants to low temperature and direct sunlight (Fig. 10.10). Hardening produces plants with short, sturdy stems, well-developed root systems, and greater dry weight. Transplanting can take place on single rows with furrow irrigation or double rows beds using plastic mulch with drip irrigation, with many variations in between (Fig. 10.7). Transplanting is best conducted under cool, moist conditions that limit water loss. In some desert areas, transplanting is done at night. A liquid fertilizer solution, sometimes in combination with insecticide, is often applied at transplanting (Swaider *et al.*, 1992). Obvious off-types, blind plants (lacking an apical meristem), and plants displaying disease symptoms are discarded prior to transplanting.

Temperature

The order of tolerance to high temperature for the major *B. oleracea* crops is cabbage (most

Fig. 10.10. Greenhouse production of cabbage plug transplants. The plastic covering was removed to harden plants before field planting. (Courtesy of Tim Loeffler.)

tolerant), broccoli, cauliflower, and Brussels sprouts (least tolerant). Low temperature tolerance depends on acclimation leading to low temperature events. After germination, plants in the juvenile phase will tolerate short-term exposure to temperature as low as 0°C to 5°C with little adverse effect. Maturing cabbage and kohlrabi plants can tolerate −3°C to −2°C (26°F to 28°F) after acclimation. Annual broccoli and summer cauliflowers, Chinese cabbage, and turnip tolerate −4° to −3°C (24° to 26°F). Brussels sprout, collards, mustard, biennial broccoli, and kale when properly acclimated by gradually decreasing temperatures can tolerate at least as low as −9°C (26°F) (Rubatzky and Yamaguchi, 1997). Prolonged periods of frozen soils combined with bright sunny days near or above freezing can severely damage plants by desiccation. During the winter following establishment, growth is slow, or plants become quiescent and are vernalized.

In spring of year two, plants resume growth, bolt, flower, and produce seeds which develop and mature during the summer of year two. Some hybrid seed crops need to be pruned to achieve proper nicking during flowering. Pruning is an intervention process to either delay or to advance flowering. Other purposes of pruning include to increase ventilation of the plant and to provide a more open frame for insect pollinators. There are associated risks since plants that are wounded can serve as an entry point for disease pathogens and a regular spray program with fungicides and bactericides is important. For broccoli, the head may be removed to stimulate the auxiliary buds to develop and produce flowers. Another technique is to remove the center one third of the head at the fresh market stage. In cabbage, removing some of the branches at beginning bloom delays flowering for four to five days. For cauliflower, the advancement of flowering starts from the perimeter of the curd and if allowed, the center of the head becomes inactive and can be a source of decay. Removing the center increases ventilation. Another technique for cauliflower is to remove the perimeter of the curd and retain the center to elongate and flower more uniformly. These techniques vary widely for each crop, production area, and some are specific for a hybrid cultivar.

To prevent lodging, the bushy plants are sometimes staked and stringed to keep them upright, especially for hybrid seed production (Fig. 10.11).

Fig. 10.11. Staking cabbage plants to prevent lodging during flowering and seed development. (Courtesy of Tim Loeffler.)

This improves air circulation, reduces disease pressure, improves pollination, and allows access for equipment to spray the crop and for seed harvest (Fig. 10.11). After maturation drying, seed harvest is conducted later summer or fall of year two of a production cycle. In hybrid production fields, the male rows are destroyed unless a reciprocal cross is being made with both parent lines having a strong SI as is done with certain cabbage and Chinese cabbage hybrids. Biennial seed production requires a location with consistently cool winters for vernalization but in a climate mild enough to avoid winter kill. In addition, the seed maturation and harvest period the second year needs to be dry to reduce disease pressure, prevent preharvest sprouting, and avoid shattering. In addition to mild winters and dry summers a suitable location for commercial Brassica seed production needs fertile well-drained soils, irrigation infrastructure, and skilled growers. Locations that meet all these criteria are limited so commercial production of biennial Brassica seed crops is limited to a relatively few places in the world.

Production areas

Large-scale commercial production of *B. oleracea* seed is concentrated in the following areas:

Broccoli and summer cauliflower production: Coastal Central California and Yuma, Arizona, USA; South Africa.
Cabbage and Chinese cabbage: Skagit Valley, Washington (Fig. 10.12) and Willamette Valley, Oregon, USA; Italy; New Zealand (South Island).
Broccoli, cabbage, cauliflower: France; Chile; Australia.
Tropical cauliflower: Northern India; Yuma, Arizona, USA; South Africa; Peru.
Hybrid radish: Western Oregon, USA; France; Italy; New Zealand.

Root crops

Root crops like rutabaga, radish, and turnip can be produced by either seed-to-seed or root-to-seed methods. For the root-to-seed system, developed roots are lifted and checked for off-types during the first year's growth before bolting occurs and is utilized primarily for basic and foundation seed production for open-pollinated cultivars as well as parent lines for hybrid seed production. Off-types are discarded, and true-to-type roots can be refrigerated for planted the following spring or biennial types can be replanted for vernalization in the field if conditions are mild. Since radish is an annual it can be immediately replanted for continued development after root inspection. For production of seed for commercial sale the seed-to-seed method is usually used. Roots remain in the ground without lifting, and rogueing is based solely on top growth evaluation. It is important that only stock seed with high genetic quality is used for seed-to-seed production. For biennial root crops produced seed-to-seed, the location should be selected where winters are mild to allow adequate vernalization and kill is unlikely such as the Treasure Valley of Idaho or the Columbia Basin of Washington state, USA.

Turnip, kohlrabi, and winter radish seed production are similar. Some cultivars of winter radish and turnip show a biennial character and flower faster after cold treatment. Spring/summer radish grows and matures rapidly, producing a seed crop in only 12 to 14 weeks. Radish roots can split with rapid growth caused by high water, high N, and high temperatures. Root splitting may increase root rots, lodging, and preharvest seed loss. Optimum average temperatures for slow quality radish growth are 13°C to 16°C (55°F to 60°F).

Seed of kohlrabi, turnip, and certain other crops can be vernalized after radicle emergence while still in the juvenile seedling stage (George, 2009). The treatment is given before spring planting. Seeds are hydrated on germination paper at approximately 20°C to 22°C (68°F to 72°F). Seventy to 90% germination (radicle emergence) occurs in the first 24 h for vigorous seed lots. Moist seeds are stored at −1°C (30°F) for 35 to 50 days (George, 2009). After the vernalization, delicate seedlings can be planted by hand or impregnating in viscous gel to protect the radicle and planted mechanically. This technique is used only for production of breeder or foundation seed and does not allow for rogueing against early bolters (George, 2009).

Fig. 10.12. Hybrid cabbage field with a good nick blooming in Skagit Valley, Washington state, USA. Beehives on a pallet at the edge of the field on the right improve cross pollination. Configuration 2F:2M rows. (Courtesy of Tim Loeffler.)

Leafy greens

Brassica greens, kohlrabi, and many of the smaller annual Asian vegetables, other than heading types of Chinese cabbage, can be established by precision direct seeding. Kohlrabi is precision-seeded to a final stand, or overseeded and thinned to 16 to 20 cm (6 to 10 in.) between plants. Adequate soil preparation is essential for precision seeding so seed to soil contact is maximized for rapid hydration. To promote best stand establishment, soil moisture and temperatures should be optimum to promote rapid germination and establishment. Other agronomic management for Brassica greens is similar cabbage. When direct-seeded in early spring, annual seed crops can mature in later summer to early fall after only 10 to 14 weeks depending on the environment. Turnip, kohlrabi, winter radish, and other biennial crops require a vernalization treatment and produce seed early during the second year after establishment unless vernalized by refrigeration and replanted in a greenhouse or warmer conditions for seed production in a root-to-seed system.

Seed Harvest

At full flower, most Brassicas will be bushy and roughly 1.2 m (47 in.) high and roughly 0.5 m (20 in.) across bearing little resembles to edible vegetables (Nieuwhof, 1969). After pollination is complete for hybrid seed crops, the male rows are destroyed. Prior to harvest, the field should be rogued for noxious and difficult-to-remove weeds as most hybrid seeds are sold in a wide range of countries that may prohibit seed movement if noxious weeds are present. Removing weeds before harvest also avoids excessive seed losses during the post-harvest cleaning. A final phytosanitary field inspection is completed for seed crops to be exported.

Seeds usually mature 50 to 90 days after fertilization. Rain can cause damage to seed crops by increasing lodging, shattering, and disease. This is why ideal production areas are dry during flowering, seed development, maturation, and harvesting. Silique size at maturity should range from 4.5 to 10 cm (2 to 4 in.) in length and 3 to 6 mm (0.12 to 0.24 in.) wide. Each pod will release 20 to 40 seeds ranging in color from dark brown to purple (Musil, 1948). Brassicaceae siliques tend to shatter easily leading up to maturity. As siliques dry, they dehisce into two halves and can fall to the ground before harvest. Harvesting seed at the proper time using the best methods can significantly reduce shattering. Silique maturation and drying is closely linked to plant senescence. Senescence is visualized by color transition from green to orange-brown foliage and silique color. Seed maturation can be confirmed by inspecting the oldest pods which will turn brown first. Mature seeds inside brown siliques will be firm, dry, and dark colored when ready for harvest.

A decision must be made when to cut plants for final drying since there will be a range of maturity stages on each plant. For the best germination and vigor, most seeds should have reached full developmental maturity with pods still intact. The seeds at this stage normally have a brown seed coat with yellow cotyledons and seed moisture is still high relatively high. Most hybrid seed production is cut by hand or with a mechanical swather (Fig. 10.13). Large fields of open-pollinated cultivars are usually mechanically swathed. Cabbage seeds, highest germination and vigor occurred when plants were cut at 30% moisture content (Gray *et al.*, 1985). Cut plants are windrowed to encourage slower, more uniform field drying to reduce shattering. For smaller scale production of hybrid seed, foundation, or breeder seed, plants may be placed on canvas or plastic sheets to final dry so any shattered seeds will be caught (Fig. 10.14).

Once dry, usually after 10 to 15 days after cutting, the plants can be fed into a stationary thresher positioned in a seed field and or combined directly from the windrows with the intake gathering reels removed to reduce shattering (Fig. 10.15). For direct combining, the roller uptake should be set as low as possible, and the drum speed should be adjusted to between 250 and 500 rpm. At the beginning of threshing, a close inspection of the seed for damage is important and to make any harvester adjustments to obtain the highest quality seed possible. A combine harvest speed of 3.2 to 4.0 kmph (2 to 2.5 mph) is recommended but this depends on the equipment. The crop should dry in the field

Fig. 10.13. Cutting senescing cabbage plants with a heavy seed load in preparation for windrowing and final drying before combining. (Courtesy of Tim Loeffler.)

Fig. 10.14. Broccoli seed stalks drying on plastic before mechanical threshing. (Courtesy of Tim Loeffler.)

Fig. 10.15. Combining of windrows precut for uniform drying to reduce shattering. (Courtesy of Tim Loeffler.)

to about 9 to 10% seed moisture before combining and will ultimately be dried to 5.5 to 7% for long-term storage.

Seed Cleaning

Brassicaceae seeds crack easily and it is therefore important to use a relatively low combine cylinder speed for mechanical threshing as described above (Fig.10.15). Higher speeds may be required if moisture contents are elevated, and tissues are not brittle. Brassica seeds are cleaned using an air screen cleaner, gravity separator, and/or spiral separator before testing for germination, seed health, and genetic quality. Spirals separate misshaped and split seeds from a lot because the round seeds move faster than non-round chaff or broken seeds down the column (Fig. 10.16). Additional upgrading processes include seed coat color sorting, chlorophyll sorters to remove immature seed, and/or magnetic separation to remove soil to meet export and market requirements. Seed lots testing positive for certain seed-borne diseases like black rot are usually disinfected using hot-water treatment and re-tested. Hybrid seeds are separated into several sizes and tested for final quality before being packaged for sale.

Seed Yield

Seed yields vary regionally and depend on the cultivar and production environment. The following numbers give a rough approximation of average seed yields that can be expected. Cabbage yields of 700 kg/ha (625 lbs/acre) are common. The 1000 grain weight of cabbage is 3.3 g (0.12 oz). Cauliflower seed yield averages about 400 kg/ha (357 lbs/acre), and the 1000 grain weight is approximately 2.8 g (0.1 oz). Brussel sprouts seed yield is 600 kg/ha (535 lbs/acre), and the 1000 grain weight is 2.8 g (0.1 oz). Kohlrabi yield is 700 kg/ha (625 lbs/acre), with 1000 grain weight of 3.2 g (0.12 oz). In general, a healthy well-pollinated *B. oleracea* plant may produce 0.2 kg (0.4 lb) of seed each and there

Fig. 10.16. Aluminum spiral separator used to clean Brassica seeds with the protective shield open. Uncleaned seeds are poured in the chute at the top. Unbroken Brassica seeds are round, roll faster, and are collected in the outer partition. Broken pieces are slower and come out the inner partition. (Courtesy of Tim Loeffler.)

are approximately 300 seeds per gram (5660 seeds per ounce) (George, 2009).

A typical open-pollinated Chinese cabbage seed yield is about 500 kg/ha (446 lbs/acre), and the 1000 seed weight is approximately 3 g (0.11 oz). For radish, seed yields of approximately 1000 kg/ha (892 lbs/acre) are satisfactory although yields as high as 1500–2000 kg/ha (1338–1784 lbs/acre) are achieved by some seed producers. The 1000 grain weight of radish is approximately 10 g (0.4 oz) (George, 2009).

Pest Management

Control of plant pathogens, insect pests, and weeds should be an integrated component of production management. Development of an integrated pest management (IPM) system should rely upon regional recommendations and be in accordance with labeling restrictions for control measures (Flint, 1992).

Brassicaceae seed crops should be grown in a three- to five-year rotation with other families to reduce soil-borne diseases. Longer rotations may be needed when soil-borne diseases are black leg (*Phoma lingam*), black rot (*Xanthomonas campestris*), or clubroot (*Plasmodiophora brassicae*). Weed control is essential to reduce competition for nutrients, water, sunlight, and to increase air circulation around plants to reducing disease. Introduction of plant pathogens, root-dwelling insects, nematodes, and weed seeds from soil-laden farm equipment, transplants, and irrigation water should be avoided. Weeds of wild Brassicas are problematic because they may cross with the seed crop and their seeds are extremely difficult to separate after harvest.

Pathogens

Black rot (*Xanthomonas campestris*) is a bacterium that affects cabbage and related crops worldwide and is one of the most serious cabbage diseases in warm climates. Seedling infection can be difficult to diagnose since only a few plants in a lot may be affected. Infected seedlings are yellow, stunted, have blackened margins on their cotyledons, and may eventually wilt and die. On older plants, the disease is easily recognized by the presence of yellow V-shaped or U-shaped areas extending inward from the margins of the leaves. As the disease progresses, the yellow lesions turn brown, and the tissue dies. The veins of affected leaves darken, and the midrib turns black as discoloration progresses toward the leaf base. Eventually, the bacteria spread into the main stem. When infected stems are cut in cross-section, the vascular ring may appear black when bacteria infect xylem vessels. This vascular discoloration extends from the stem to the upper leaves and down into the roots. In later stages of infection, all central tissues of the main stem turn black. The vascular symptoms of black rot should not be confused with Fusarium yellows disease which causes a similar brown discoloration (Kemble *et al.*, 1999).

The black rot bacterium can overwinter on infected cabbage seeds, in weeds from the family Brassicaceae including black mustard, field mustard, wild turnip, wild radish, shepherd's purse, and pepperweed, or in infected plant material in the soil. The bacterium can persist in plant residue for one to two years or as long as the plant debris remains intact. Black rot is spread on seed and seedlings, by movement of contaminated plant material, in irrigation water or splashing rain, by insects, by cultivation equipment, and by field workers. Post-harvest eradication treatments have been developed, but it is best to keep this disease out of seed fields. Aphid control is critical to prevent spreading as well (Kemble *et al.*, 1999).

Black leg (*Phoma lingam*) causes oval, sunken, light brown cankers, often with black or purple margins near the stem base. The fungal canker enlarges until stems are girdled causing wilting and eventual death. Pale, irregular-shaped leaf spots gradually enlarge, becoming circular with gray centers. Small, black, pepper-like spots (pycnidia) are visible within the leaf spots and produce spores. Severely infected plants may look blue, are stunted, wilted, and eventually collapse (Kemble *et al.*, 1999).

Black leg can survive in plant residue for two to three years and may be seed-transmitted. Plants are often infected during transplant production and field planting, where water spreads fungal spores among plants. Only transplants free of black leg should be planted. Crop rotation is an effective control. Remove plant debris at the end of the season to prevent overwintering and spread to adjacent fields. Fungicide seed treatments, soil fumigation, or solarization can effectively control black leg (Kemble *et al.*, 1999).

Club root (*Plasmodiophora brassicae*) is another common fungal disease of Brassicas (Cornell, 2022). Roots can be infected before leaves show symptoms. Infected roots are unable to absorb water and nutrients, so top growth is stunted, and lower leaves may yellow and die. Affected plants may wilt during the day but recover at night. Secondary infection may kill plants by invading root galls. Control measures include maintaining a high pH which inhibits the fungus. Foliar applications of micronutrients such as boron may be necessary if deficiency symptoms occur due to high soil pH. Long rotations with other Brassica crops can provide effective control as well as improved soil drainage, preventing spread from other infected fields, and use of uncontaminated irrigation water (Rimmer *et al.*, 2007).

Alternaria leaf spot (*Alternaria brassicae*) causes small, circular, dark spots on older leaf surfaces. As spots enlarge, concentric rings develop within lesions surrounded by a yellow halo. The tan-colored centers of lesions may eventually fall out, producing a hole. Under wet conditions, lesions may become covered with masses of black spores. Weeds from the family Brassicaceae may also transmit the fungus. Spores of Alternaria may spread by wind and water. The disease is most damaging under wet, warm conditions of 20°C to 30.5°C (68°F to 81°F). Other Brassica crops should not be grown in the same field for at least three years. Infected plant debris should be removed or destroyed after the season. Only disease-free transplants should be used. A fungicide spray program at seven- to ten-day intervals after the first sign of the disease may be effective (Kemble *et al.*, 1999).

Wirestem (*Rhizoctonia solani*) is a fungal disease that affects transplant beds and production fields. Affected plants show reddish-brown discoloration and constriction of the stem near the soil line causing distorted and twisted stems from which its common name is derived. Surviving plants are weak and stunted. The fungus may be present in soils where infected plant debris persists. Prolonged, overly moist soil favors the disease. Wirestem is greatly influenced by recent cropping history. Crop rotation is an effective control and related Brassicas should not be grown in the same field for three years in a row. Soil fumigation, use of raised beds, and careful watering may reduce this disease. Fungicide treatments applied at transplanting may provide effective control (Kemble *et al.*, 1999).

Downy mildew (*Peronospora parasitica*) is a fungus spread by wind-blown spores from infected plants. In moist weather, a white, fluffy, fungal growth develops mostly on the undersides of leaves. With time, yellow to tan spots develop on the corresponding upper leaf surface. Infected leaves on young plants may drop and the plants eventually die. On older leaves, the infected areas turn tan with a papery texture and gray spore masses develop on the undersides of infected leaves. Plants infected with downy mildew are predisposed to infection from bacterial soft rot. Mildews overwinter on seed, in weeds from the family Brassicaceae, and possibly in soil. Heavy fogs, light rains, prolonged dews, overhead irrigation, and night temperatures between 8°C and 16°C (46° and 61°F) with day temperatures below 23.8°C (75°F) favor downy mildew. Spores of the fungus drift long distances in cool, moist air (Kemble *et al.*, 1999).

Fusarium wilt (*Fusarium oxysporum* f. sp. *conglutinans*), also called "yellows", infects all members of Brassicaceae and is closely related to Fusarium diseases in other vegetables. Infected plants turn yellowish-green two to four weeks after emergence or transplanting. The lower leaves turn yellow first, and then symptoms move to the upper leaves. With time, the yellow tissue turns brown becoming dry and brittle. The vascular tissues in the stems and leaves turn dark brown, expressing symptoms similar to black rot. However, Fusarium wilt develops internally and appears in the lower portions of the plant first. Fusarium grows in the soil and on plant debris. It also produces two types of spores, one of which is short lived, and the other, heavy-walled and capable of withstanding long periods of low temperatures and drought. Fusarium can prosper in soil for many years even without Brassica host plants. The Fusarium fungus grows poorly at temperatures below 16°C (60°F) and reaches its maximum growth rate between 27°C and 32°C (80°F to 90°F). At 35°C (95°F) and above growth is inhibited. Since Fusarium can live free in the soil without a host, conventional controls such as rotation, fungicide treatments, and destruction of crop refuse are not effective. Some cultivars are resistant (Cornell, 2022).

Turnip mosaic virus (TuMV) is one of the most important and widespread viruses infecting Brassicas. This virus infects radish, turnip, lettuce, endive, and spinach and other Brassicas. TuMV is efficiently transmitted in a non-persistent manner by several aphid species, most notably the green peach aphid (*Myzus persicae*) and the cabbage aphid (*Brevicoryne brassicae*). TuMV causes mosaic and black necrotic ring spots in cabbage, cauliflower, and Brussels sprouts. Spotting may be several layers deep within the head and appears on the midribs, the side veins, and in the interveinal areas where spots often coalesce. TuMV also causes mosaic with leaf distortion and necrosis on lower leaves of turnip, radish, and mustard plants (Cornell, 2022).

Cauliflower mosaic virus (CaMV) is another virus that infects this family. Symptoms have often been confused with TuMV infections. Like TuMV, cauliflower mosaic is transmitted by the same aphid species in a non-persistent manner. CaMV has a host range limited to Brassicas and is distributed mainly in the temperate regions of the world. The virus induces mosaic and a striking veinal chlorosis in most hosts. A masking of symptoms may occur in chronically infected plants, particularly at high temperatures (Cornell, 2022).

Insects

The insect pest management program should keep pest populations below damaging levels, i.e. the economic threshold, while protecting natural enemies and competitors (Rimmer *et al.*, 2007; Cornell, 2022). Negative effects on pollinizers should be considered before using systemic or contact insecticides and other pesticides.

Diamondback moth (*Plutella xylostella*, DBM) caterpillars have become a serious worldwide pest of Brassicas because of insecticide resistance (Cornell, 2022). The caterpillar's name comes from the diamond-shaped markings on the adult moth. DBM caterpillars are most active in spring and early summer, producing up to ten generations per growing season. DBM larvae are smaller than larvae of other caterpillar pests of Brassica crops. Mature larvae are green, 10 to 15 cm (1/3 to 5/8 in.) long, and tapered at both ends. A larva wriggles violently when disturbed and often dangles from a leaf by a silken thread.

Because DBM populations can increase rapidly at temperatures above 27°C (80°F), atleast twice-weekly scouting is recommended. Heavy rains may reduce populations dramatically. *Bt* insecticides are usually effective and should be applied when worm activity is noticed. Sprays may need to be applied at five-day intervals when populations are high. A combination of *Bt* and pyrethroid insecticides may be used as well. DBM pupae suffer natural mortality from native parasitoids, and parasitized pupae have a broad white stripe around the pupa. Removing or destroying crop debris will prevent buildup of DBM and migration to adjacent fields.

Cabbage loopers (*Trichoplusia ni*, CL) are most active in early summer and fall. Mature larvae grow as large as 3.8 cm (1.5 in.) in length and have three pairs of prolegs in the rear. The larvae loop their bodies when they move and grasp plants firmly when handled. Their green color blends well with the foliage, and their dark excrement or frass is often visible before the larvae. CL larvae chew holes in leaves, and large, mature larvae consume large amounts of plant tissue. Because large larvae are less susceptible to insecticides than young, sprays should be applied when larvae are small. *Bt* insecticides are moderately to highly effective and may be used separately or in combination with pyrethroid insecticides (Kemble *et al.*, 1999).

Imported cabbageworm (*Pieris rapae*, ICW) is a white to yellowish-white moths with black spots on its wings. Unlike nocturnal CL moths, ICW moths fly during the day. The green larvae have a velvety appearance because of their dense coating of fine hairs. The mature larva has a faint orange stripe down its back and chews irregular holes in the leaves, bores into heads, and contaminates leaves and heads with frass. Damage is similar to that caused by cabbage loopers (Kemble *et al.*, 1999).

Cabbage webworm (*Hellula undalis*) larvae are 2 cm (0.75 in.) long when mature and gray in color with five dark stripes. The head capsule is black with a distinct white V-shaped mark. Cabbage webworm is an occasional pest of the Brassicas feeding primarily around buds. The caterpillars produce webbing that can protect them against insecticide contact (Kemble *et al.*, 1999).

Cross-striped cabbageworm (*Evergestis rimosalis*) has black and white transverse stripes on its back. It is controlled with *Bt* and other insecticides recommended for caterpillar control (Kemble *et al.*, 1999).

Beet armyworm (*Spodoptera exigua*) may infest fall plantings of cabbage and related crops. Armyworms migrate to cabbage in large numbers as other crops die or are harvested, so it is important to carefully monitor late summer/fall plantings and apply insecticides when infestations are first detected. The moths lay egg masses on the undersides of leaves and cover the eggs with white, fuzzy scales. The larvae are light green to dark olive green and may have longitudinal stripes on the back or sides of their body. Control measures should be applied if egg masses or larvae are found on 2 to 3% of the plants (Kemble *et al.*, 1999).

Cutworms (*Agrotis* spp.) are caterpillars that rest beneath the soil during the day and feed on stems and foliage at night. They are dark gray-brown with an oily appearance. Cutworms often curl into a C-shape at rest or when disturbed. Cutworm larvae overwinter in fields and therefore may be present at planting, particularly in fields with high organic matter and previous crop residue. Before planting, remove weeds and plant debris to starve developing larvae. Beneficial nematodes may attack and destroy cutworms. Release of trichogramma wasps over three consecutive weeks will parasitize cutworm eggs. Diatomaceous earth sprinkled around the base of plants can be an effective control. A mixture of bran or corn meal and Dipel dust (*Bt*-kurstaki) and molasses sprinkled on the soil surface will reportedly kill caterpillars. After harvest, organic debris should be removed, and deep ploughing may prevent larvae from overwintering. A preplant or at-plant soil insecticide can kill cutworms. Insecticide sprays can be applied to the base of the plants if cutworm damage is observed after planting (Kemble *et al.*, 1999).

Seed and root maggot (*D. platura, D. radicum*) larvae develop into flies and are attracted to decaying organic matter. They feed on young roots and stems, severely reducing Brassica stands. Damage usually occurs during cool, wet conditions when plants grow slowly. Delaying planting until warm dry weather may reduce the risk of maggot damage (Kemble *et al.*, 1999).

Several aphid species, including cabbage aphids (*Brevicoryne brassicae*), may infest Brassicas. Cool, dry weather is most favorable for aphid development. Aphids do not usually infest transplants but may build up in fields. Large numbers of aphids may kill small plants. Their feeding can cause leaf curl on older plants. Often, the most serious problem associated with aphids is contamination. Aphids that colonize deep inside buds or heads are almost impossible to remove. Natural enemies can control light infestations but may not be effective against heavy populations. Broad-spectrum insecticides like pyrethroids can increase aphid numbers by eliminating natural enemies; therefore, these insecticides should be used only when necessary (Kemble *et al.*, 1999). Insecticidal soaps provide biological control.

Sweetpotato or silverleaf whitefly (*Bemisia tabaci*) adults are tiny, winged insects about 2 mm (1/16 in.) long with white bodies and yellow heads. They are easy to detect on the undersides of leaves or as they fly when plants are disturbed. The adults lay eggs on the undersides of leaves where immobile nymphs develop. The nymphs are scale-like in appearance. Like aphids, whitefly adults and nymphs suck plant sap with piercing mouthparts. They produce sticky honeydew on leaves where they are feeding. Black sooty mold often grows on the honeydew. Insecticides are not highly effective against whiteflies, so prevention is the best approach. Transplants should be free of whiteflies, and fields planted as far as possible away from other whitefly-infested crops. Destroying weeds and previous crop residues that may harbor whitefly populations is also an important pest management technique (Kemble *et al.*, 1999). Whiteflies tend to be a pest predominantly in long-season areas with mild winters.

The harlequin cabbage bug (*Murgantia histrionica*) is closely related to the common stinkbug. Harlequin bug is a brightly colored, shield-shaped insect with red and black markings and piercing-sucking mouthparts that feeds on the veins of cabbage and other Brassica leaves. Harlequin bug eggs are barrel-shaped and laid in clusters on leaves. The eggs are white, and each has two black bands around the circumference (Kemble *et al.*, 1999).

Flea beetles (*Altica* spp.) are most common in the spring and in fields that are weedy or surrounded by weeds. The adults are small, dark-colored beetles with enlarged hind legs enabling them to jump great distances. Flea beetles feed on the undersides of leaves, causing numerous small, round, or irregular holes. Flea beetles are of greatest concern during seedling development when they can defoliate small plants (Kemble *et al.*, 1999). Natural predators can successfully control flea beetles including braconid wasps and tachinid flies. In both cases, the larval stage feeds on the flea beetle, while the adults feed on nectar and pollen. To encourage braconid wasps and tachinid flies, caraway, herbaceous fennel, coriander, California poppies, and yarrows can be planted near seed fields.

The vegetable weevil (*Listroderes difficilis*) adult is 6 to 10 mm (1/4 to 3/8 in.) long, brownish gray with two light-colored marks on its wing covers and a long snout. Larvae are legless grubs. Adults and larvae feed on the foliage and stems of plants causing the most damage to seedlings in early fall and spring. Formulated mixtures of the botanical insecticides pyrethrin or azadirachtin may provide control. Spot treatments may be all that is necessary since vegetable weevils do not fly. Treatment is only recommended when more than 5% of the stand is damaged (Kemble *et al.*, 1999).

References

Akobundu, I.O. (1980) Live mulch: a new approach to weed control and crop production in the tropics. In: *Proceedings 1980 British Crop Protection Conference – Weeds*, Vol. 2. British Crop Protection Council, Thornton Heath, UK, pp. 377–382.

Bratsch, A. (2009) *Specialty Crop Profile: Horseradish*. Virginia Cooperative Extension, Publication 438–104. Virginia Polytechnic Institute and State University, Virginia. https://vtechworks.lib.vt.edu/handle/10919/55168 (accessed 11 September2023).

CABI (2023) *Crop Protection Compendium*. CAB International, Wallingford, UK. www.cabi.org/cpc/ (accessed 11 September 2023).

Collins, R. (2003) Growing wasabi in western North Carolina. https://newcropsorganics.ces.ncsu.edu/specialty-crops/specialty-crops-research/collins2/ (accessed 11 September 2023).

Cornell (2022) Integrated crop and pest management guidelines for commercial vegetable production. https://cropandpestguides.cce.cornell.edu/Preview/2022/22_Veg_Crops_Preview.pdf (accessed 11 September 2023).

Davis, J., Evans, R., Grabow, G., Jester, B., Little, B., *et al.* (2007) *Offsetting Drought for Small-Scale Vegetable Production in North Carolina*. Publication AG-519-04. North Carolina State Extension. https://content.ces.ncsu.edu/offsetting-drought-for-small-scale-vegetable-production-in-north-carolina (accessed 20 September 2023).

Fenwick, G., Heaney, R.K., Mullin, W., VanEtten, J. and Cecil, H. (1982) Glucosinolates and their breakdown products in food and food plants. CRC *Critical Reviews in Food Science and Nutrition* 18, 123–201.

Flint, M.L. (1992) *Integrated Pest Management for Cole Crops and Lettuce*. Agriculture and Natural Resources, University of California. https://ipm.ucanr.edu/IPMPROJECT/ADS/manual_colecropslettuce.html (accessed 11 September 2023).

Fu, I., Shennan, C. and Welbaum G.E. (1991) Evaluating Chinese cabbage cultivars for high temperature tolerance. In: Janick, J. and Simon, J.E. (eds) *Proceedings of the Second National Symposium on NEW CROPS, Exploration, Research, Commercialization*. Timber Press, Portland, Oregon, pp. 570–573.

George, R.A.T. (2009) *Vegetable Seed Production*, 3rd edn. CAB International. Wallingford, UK.

Griffiths, M. (1994) *Index of Garden Plants.* Timber Press, Portland, Oregon.

Iwano, M. and Takayama, S. (2012) Self/non-self discrimination in angiosperm self-incompatibility. *Current Opinion in Plant Biology* 15, 78–83.

Jett, L.W., Welbaum, G.E. and Morse, R.D. (1996) Effects of matric and osmotic priming treatments on broccoli seed germination. *Journal of the American Society for Horticultural Science* 121, 423–429.

Kemble, J.M., Zehnder, G.W., Sikora, E.J. and Patterson, M.G. (1999) *Guide to Commercial Cabbage Production*. Publication ANR-1135. Alabama Cooperative Extension System, Alabama A&M University and Auburn University, Alabama.

MBGP (2023) The Multinational Brassica Genome Project. https://www.brassica.info/home/about_mbgp.html (accessed 7 September 2023).

Muñoz-Sanz, J.V., Zuriaga, E., Cruz-García, F., McClure, B. and Romero, C. (2020). Self-(In)compatibility systems: Target Traits for Crop-Production, Plant Breeding, and Biotechnology. *Frontiers in Plant Science* 11, 195. doi.org/10.3389/fpls.2020.00195 (accessed 21 September 2023).

Musil, A.F. (1948) *Distinguishing the species of Brassica by their seed*. Circulation 857. US Department of Agriculture, Washington DC. https://searchworks.stanford.edu/view/11555182 (accessed 11 September 2023).

Myers, C. (1991) *Specialty and Minor Crops Handbook*. Publication 3346. University of California, Small Farm Centre, University of California.

Myers, J.R. (2006) *Outcrossing Potential for Brassica Species and Implications for Vegetable Crucifer Seed Crops of Growing Oilseed Brassicas in the Willamette Valley*. Report 1064. Oregon State University, Oregon. https://www.semanticscholar.org/paper/Outcrossing-potential-for-Brassica-species-and-for-Myers/8121956d534c481891e429595836b4c6bd0fbdb1 (accessed 11 September 2023).

Nieuwhof, M. (1969) *Cole Crops*. World Crops Books Series. Leonard Hill, London.

Peirce, L.C. (1987) *Vegetables: Characteristics, Production, and Marketing*. Wiley, Hoboken, New Jersey.

Ridge, S., Brown, P. H., Hecht, V., Driessen, R.G. and Weller, J. L. (2015) The role of BoFLC2 in cauliflower (*Brassica oleracea* var. *botrytis* L.) reproductive development. *Journal of Experimental Botany* 66, 125–135.

Rimmer, S.R., Shattuck, V.I. and Buchwaldt, L. (eds) (2007) *Compendium of Brassica Diseases*. American Phytopathological Society, St Paul, Minnesota.

Rubatzky, V.E. and Yamaguchi, M. (1997) *World Vegetables: Principals, Production, and Nutritive Values*, 2nd edn. Chapman & Hall, New York.

Schmidt, R. and Bancroft, I. (2011) Perspectives on genetics and genomics of the Brassicaceae. In: Schmidt, R. and Bancroft, I. (eds), *Genetics and Genomics of the Brassicaceae*. Springer, New York, pp. 617–632.

Sehgal, N. and Singh, S. (2018) Progress on deciphering the molecular aspects of cell-to-cell communication in Brassica self-incompatibility response. *3 Biotech* 8, 347. doi:10.1007/s13205-018-1372-2 (accessed 20 September 2023).

Swaider, J.M., George, W. and McCollum, J.P. (1992) *Producing Vegetable Crops*, 4th edn. Interstate Printers and Publishers, Danville, Illinois.

Takasaki, T., Hatakeyama, K., Suzuki, G., Watanabe, M., Isogai, A. and Hinata, K. (2000) The S receptor kinase determines self-incompatibility in Brassica stigma. *Nature* 403, 913–916. doi:10.1038/35002628 (accessed 21 September 2023).

USDA (2023) Plants Database. US Department of Agriculture. https://plants.usda.gov/home (accessed 21 September 2023).

Welbaum, G.E. (2015) *Vegetable Production and Practices*. CAB International. Wallingsford, UK.

Wittwer, S. (1987) Chinese cabbage-year-round. In: Wittwer, S., Yu, Y., Sun, H. and Wan, L. (eds) *Feeding a Billion: Frontiers of Chinese Agriculture*. Michigan State University Press, East Lansing, Michigan, pp. 271–277.

Yasuda, S., Wada, Y., Kakizaki, T., Tarutani, Y., Miura-Uno, E., *et al.* (2016) A complex dominance hierarchy is controlled by polymorphism of small RNAs and their targets. *Nature Plants* 3, 1–5. doi:10.1038/nplants.2016.206 (accessed 22 September 2023).

11

Family Amaranthaceae Subfamily Chenopodioideae

Overview

Historically, garden or table beet (*Beta vulgaris* subsp. *vulgaris*) and spinach (*Spinacia oleracea*) were classified in the family Chenopodioideae (Chenopods). More recently, beet, spinach, and other related crops have been redesignated in Chenopodioideae, a subfamily within the family Amaranthaceae. The family Amaranthaceae contains many widely distributed dicot, annual, biennial, and perennial plants, including many weeds (USDA APHIS, 2023). Weedy species include lambsquarters (also called annual goosefoot, *Chenopodium album*), and red-rooted pigweed (*Amaranthus retroflexus*) and other related weeds are members of this family.

Beets are an ancient crop dating back to the second millennium BC and were likely domesticated in the Mediterranean region (Zohary and Hopf, 2000). Spinach was domesticated later somewhere in southwestern Asia, Iran, and nearby areas of the Middle East. Spinach was described in Chinese literature around the year 650 but was unknown to ancient Greeks and Romans. Spinach became popular as a leafy vegetable and preferred by many over beet. Chard originated around the 13th century. Beets, spinach, and chard were introduced to the western hemisphere from Europe. Due to space limitations, this chapter will emphasize production of garden beet and spinach seeds. In addition to beet and spinach, other important vegetables in this subfamily include: Swiss chard (*Beta vulgaris* subsp. *vulgaris*), which is a leafy green or foliage beet; orach (*Atriplex hortensis*), a warm-weather leafy annual alternative to spinach; and Good-King-Henry or perennial goosefoot (*Blitum bonus-henricus*, syn. *Chenopodium bonus-henricus*), a perennial leafy green. Non-vegetable related crops include mangelwurzel or fodder beet and sugarbeet (both *Beta vulgaris* subsp. *vulgaris*), and quinoa (*Chenopodium quinoa*), a small grain. New Zealand Spinach (*Tetrogonia expansa*) is a member of the family Tetrogoniaceae and is not true spinach.

BEETS

Botany

Beets and chard have been historically subdivided into two distinct Groups: Flavescens or Crassa and Cicla (Hanelt *et al.*, 2001). Cicla is considered the Leaf Beet Group within *Beta vulgaris* subsp. *vulgaris*. This Group designation includes Swiss chard and foliage beet cultivars grown primarily for their leaves (Welbaum, 2015). Similarly, the former botanical variety Crassa, now designated as either Group Flavescens or by some Group Crassa, includes garden, fodder, and sugarbeets, consumed for their roots (Welbaum, 2015; USDA ARS, 2023b).

DOI: 10.1079/9781789243260.00011

Beets are a cool-season, frost-tolerant herbaceous biennial. Beets germinate well over a range from 8°C to 30°C (46°F to 86°F). After germination, seedlings form a crown, consisting of an unelongated stem that forms a rosette of leaves. Leaf color varies from light green to dark red. Petioles vary in length and tend to be thin especially when plants are crowded. Leaf shapes vary from oblong to triangular, leaf margins may be curved or straight, and leaf-blade surfaces smooth or savoy (wrinkled).

Beets are a qualitative long-day biennial with a cold requirement for bolting and flower initiation (George, 2009). After mature plants are vernalized, flower stalks develop rapidly usually during spring of the second year with little additional vegetative growth. Seed stalks reach a height of up to 1.5 m (4.9 ft) with extensive branching to form a large, compound, open inflorescence with small, greenish, almost sessile flowerers that form in clusters of two or three, sometimes singly, in axils of bracts on open spikes (Rubatzky and Yamaguchi, 1997). Beet flowers are perfect having both male and female parts on the same plant but protandrous. This means that on any flower anthers develop first and dehisce pollen for two or three days before shriveling and senescing. Stigmas develop on the same flower and are receptive for about two weeks after anther senescence. Protandry ensures that cross-pollination occurs. Pollen is light and carried by wind as the primary transfer mechanism, although insect pollination has been known to occur sometimes.

Fruit called utricles are irregularly shaped, hard seed balls containing multiple embryos that develop because multiple flowers grow together and are fused by corklike bract tissues (Fig. 11.1; Rubatzky and Yamaguchi, 1997). Polygerm utricles can produce several seedlings giving the appearance of greater than 100% germination. Monogerm cultivars produce single flowers and utricles containing only a single embryo (Fig. 11.1). Monogerm cultivars produce stands with greater uniformity and do not need thinning when precision-planted.

Fig. 11.1. Irregularly shaped beet utricles. Most garden beet cultivars are polygerm but a few are monogerm with only embryo per utricle.

Maturing beet plants are composed morphologically of the leaves, crown, a neck, enlarged hypocotyl, and root. An enlarged hypocotyl–root axis forms near the soil surface along with a narrowing, tapered, true root that extends deeper into soil as an anchor. Under favorable growing conditions, accumulation of photosynthates causes the hypocotyl/roots to increase in diameter. Cambial growth increases root thickness due to division and enlargement of parenchyma tissues. The neck is exposed extending from below the crown to near the soil line. The broadest portion of the beet lies below the neck, has no lateral roots, and is completely smooth. The mature beetroot (hypocotyl and upper root) grown for commerce is a spherical or elongated fleshy axis of varying shape depending on the cultivar. This enlarged hypocotyl–root axis is eaten as food and frequently referred to in the vegetable trade as a root or beetroot. The root interior consists of alternating zones of vascular and storage tissues that are uniformly colored or contrasting rings in other cultivars. Below the hypocotyl–root axis is a "true" root. The root tapers to a slender taproot, which may be quite long. Lateral or secondary roots arise along the length of the taproot and are important for nutrient uptake.

Garden beets occur in a wide range of colors ranging from deep red to purple to white to deep yellow. Purple, typically referred to as dark red in the seed trade, is the preferred color of commercial garden beets in North America and much of Europe. Interior color differences are due to different levels of pigmentation. The red and purple colors of beets are caused by betacyanin, a pigment with properties similar to anthocyanin, which is the source of purple color in many vegetables. Beets also contain betaxanthin, a yellow pigment. The ratio of betacyanin and betaxanthin varies among cultivars and with environmental conditions. White cultivars lack both pigments. Intense red roots are an indication that betaxanthin is present. Yellow color means there is little betacyanin.

Beet types: vegetable and agronomic

There are three types of *Beta vulgaris* grown: fodder beets, sugarbeets, and garden (or table) beets. All three have a common ancestor and were developed by human selection and plant breeding. Since they are all members of the same species, sexual crossing will occur when they are grown in proximity. Fodder beet, or mangelwurzel, is an agronomic crop grown best under cool-temperate conditions. The enlarged roots are stored and later fed to livestock. Average fodder beet maturity is approximately 200 or more days depending on location. Sugarbeet is a comparatively new crop created through breeding and selection of fodder beet for high sugar content (Hill and Langer, 1991). Over the years, plant breeding increased root sugar content to between 15 and 21% of the total weight.

Vegetable beets are commonly referred to as red, garden, table beet, or beetroot. Vegetable beets are smaller and develop more rapidly than fodder or sugarbeets. Garden beets develop from 50 to 80 days after direct seeding and tend to develop fiber with age especially under stressful growing conditions. The dark green leaves and stems of some cultivars are eaten raw in salads, steamed or boiled, or sometimes stir-fried. Garden beets are also pickled in vinegar and then eaten cold as a side dish or condiment. Beets may also be peeled, shredded raw, and then eaten as a salad or soup. Beet juice, obtained from cutting or crushing the hypocotyl axis (root), contains betanins used as natural red food colorant.

Swiss chard

Swiss chard is a biennial closely related to garden beet but without an enlarged hypocotyl or root. Chard is essentially a foliage beet. Plants form a tight upright rosette of large, elongated leaves with broad crisp petioles (Schrader and Mayberry, 2002). Cultivars with flat leaf blades lack undulation between veins while savoy-type leaves are cupped with a convoluted texture. Many chard cultivars have large, savoy, dark green leaves with broad, white, red, or yellow midrib or petiole. Red-veined cultivars are often called "rhubarb" chard type although unrelated to the vegetable rhubarb in the family Polygonaceae (Welbaum, 2015). Some multicolored cultivars are grown as edible summer ornamentals. Swiss chard is more heat tolerant than spinach and other cool-season greens. Chard grown as an annual for edible production. Whole plants can be harvested, or larger outer leaves can be collected from multiple plants and bundled for sale. Chard leaves are cooked before eating. More recently, chard is planted at high populations like spinach and harvested at the baby leaf stage for salad mixes. Popular cultivars of Swiss chard include Lucullus, White Silver, Large White Ribbed, Fordhook Giant, and Bright Lights (Schrader and Mayberry, 2002).

Cultivars

Variation exists among garden beet cultivars with respect to a wide range of horticultural characteristics that should be checked for trueness-to-type during seed production. Beet root shapes may be globe, cylindrical, top-shaped, or flattened. Shape may also be affected by environment conditions and particularly spacing. The size of storage roots is quite variable, ranging from 2 to 20 cm (0.8 to 8 in.) in diameter depending on the stage of development, environment, spacing, and cultivar. Leaf size, shape, and color are other traits that define cultivars. Leaf venation and root color may vary from green to red to purple to yellow with green or purple leaf blades. Other characteristics that differentiate beet cultivars include growth rate and foliage size (often referred to as top size) with small, medium, or large cultivars available. Leaf strength of attachment, determined by neck and crown size, petiole thickness, and petiole length, varies and is important for beets grown for sale as bunches with leaves attached.

Red garden beet cultivars are often divided into distinct groups based on root shapes and internal characteristics. Early Wonder, Crosby Egyptian, or often simply Egyptian cultivars are early maturing, flattened round, and have smooth clean shape with a small taproot without much transition taper. Egyptian cultivars have prominent internal rings of alternating deep red and lighter pinkish red to white zones corresponding to regions of xylem and phloem tissues. Egyptian-type cultivars are well suited for bunched display for fresh market, whole use, or pickling due to their attractive smooth root shape. However, internal characteristics are less attractive because of color variation.

The second distinct group is Detroit Dark Red sometimes referred to as simply Dark Red. Cultivars in this group mature later, shapes are elongated to round with a rougher appearance, and they have a prominent tapered taproot. Interiors are uniformly solid red with little to no zoning. Detroit Dark Red cultivars have less attractive rougher exteriors but uniform and intensely colored interiors that are better suited for processing into cubes or slices. While many markets prefer beets that are dark purple, some cultivars may have white, red, and yellow coloration (Grubben and Denton, 2004).

A third type, Ruby Queen and related cultivars, was originally developed by Northrup King Seed Company in the 1950s from a selection of Special Crosby combining many favorable characteristics of Egyptian and Detroit types, e.g. smooth globe-shaped roots similar to Egyptian with deep red interior color and fine texture like Detroit types. Ruby Queen and related cultivars have short to medium tops with high dry matter content and are popular for processing and with home gardeners.

Another distinct type has a cylindrical root shape and is grown for processing and specialty markets. Cultivars in this group include Cylindra, Formanova, and others. Cylindrical beets have the benefit of providing a higher percentage of uniform slices for processing.

Consumers are often attracted to vegetables that are harvested immature because they are perceived to be more tender, have higher nutrient content, require less cooking, and perceived to be free of pesticides. "Baby" beet whole seedlings with only three to five true leaves are harvested and sold as microgreens. Immature foliage without roots may be marketed as microgreens as well. Leaves from immature beet plants can be mixed with other immature salad vegetables and herbs to create microgreen mixes. The leaves of beets are also harvested at a "baby leaf" stage and are planted at high populations like baby leaf spinach and used in salad mixes. Such baby vegetables and microgreens often sell at a premium price because of their perceived greater value especially when sold certified organic. Golden Beet, Bull's Blood, and Early Wonder Tall Top are cultivars popular for microgreen production. Baby beet roots are another variation of immature production. Baby is not a specific cultivar but rather beet roots that are harvested at a small $<$ 4 cm (1.6 in.) diameter and sold for a premium (Schrader and Mayberry, 2002).

Standard beet cultivars are sometimes planted at close in-row spacing and harvested early to reduce root size rather than growing special small cultivars.

Production

Site selection and preparation

Beet seed crops should be grown on well-drained and friable soils to ensure uniform stands. Sandy loam to silt loams are well suited for beet seed production. Deep plowing (mold-board or chisel), with rotovators or disks creates finely tilled soil free of rocks, clods, and heavy organic debris leading to complete stands. Good soil structure favors straight uniform root development for easy rogueing. Beet seed crops are also planted on raised beds in some areas to improve drainage and facilitate furrow irrigation in some cases.

Establishment

Beet cultivars can be monogerm or polygerm, which are referred to as multigerm in the seed trade as described above. Most cultivars are multigerm and only a few cultivars of monogerm garden beet are available and these are F1 hybrids.

Beet seed crops are either direct-seeded or transplanted depending on the requirements and climatic conditions for the growing area. Since seed production is biennial, this means that concurrent seed crops will overlap with future seed crops to be harvested the next season. In some areas such as the Skagit Valley of Washington state, USA, direct seeding of beets is not allowed due to the "green bridge" and the cross-contamination of plant pathogens from existing production which impacts seed yields. In the Skagit Valley region, beet roots or seedlings are planted in an isolated nursery and later transplanted to the production field the following spring (Fig. 11.2). Establishment using

Fig. 11.2. Uniform plant stand in a beet seed production field after transplanting. (Courtesy of Tim Loeffler.)

direct seeding or from transplanting will be different. In a direct-seeded production field, cultivars or parent lines can be precision-seeded to a higher plant population than transplanting to account for stand loss during emergence, crop establishment, and overwintering. Polygerm embryos may germinate all at once or at different times, thus may require hand or mechanical thinning to ensure the desired uniformity needed for seed crop production. Transplanting has the advantage of precision planting to a final stand. This is beneficial because planting to a stand for transplanted production promotes uniform development for good nicking of inbred lines for F1 hybrid crops. Uniform stands also reduce variability for once-over mechanical harvest.

Beets are direct-seeded 1.3 to 2.6 cm (0.5 to 1 in.) deep. For direct-seeded production, garden beets are often spaced in rows from 50 to 100 cm (20 to 39 in.) apart with final in-row spacing from 30 to 50 cm (12 to 20 in.) depending on plant size. The direct-seeded crop is planted mid to late summer depending on the growing area. The final optimum plant population should be from 150 to 225 plants/m^2 (14 to 21 plants ft^2) (George, 2009). For seed production to be transplanted from roots, the crop is planted in early summer in a plant bed under dense production like vegetable production to allow the roots to develop to vegetative maturity before they are harvested in the fall (Fig. 11.2).

Growth and Development

The optimum temperature range for beet seed crop vegetative growth is 15°C to 27°C (59°F to 81°F). In contrast to spinach, beets are more tolerant of higher temperatures and can be grown at warmer temperatures particularly during seedling growth. Root growth is more heat sensitive and develops lignin and fibers faster with increasing temperature. Once established, beets are considered hardy vegetables because they can tolerate subfreezing temperatures as low as –9.4°C (15°F) for brief periods (Rubatzky and Yamaguchi, 1997). However, extended exposures of several days are lethal. Freeze tolerance varies somewhat with cultivar, planning depth, and soil type. Beet root and leaf quality are highest when grown under cool night temperatures that slow respiration, enhancing retention of stored carbohydrates. Beetroots require 50 to 80 days to reach suitable size for thorough evaluation near vegetative maturity.

Most beet cultivars grown in northern latitudes are biennials. Like other biennials, post-juvenile mature beet seedlings require vernalization to flower. Mature seedlings have a crown diameter approximately the size of a small pencil or about 3 mm (0.12 in.), and five to eight true leaves. Vernalization requirements vary among cultivars and can be quantified by the length of exposure to low temperature. Requirements for garden beet vernalization are generally based on data from the more widely studied sugarbeet. Many beet cultivars must be vernalized prior to 14 h long-day exposure, a process called photothermal flower induction (Owen *et al.*, 1940; Owen, 1954). The optimal temperature for sugarbeet vernalization is about 8°C (46°F). At the threshold temperature of 15°C (59°F), vernalization requires longer exposure times (Smit, 1983). Many beet cultivars need at least two months below 15°C (59°F) to be vernalized, although some cultivars developed for subtropical areas may bolt after exposure to as little as three weeks below 15°C (59°F) (see Genetics section below) (Rubatzky and Yamaguchi, 1997). The chilling required for vernalization of many beet cultivars is usually greater than for other biennial vegetables. In practical terms, early seeded beet seed crops are at less risk of premature seed stalk development compared to onion and certain other biennial vegetables (Goldman and Navazio, 2007). Rapid seed stock elongation, called bolting, and flowering occur after vernalization when favorable growing conditions return.

Beet seed crops can be established using direct seeding (also known as seed-to-seed) or transplanting (also known as root-to-seed) systems. Seed-to-seed production is simpler because plants remain in the same field over the entire production cycle. A major disadvantage of seed-to-seed crops is that it overlaps with existing maturing seed crops and adequate distance between fields should be considered to minimize the effects of the "green bridge". Planting dates are somewhat dependent on the climate and length of the growing season. For the seed-to-seed method planting can be later in the summer, since the plants remain in one location except when transplanting is used. Most garden beets cultivars are polygerm and precision planting is based on the "sprout count" per 100 seeds to obtain to a final stand like monogerm types. An advantage of polygerm seed is that it can compensate for lower germination rates. Alternatively, seeds may be sown at 12 kg/ha (10.7 lbs/acre) in a seedling nursery, uniform seedlings pulled, and transplanted to a final production stand to ensure optimal spacing. Transplanting is often conducted early the following spring as soon as weather permits and seedlings should be small, about 3 cm (1.2 in.) in diameter and weighing 40 to 45 g (1.4 to 1.6 oz), to reduce transplant shock (George, 2009). Roots should not be severely damaged during digging and transplanting. For transplanting, garden beets are often spaced in rows from 50 to 100 cm (20 to 39 in.) apart with final in-row spacing from 30 to 50 cm (12 to 20 in.) similar to direct seeding depending on plant size. The final optimum plant density is approximately 150 to 225 plants/m^2 (14 to 21 plants ft^2) depending on the cultivar. Environmental conditions must be well adapted to beet seed-to-seed production, enabling successful vernalization without excessive winter kill. Summer conditions should be dry and mild. Identifying satisfactory locations with consistently mild winters and dry summers where irrigation is available for beet seed-to-seed production can be challenging. Seed-to-seed production occurs in Oregon in the USA, southern Chile, Denmark, France, and New Zealand. The seed-to-seed system produces more genetically variable crops because less inspection and rogueing are performed to remove off-types. Therefore, it is critical that only genetically high-quality stock seed is used in the seed-to-seed production system. Plants bolt and flower the second spring following successful vernalization. Utricles mature and are harvested in late summer or early fall of the second year.

As mentioned above, root-to-seed production occurs in Western Washington in USA. Root-to-seed production occurs in two stages and requires more labor. This production system results in greater uniformity and less genetic variation because roots are lifted and inspected for off-types at the end of the first-year's vegetative growth (Fig. 11.3). Initial field seeding occurs in

Fig. 11.3. Digging (lifting) beet roots with tops attached in fall for inspection as part of the root-to-seed production method. (Courtesy of Tim Loeffler.)

early summer, June in the northern hemisphere, or later in areas with longer growing seasons. Monogerm and polygerm cultivars are both precision-planted to a final stand using the same equipment based on the standard germination of each seed lot. Polygerm cultivars are not thinned to a final stand. Sorting at harvest is based mostly on size and gross root differences. There is a heavy reliance on high-quality stock seed. Foundation seed production is more critical at the sorting stage

Roots are dug using a modified potato digger or similar device after the tops have been removed (Fig. 11.3). Severe root damage should be prevented during top removal and digging. Removal of the tops is often accomplished with a flail defoliator or similar device before digging. Some of the traits examined during root inspection are listed under Cultivars above, in conjunction with the breeder's description of cultivar characteristics. After digging and visual inspection for off-types, roots must be vernalized to produce seed the following summer. To ensure vernalization and reduce winter kill from excessively low temperatures, dug roots are placed in windrows and covered with soil in the field or placed in a cold storage facility overwinter and replanted the following spring (Fig. 11.4A, B). The optimum temperature for beet root storage is 4°C to 5°C (39°F to 41°F) and 80 to 90% relative humidity to reduce shriveling (George, 2009). Roots stored indoors are placed in palleted bins and may be briefly aerated to remove surface moisture before placement inside temperature- and humidity-controlled storage.

After winter storage, roots are again inspected for root quality during spring digging in the field or removal from cold storage before they are transported and replanted in the seed production areas (Fig. 11.5). As mentioned above, the seed production area is usually in a different region better suited to manage diseases and other pests to promote better beet seed maturation and harvest. Replanting occurs the following spring as early as weather conditions permit. Roots are spaced in rows 100 cm (39 in.) apart with in-row spacing of approximately 30 cm (11.8 in.) depending on the cultivar (Fig. 11.5). Mechanical transplanters can be used if they accommodate the diameter of beet roots. Carousel planters can be adjusted for beet roots if the diameter is not too large. Press wheels on the transplanter help to ensure good root–soil contact. Starter fertilizers can be applied by the transplanter as well. Planting depth is important. Roots should have the crown facing upward just above soil level. Soil structure should be loose and friable at transplanting (Fig. 11.5). In drier areas, sprinkler irrigation aids establishment by settling soil around the roots. Seed harvest occurs in late summer or early fall of the second year.

Male sterility genes are incorporated by plant breeders to ensure there is no selfing of the female parental line during F1 hybrid seed production. A male-fertile maintainer line can be crossed with male-sterile female parent inbred line to produce seed. However, restoring fertility to the female parent is not necessarily needed since only roots are harvested in commercial beet production. In a crop like corn or sorghum, restoring fertility with a maintainer line is required. The male inbred line used for hybrid seed production is 100% fertile to supply pollen to the female parent. After pollination is complete, the male-fertile line is destroyed by mowing or tillage or a combination of both. Seed is only harvested from the male-sterile rows (female parent) (Fig. 11.6). A typical ratio is 4 male to 12 female rows and depends on the parent lines.

Fertilization

Beets are relatively tolerant of salinity and drought like other members of the subfamily Chenopodioideae but are sensitive to soil acidity (Rubatzky and Yamaguchi, 1997). Beets grow best at pH 6.4–6.8. Nitrogen is often a limiting nutrient, and its management is critical for maximizing growth. The most efficient nitrogen fertilizer applications should be based on total biomass production including reproductive tissues, minus the contribution for residual soil nitrogen in the soil root zone, as well as nitrogen mineralized from soil organic matter decomposition. Soil testing is needed to determine contributions from residual nitrogen and organic matter. Nitrogen in soil organic matter becomes available through mineralization. This complex process is dependent on both moisture and temperature, complicating calculations of mineralization rates. Research shows that approximately

Fig. 11.4. (A) Roots stored in narrow beet pits (60–90 cm wide × 15–30 cm deep) overwinter at temperatures (from 7°C to 10°C) required for vernalization. (B) Beet roots are dug from pits in March using a potato digger. (Courtesy of Tim Loeffler.)

33.6 kg/ha of N (30 lbs/acre) will be mineralized in a growing season for each 1.0% organic matter in soil surface (Davis and Westfall, 2015). If uncomposted animal manure is applied, the nitrogen will likely be released as NO_3^- later in the season without benefiting early development (Davis and Westfall, 2015). Composted manure tends to release nitrogen faster. Fertilizer application should be precision-placed, often by row banding, in the top 40 to 60 cm (16 to 24 in.) of the soil since secondary roots, important for nutrient uptake, develop primarily in this profile.

Fig. 11.5. Replanting beet roots in spring using the root-to-seed method. (Courtesy of Tim Loeffler.)

Fig. 11.6. Hybrid beet seed field after cross-pollination between the two inbred lines. Male plants (four rows) have been removed, male-sterile female (12 rows) remain for seed production and harvest. (Courtesy of Tim Loeffler.)

Overall, beets require what would be considered moderate amounts of N-P-K. For example, a beet seed crop yielding 1000 kg/ha (892 lbs/acre) removes via harvested seed 70, 9, 90, 8, and 13 kg/ha (60, 8, 80, 7, and 12 lbs/acre) of N, P_2O_5, K_2O, CaO, and MgO, respectively, while plant biomass removes roughly 96, 0, 60, 108, and 102 kg/ha (86, 0, 54, 96, and 91 lbs/acre) of N, P_2O_5, K_2O, CaO, and MgO, respectively. These values vary with both cultivar, soil type, and environmental conditions.

Phosphate, potassium, sulfur, and micronutrient fertilizers are applied based on soil test results because levels vary greatly with soil type and cropping history. Beets require high levels of some trace elements, especially boron and zinc, for normal growth. Cavity spot is a physiological disorder caused by boron deficiency, in which black necrotic lesions develop on the outside of roots and heart rot on the interior.

Pollination

Beet flowers exhibit protandry with the anthers shedding pollen before stigmas are receptive. Protandry ensures cross-pollination. The pollen is light and can travel significant distances in the wind without assistance from insects.

To ensure the best cross-pollination, seed fields rows should be planted in a block configuration, so pollen is equally distributed across a field (Fig. 11.7). For F1 hybrid production, when practical it is best to orient the planting, so the prevailing wind blows across the male parent rows first, distributing pollen across the adjacent female rows.

Isolation

Beta vulgaris pollen is wind-borne and may travel over relatively long distances. Sufficient isolation is important to prevent crossing with other seed fields or weeds. Garden beets are cross-compatible with the other *B. vulgaris* crops (i.e. sugarbeet, fodder beets, spinach beet, and Swiss chard) so isolation is essential to prevent crossing. Seed companies work together to establish growing areas and isolation distances for each type of crop of *B. vulgaris* to maximize high-quality seed production. Separate production areas are designated for table beets, Swiss chard, fodder beet or

Fig. 11.7. A beet field at full bloom. Plants bolt to initiate reproductive development after vernalization and then develop small green flowers that are andromonoecious and cross-pollinated by wind. (Courtesy of Tim Loeffler.)

mangel, and sugarbeets. Isolation distances vary from 800 m (0.5 miles) for similar types up to 3200 m (2 miles) for differences for among type, color, and shape of roots of table beets as well as between fields of F1 seed production. Up to 8 km (5 miles) is required between table beets and Swiss chard, fodder beet, and sugarbeets. In some seed production areas, fields are "pinned" on an isolation map to assure isolation distances are maintained between seed growers. Small lots of breeder or foundation seed can be grown in pollen-proof screen houses or greenhouses to control pollination if there is sufficient air movement to adequately distribute pollen.

Irrigation

Beets have a shallow root system and must receive a consistent supply of water to maintain active growing. Spring plantings may not require irrigation in areas with adequate rainfall. Some areas have adequate rainfall and supplemental irrigation is not needed. In spring, field water loss from evapotranspiration is relatively low. Overhead irrigation after spring transplanting reduces transplant shock and settles soil around plants. However, during the summer months, evapotranspiration is higher, so irrigation is often needed to maintain constant growth and reduce embryo abortion. Drip lines are water efficient but may interfere with seed harvest if not removed. Use of subsurface drip is increasing in some areas because if its greater efficiency. Furrow irrigation is not commonly applied to beet seed fields because it requires level soils, more water, and may negatively affect root development on heavy soils. Beets need at least 2.5 cm (1 in.) of water per week to ensure continuous growth and a minimum of 350 mm (14 in.) is required to produce seed crop depending on the weather and soil type (Rubatzky and Yamaguchi, 1997).

Rogueing

Removing off-type, diseased, and low vigor plants is essential for high-quality beet seed production. In addition to obvious off-types, subtly different plants that are not "true-to-type" and lack traits the developing breeder selected should also be removed. This simply means that each cultivar was developed with specific characteristics of color, size, shape, speed of growth, etc. by the cultivar developer that should be passed along to succeeding generations to prevent genetic drift. Careful detailed rogueing helps ensure the seed sold will produce plants with all the characteristics described to the buyer. Having the breeder's description of traits for each cultivar is important for thorough, consistent rogueing. More inspection and rogueing are performed in the root-to-seed system. Extra rogueing decreases off-types and genetic variation although it requires more labor and time.

Primary rogueing for root traits occurs at lifting in the fall and after storge just prior to transplanting in the root-to-seed system. Before cutting tops prior to lifting, leaf characteristics should be inspected and plants with incorrect leaf color and morphology, early bolting, and seed-borne pathogens should be removed. After lifting, roots showing incorrect color, signs of premature bolting, incorrect root shape, crown defects, surface corkiness, diseased, or other root defects should be discarded. Replanting also provides another opportunity to again identify and remove these same abnormalities before seed is produced.

Seed-to-seed systems can also be rogued if smaller transplants are set in final production fields. Leaf shape and color and general root/hypocotyl shape exposed above the soil in a seed-to-seed field can be inspected toward the end of the first year's growth. A second rogueing based primarily on above ground traits and disease symptoms may be conducted during regrowth and during bolting in the second year.

Genetics

Both open-pollinated and F1 hybrid cultivars are available. Even though presence of F1 hybrids has increased, open-pollinated cultivars are still popular in the seed market. Special considerations for producing hybrid seed and maintaining inbred lines for F1 production were described above. A population in the range of 120–200 plants is needed to produce seed and maintain the genetic health of a population (Organic Seed Alliance, 2018). Genetic drift is a concern when seeds are produced in areas where the climate is different than where they will ultimately be grown for food. There is a genetic linkage between shape and zoning and maturity. This has allowed Crosby Egyptian and Detroit Dark Red cultivars to remain distinct despite crosses that have attempted to combine the best characteristics of each.

The dominant *B* gene (B-allele) controls bolting in beets (Abegg, 1936). All cultivated biennial beets grown commercially in northern Europe and North America are genotype *bb* and require extensive vernalization to induce flowering under long days.

Seed Harvest

The harvesting of beet seed commences when the utricles at the bases of inflorescent side shoots mature. At this stage the fruits turn from green to brown. To verify maturity, transversely cut an utricle. Cut immature utricles are wet and milky while ripe fruits are drier and with mealy texture. Seeds ripen successively from the bottom branches to the top of the inflorescence. The terminal point of the compound inflorescence matures last. It is important to determine the optimum time for cutting the flower stalks for final drying.

In dry areas, large-scale fields can be combined directly without first cutting and windrowing senescing plants. The combine gathering reels should be removed to reduce shattering and screen sizes appropriate for beet utricles should be installed. However, direct combining may lead to considerable loss from shattering. Birds may feed on seeds if plants remain upright until full maturity as well. A common practice in many areas is to cut and windrow flower stalks to facilitate crop desiccation and reduce shattering. As soon as the lower branches are mature and begin to dry, the earliest seeds are also mature. Immature seeds near the top of the plant may shrivel if plants are cut prematurely. If cutting is delayed, some utricles will be lost to shattering as well. Determining the best stage for cutting is important for combining best seed quality with maximum yield. The best harvest stage occurs when later ripening plants begin to lose chlorophyll and turn yellow. Maturing beet flower stalks tend to lodge and be prostrate with the onset of maturity rather than remaining vertical until fully dry. The method of cutting depends on the size of the operation. Large fields are cut using self-propelled swathers to place seed stalks into windrows (Fig. 11.8). Smaller-scale production for breeder, foundation seed relies on hand cutting with knives or scythes and manual placement of stalks in windrows or on drying sheets (George, 2009).

Windrows are carefully turned as needed depending on the humidity or after rainfall. For small-scale production the cut stalks may be bundled and suspended to aid drying or placed in protected structures especially in humid areas. The cut material can take from three to 14 days to dry depending on the maturity at cutting, air movement, humidity, and temperature. After cutting and field drying, windrows are combined or run through a stationary threshing machine brought to the field (Fig. 11.9). On some occasions when stacks of stalks are left in the field to final dry, they are placed on sheets to collect seeds that shatter. Threshing of the crop can begin when the utricles easily separate from the stem and seed moisture content should be below 15% (Jarmin and Thornton, 1985). The dry straw of beetroot seed can be extremely brittle, and it is therefore important to use a relatively low cylinder speed and air blast to remove chaff and other debris. Concave openings must be wide enough to avoid producing too many small pieces of straw because they can be difficult to separate during seed cleaning. The seed crops harvested at higher seed moisture contents should be dried to 7.5% or lower for safe storage (Organic Seed Alliance, 2018).

Post-harvest Seed Conditioning

Beet seed is initially scalped after harvest to remove sticks, weed seeds, and crop trash prior to the seed conditioning process. A debearder is usually placed ahead of the air screen machine to reduce the amount cork material of the utricles. Other seed conditioning equipment includes the gravity table and the inclined draper which can be adjusted for the final cleaning and conditioning of the seed crops. Most table beets are sized into several seed size categories to allow precision planting in the seed market. There is relatively little chaff from

Fig. 11.8. Windrowing of mature beet plants with a vertical and horizonal cutter. Plants are cut and dried in the field before full senescence to reduce shattering. (Courtesy of Tim Loeffler.)

beetroot material, but a large amount of dust is created during the seed-cleaning process.

Seed Yield

Utricles vary greatly in size depending on how many seeds they contain. The range varies from 2 to 6 mm (0.08 to 0.23 in.) for fruits containing from one to multiple embryos, respectively. Individual beet seeds are reddish brown and measure 1.5 to 3.0 mm (0.06 to 0.12 in.) across. The weight of 1000 multigerm utricles is roughly 17 g. About 55 to 60 beet seeds weigh 1 g (0.04 oz) (Rubatzky and Yamaguchi, 1997). Satisfactory seed yields vary over a wide range with location and can range from 1000 to 2200 kg/ha (892 to 1963 lbs/acre). Utricles are sometimes milled to remove excess pericarp and improve singulation. One thousand milled and conditioned "monogerm seed" generally weigh less, in the range of 10 g. (George, 2009).

Pest Management

Weeds

Weed control is another challenge for producing beet seed. Beets germinate slowly in cold soils planted in early spring. Beet transplants take time to resume growth and may not compete well with vigorous annual weeds. Common invasive weeds like lambsquarters (*Chenopodium album*) and goosefoot (*Chenopodium albescens*) are related to beet and are prolific seed producers. Herbicides labeled for beet will not be effective against these weedy relatives. If allowed to produce seed, these weeds and others contaminate beet and spinach seed lots, requiring additional cleaning steps (Hembree and Norris, 2010). As pointed out in the Pathogens and Insects management sections below, certain weeds and escaped beet plants and their relatives can serve as alternative hosts for pests. Monitoring the areas surrounding for seed fields for these

Fig. 11.9. Harvesting dried beet seed (utricles) with a combine from windrows. The combine gathering reel has been removed to reduce shattering. (Courtesy of Tim Loeffler.)

weeds, insects, and diseases is essential for an effective pest management strategy.

It is wise to establish seed crops in fields without a history of weed problems, especially troublesome perennial weeds. Controlling weeds in areas surrounding seed fields goes a long way toward preventing seed from entering. Mowing, hand removal, or targeted use of herbicides can effectively control weeds in and around seed fields.

Non-chemical management techniques are effective for controlling weeds. If a field is relatively weed free, hand removal at the time of rogueing can be effective before weeds produce seed. Mechanical cultivation can remove weeds as well. Development of weed-removing "intelligent" robotic cultivators offers promise for improved weed control while reducing labor costs. Avoiding deep tillage prevents buried seeds from germinating in some areas. The stale bed technique allows weed seed to germinate after soil is prepared for planting. If a flush of weed growth follows soil preparation, an additional shallow tillage removes weeds before the seed crop is planted.

Pathogens

Powdery mildew (*Erysiphe polygoni*) can be a serious fungal pest of beet seed crops. It can be serious for seed-to-seed production when plants overwinter, particularly in mild climates. Second year growth after transplanting in the root-to-seed system is also susceptible when plants are exposed to cool, damp conditions and overwintering spores on plant debris. Conidia (asexual spores) of *E. polygoni* are short-lived and unable to overwinter in climates with harsh winters. Ascospores (sexual spores) that develop later in the season may overwinter on plant debris. Powdery mildew fungus is reintroduced annually from wind-borne conidia. Symptoms are small, white, powder-like spots that appear first on the under surface of older leaves, and under favorable conditions rapidly spread over the entire leaf surface. Powdery mildew is favored by warm, dry weather, with optimum temperatures between 15°C and 30°C (60°F and 86°F). Excessive nitrogen fertilization should be avoided. Balanced irrigation, and fungicide applications at first appearance may provide control. Downy mildew (*Peronospora parasitica*) is transmitted by seed and may cause red-rimmed spots on the leaves of adult plants (UC IPM, 2023).

Phoma leaf spot (*Phoma betae*) is a seed-borne fungal pathogen and is the asexual stage for black leg (*Pleospora bjoerlingii*). The conidia are spread between seedlings during rain or irrigation events. Infected seedlings can develop systemically, and the pathogen can form perfect stage pseudothecia that can overwinter in the roots. These pseudothecia produce air-borne ascospores that can infect the spring planted crops. A standard practice is to use fungicidal seed treatments and hot-water treatments are also effective. Crop rotations and proper destruction of crop residues are important control measures.

Fungal foliar disease *Cercospora beticola* causes circular leaf spots that are tan to light brown with reddish purple borders. As the disease spreads, individual spots coalesce. Affected foliage turns yellow, eventually brown, and necrotic (UC IPM, 2023). This disease was primarily a problem in areas where warm nights combine with high humidity (CABI, 2023; UC IPM, 2023). Optimum daytime temperatures for *Cercospora* development are 25°C to 35°C (77°F to 95°F) with night temperatures above 16°C (61°F) and a relative humidity of 90 to 95%. The fungus spreads on infected plant residue, can be seed-transmitted, and numerous weeds can serve as alternate hosts. *Cercospora* is controlled by removing inoculum from a field, plowing to incorporate crop residues between seed crops, and using three-year rotations with non-host crops. *Pleospora betae* is a fungal pathogen responsible for a seed-transmitted, damping-off disease similar to Pythium and Rhizoctonia (CABI, 2023).

Bacterial leaf spot (*Pseudomonas syringe* pv. *apata*) is seed-borne and this bacterial disease causes leaf spot symptoms in the early stages of plant growth. The pathogen has a wide host range that includes other Chenopods such as spinach, sugarbeet, and Swiss chard as well as beans, Cucurbits, and lettuce. The pathogen can spread in a field by infected seed, rain, or irrigation. This is important for baby leaf production and seeds should be tested prior to use. Crop rotation and tillage of crop refugia to reduce inoculum should be practiced (Sharma *et al.*, 2021).

Black spot or cavity spot is a non-pathogenic physiological disorder caused by boron deficiency that appears as necrotic black spots or pits on root surfaces. This is mainly cosmetic, and unless severe, doesn't adversely affect seed production.

Some of the most common viral diseases that affect beet include: curly top virus, mild yellowing virus, mosaic virus, pseudoyellows virus, and soil-borne mosaic virus. Common symptoms for these beet virus diseases include dwarfed leaves and distorted leaf blades that are rolled upward and inward. Lower leaf veins are irregularly swollen with bumps. In cross-section, vascular tissue has dark rings in large roots. Young roots of virus infected plants are stunted, twisted, distorted, and will eventually die in severe infections. Death of small roots is followed by production of new ones leading to a "hairy root" phenotype (UC IPM, 2023).

Beet curly top virus is vectored by the beet leafhopper, *Circulifer tenellus*, which has an extensive host range, a high reproductive capacity, and can migrate long distances from its breeding grounds. Leafhoppers acquire the virus after feeding on infected plants. They overwinter on many different annual and perennial weeds. Once leafhoppers become a vector, they can often transmit the virus until they die. The severity of the curly top virus on beet depends on climatic factors. Plants showing any of the viral symptoms listed in this section should be removed from seed fields or surrounding areas.

Insects

Beet armyworms may cause severe injury to seed crops in summer and fall (Natwick *et al.*, 2010). Beet armyworm (*Spodoptera exigua*) and yellow striped armyworm (*Spodoptera praefica*) are small, mottled gray- or dusky-winged moths that deposit pale greenish or pinkish striated eggs on leaves. Eggs are small or large masses covered with white cottony matter visible to the naked eye. Eggs hatch quickly in a few days, and tiny caterpillars begin feeding. Olive-green caterpillars have a yellow stripe on each side of their bodies. They are fully developed in only about two to three weeks, when 3.2 cm (1.3 in.) long. Armyworms skeletonize leaves, leaving only veins. In severe infestations, armyworms can consume veins, petioles, and exposed roots. Fields should be closely monitored for armyworms because of the severe damage they can cause. Controls include killing larvae that survive in plant debris using cultivation, attracting beneficial insects or releasing them, pheromone trapping, or periodic insecticide applications.

Lygus (western tarnished, *Lygus hesperus*) bugs attack many different plant species including sugarbeet as well as Swiss chard. A discussion of this important pest and its control are covered under the spinach Insects section below.

Larvae of beet webworms (*Hymenia perspectalis, Loxostege sticticalis, Spoladea recurvalis*) produce webs and feed on beet foliage, stunting growth and reducing yields in severe cases. Sometimes aphids, flea beetles (*Chaetocnema confinis*), and beet leafminers (*Pegomya hyoscyami*) are a problem on garden beets. Meloidogyne root-knot and other nematodes may adversely affect beetroots, causing distorted development and reduced yield (Natwick *et al.*, 2010). Other notable insect pests of beet include vegetable weevil (*Listroderes costirostris*), beet bug (*Piesma quadratum*), and green looper caterpillar (*Chrysodeixis eriosoma*) (CABI, 2023).

SPINACH

Botany

Spinach is an annual grown for its leaves which are eaten raw or cooked. Spinach has a distinctive vegetative phase. It forms a rosette of leaves on a short-compressed stem. Both the leaf petiole and blade are harvested during this stage as a vegetable. Plant spacing and weather conditions interact to determine both leaf size and number. Spinach grows to a vegetative height of up to 30 cm (12 in.). The leaves are alternate, simple, and vary in length from about 2 to 30 cm (0.8 to 12 in.) and 1–15 cm (0.4 to 6 in.) across, with larger leaves at the base and small leaves higher on the flowering stem. Spinach petioles are about the length of the leaf blade and hollow at maturity. Spinach is sometimes called a winter annual because it will overwinter at temperatures above −12°C (10°F).

Flowering is initiated in response to environmental cues. Seed stalk elongation and flowering are induced by long days, while temperatures above 20°C (68°F) accelerate bolting. Flowering occurs without vernalization but is hastened by previous chilling (George, 2009). Seed stalk elongation disrupts the rosette severely restricting new leaf development. The additional leaves that do form tend to be narrow and small and can appear distinctly different from the leaves at the vegetative stage. The critical photoperiod that initiates flowering varies with cultivar and ranges between 12.5 and 15 h (Rubatzky and Yamaguchi, 1997). Spinach has an indeterminate flowering pattern. Flowering and new seed set continue if conditions are favorable. Seeds developing on the lowest branches ripen first, and maturity progresses up the stalks. Lateral branches form as the central stem elongates above the rosette. Branches contain clusters of up to 20 small, inconspicuous, green wind-pollinated flowers 3–4 mm (0.13 to 0.16 in.) diameter that mature in clusters of small, hard, dry, lumpy fruit 5–10 mm (0.2 to 0.39 in.) across, each containing several seeds (Fig. 11.10). Spinach plants have an enlarged taproot with little branching. Many shallow fibrous laterals develop from the taproot.

The sex expression of spinach plants is difficult to determine solely based on early vegetative growth, although generally male plants tend to be smaller and produce fewer leaves. Spinach sex expression is most often dioecious with separate male and female plants, although a diversity of other types may occur (Figs 11.10, 11.11). The common plant types include male, female, or monoecious but plants with perfect flowers (hermaphrodites) may also occur. Extreme males are another sexual type that produce only staminate flowers, minimal foliage, flower early, and often die shortly after flowering.

Vegetative males produce only staminate flowers but have more foliage and flower later than extreme males. Female plants produce only pistillate flowers, have well developed foliage, and flower later (Fig. 11.10). Monoecious plants have both staminate and pistillate flowers on the same plant, well developed foliage, and are slow to flower (Rubatzky and Yamaguchi, 1997).

Fig. 11.10. Clusters of pale green seeds forming at leaf nodes of a female spinach plant. (Courtesy of Tim Loeffler.)

Fig. 11.11. Male spinach plant has fewer leaves and flowers earlier than the female plant on the right which flowers later and has more leaves. (Courtesy of Tim Loeffler.)

The apetalous staminate flowers are clustered on spikes. Pistillate flowers lack petals and are attached to the base of the calyx that encloses the ovary (Fig.11.10). Following wind pollination, the ovary develops into a single-seeded fruit or utricle often referred to as simply a seed. Spinach seeds may be either smooth and round or irregularly shaped with pointed spines depending on the cultivar.

Types and Cultivars

Spinach is a long-day plant and flower induction is more difficult in some cultivars than others. This is because for food production, breeders select for increasing bolt resistance for cultivars to be grown from spring to summer to prolong the harvest longer days. Cultivars grown during the shorter days of winter are developed to have less bolt resistance, so they grow faster under shorter days and cooler climatic conditions. Seed crops are timed to mature during long days and cool to moderately warm temperatures to promote bolting. Seed production of cultivars with high bolting resistance often requires to be produced in regions with higher latitudes. As a rule, older or heirloom cultivars tend to bolt early under long-day conditions and can be grown at lower latitudes. Modern F1 hybrid cultivars for commercial production tend to exhibit heterosis and grow more rapidly but to resist bolting. Spinach cultivars are divided into four types based primarily on leaf characteristics:

Savoy cultivars: dark green, puckered, and curly leaf blades. Leaf blades range from ovate to elongated to almost triangular and narrow arrowhead shapes in some older cultivars. Leaves are savoyed because of differential growth of parenchyma tissues between leaf veins (Rubatzky and Yamaguchi, 1997). Leaf growth habit varies from prostrate to upright with plant spacing having an effect. Leaf margins are smooth or curved. Because of their showy appearance, savoy types are preferred for fresh-bunched spinach sales. An example of savoy spinach is the heirloom cultivar Bloomsdale, which has been replaced commercially by higher yielding predominately female F1 hybrid cultivars that are sometimes collectively referred to as Bloomsdale types.

Flat- or smooth-leaf spinach: broad, smooth leaves that are easier clean free of soil particles compared to savoy. However, flat leaves tend to stick together when packaged. Smooth-leaved cultivars are canned, convenience packaged in bags as ready-to-eat fresh spinach, frozen, and processed as baby foods, etc.

Semi-savoy leaf blades: intermediate between flat and savoy. Semi-savoy leaves exhibit some cupping of blades but are not as difficult to clean. Semi-savoy leaves resist compression and clumping during packaging and transport. They have many of the same uses as flat-leaf types.
Oriental spinach: heart-shaped pointed leaves with longer petioles than other types and are faster growing. Oriental spinach has an upright growth habit, has strong petioles, and is used primarily for bunching. The cultivars are most popular and are dominant in the Asian markets. This type also bolts quicker especially during longer day lengths and recent breeding efforts have incorporated bolting resistance from other types to reduce bolting for summer plantings.

Baby spinach is popular in some markets because immature plants are tender with less bitterness and are perceived by consumers to be of higher quality. Although special "baby" cultivars with smaller plant architecture are sometimes grown, conventional types are often seeded at higher densities and harvested with only five to seven leaves to produce baby spinach. Baby spinach leaves can be flat, savoy, or semi-savoy.

In addition to leaf-blade surface variations described above, there are also a wide variety of leaf shapes, ranging from round-oval and heart shaped to pointed. Cultivars are available with genetic resistance to both downy mildew and cucumber mosaic virus, major diseases that affect spinach and described in more detail under the Pest Management, Pathogens section below. Characteristics of commercial importance that vary among cultivars in addition to those mentioned above include plant uniformity, yield, size, high temperature resistance, cold resistance, leaf size, color, shape, and petiole length. High-quality, modern spinach cultivars generally have short petioles, medium to dark green leaves, are free of bitterness, tender thick leaves, and disease and bolting resistance (Rubatzky and Yamaguchi, 1997).

Cultivars that have seed with spines, also known as prickly seed, are often called winter types, while smooth-seeded types are often referred to as summer types. Spiny-seeded cultivars are often considered heirloom and are less frequently grown commercially. Prickly seeded types are likely a progenitor of smooth-seeded types which are the result of continuous selection.

Production

Site selection and preparation

Crop rotation is one of the major factors for selecting fields for seed production to reduce disease pressure and to avoid volunteer plants from previous spinach seed crops. Planting a seed field next to where a previous seed crop existed should be avoided. Residues from previous crops in the same or related species may harbor diseases that can quickly spread to a new crop. Due to high susceptibility to soil-borne Fusarium wilt disease, long rotations away from infected fields for typically 6 to 10+ years depending upon the levels of infection are recommended.

Since spinach is a long-day plant, seed production has exacting environmental requirements and depends on latitude. The day-length requirement varies depending on the type of cultivars or parent lines considered. Faster bolting cultivars may only require 13 to 14 h of day length whereas slower bolting cultivars may require as much as 16–17 h of day length to initiate bolting. There are few geographic locations in the world ideally suited for spinach seed production. Spinach grows best at 18°C to 20°C (64°F to 68°F) and prefers adequate moisture during the growing season and cool dry summers. Most spinach seed production is planted in early spring and harvested in late summer or early fall. In some areas, seed production can be planted in the fall, overwintered, and then harvested the following summer. Seed crops grow vegetatively under cool conditions often during winter and early spring. It is important that winter climate is sufficiently mild for plants to survive with little or no winter kill. Plants must attain adequate size to support flowering and seed production when the inducing day length is reached in late spring or early summer. High temperatures are often incorrectly believed to cause bolting, but higher temperatures only speed flower stalk growth after it is initialized by the inducing day length. A third requirement for spinach seed production location is dry conditions in late summer as seeds mature to reduce disease pressure and shattering. Add to these climactic requirements a need for fertile well-drained soils, skilled growers, and adequate infrastructure and there are few places on earth that meet all requirements. The majority of the world's commercial spinach seed is produced in Denmark and the Pacific Northwest region of the USA; lesser seed production areas include southern Chile, France, the Netherlands, Italy, and New Zealand among others. China is also a large seed producer and mostly for its domestic market.

Soil and nutrient requirements for spinach seed production are similar to beet. Spinach can be grown on a wide range of soil types but very sandy or poorly drained soils prone to waterlogging should be avoided. Soils best suited to spinach seed production are well drained but with adequate moisture-holding capacity. Spinach is sensitive to acidic soils and pH should range between 6.5 and 7.0. Soils with and pH less than 6.0 should be limed or avoided. Raised beds are used in some areas to improve drainage on heavier soils. Deep plowing is a practice used to eliminate heavy residue from previous crops. Deep plowing also reduces overwinter survival of diseases and insects on plant debris. Finely tilled soils, using rotovators or disks, help ensure uniform emergence and good stand establishment (Hoyt *et al.*, 1994).

Establishment

Due to differences in plant architecture, spinach seed crops can be planted at higher densities compared to beets. Spinach seed crops are most often established by direct seeding and plant spacing should allow individual plants to develop for easy rogueing. Rows are spaced a minimum of 50 cm (20 in.) apart and seeds are precision-planted to achieve final stands of 200,000 to 300,000 plants per hectare (81,000 – 121,400/acre) (Fig. 11.12). This equates to a seeding density of approximately 2.5 to 3.75 kg/ha (2.3 to 3.4 lbs/acre) depending on the size of the seed. Seed spacing varies among cultivars. Unlike beet seed production, transplanting is rarely practiced and typically only for seed production of parent lines.

Fig. 11.12. Hybrid spinach seed production field. Dark green rows are male rows and light green rows are female. The male-to-female row ratio is 4:8. (Courtesy of Tim Loeffler.)

Growth and Development

Spinach germinates at soil temperatures above 1.7°C (35°F). Seedling emergence and growth are extremely slow as temperatures approaches 0°C (32°F). Germination optimum for spinach is 15°C to 20°C (59°F to 68°F). Germination at 5°C to 10°C (41°F to 50°F) is faster than at 25°C (77°F) (Rubatzky and Yamaguchi, 1997). Thermoquiescence may reduce emergence and has been reported at temperatures exceeding 30°C (35°F) (Harrington and Minges, 1954). Acid treatment or presoaking can stimulate germination by softening pericarp tissues and removing inhibitors (Atherton and Farooque, 1983a,b).

Temperatures for optimum spinach growth and best development are 15°C to 20°C (59°F to 68°F) with a minimum growth temperature of approximately 5°C (41°F) and a maximum of 32°C (90°F). Temperature affects leaf qualities, with low temperature increasing leaf thickness but decreasing size and smoothness (Rubatzky and Yamaguchi, 1997). High temperatures reduce foliage color, and savoying, and increase petiole length Spinach is moderately winter hardy and can survive temperatures as low as −12°C (10°F) for short periods of a couple of days particularly if preceded by an acclimation period. Prolong temperatures of −12°C (10°F) or lower for several days will likely result in winter kill. Spinach seed crops are direct-seeded in summer or early fall, overwintered, and harvested the following summer.

Spinach plants attain full vegetative maturity approximately 40 to 70 days after direct seeding if grown during fall, summer, or spring. Depending on temperatures and sunlight, winter growth is considerably slower. During winter seed production, vegetative spinach plants remain quiescent and resume growth with warmer temperature and increasing sunlight in late winter and early spring. Spinach is more tolerant of low light conditions than many other vegetables. Plant development is also related to growth rate. Early cultivars tend to grow faster than later maturing ones (Rubatzky and Yamaguchi, 1997).

Fertilization

Spinach plants are sensitive to acidity. Spinach is less tolerant of saline soils compared to beet. Spinach is moderately sensitive to salinity at electrical conductivity (EC) levels greater than 2.0 ms/cm. Spinach does well in slightly alkaline soils and raising soil pH through liming can improve plant vigor and help keep the fungal disease Fusarium wilt in check. The pH optimum is in a range from 6.2 to 7.5. Soils should be limed if pH is below 6.0 before establishment. All fertility inputs should be based on soil tests before planting and foliar tests during the season to monitor crop nutrition. Spinach is not an especially heavy feeder. If soil test recommendations are unavailable a rule of thumb is an application of 1:2:2 of N-P-K applied when the seed bed is prepared (George, 2009). Row banding places nutrients in the root zone, reduces losses to leaching, and is preferred to broadcast applications. On fertile alluvial soils, usually only modest fertilization is needed in conjunction with soil test recommendations. On lighter sandy soils, higher quantities of N-P-K may be required. Nitrogen inputs increase vegetative growth. During winter months soil N remains constant as little nitrification occurs at low soil temperatures in temperate regions. Fertilizer is applied as a split application preplant and at or just before bolting to satisfy requirements of reproductive growth. Excessive N or late applications may cause seed stalks to lodge complicating harvest (George, 2009).

Pollination

Breeding populations of spinach plants tend to be primarily male, female, monoecious, or occasionally hermaphroditic. Male plants tend to flower and shed pollen before female plants. It is impossible to tell the sex of the seeds or small plants; however, male plants tend to produce fewer smaller leaves in the rosette stage and form a distinct compressed button that rapidly expands before flowering. A block planting configuration ensures distribution of pollen as equitably over a field as possible since for open-pollinated dioecious cultivars the ratio of male to female is 1:1. Long narrow rows may result in poor pollination of some female plants.

In an F1 hybrid seed field the male-to-female ratio depends on the pollen production potential of the male parent. Male parents may be either dioecious or monoecious. Common male-to-female planting ratios to ensure adequate pollination vary greatly depending on the amount of pollen produced by the male. One best practice is to plant the field with the male line on all sides of the field. Seed companies typically conduct nicking trials for each parent line to determine the timing of planting, amount of pollen, and duration of pollen production to determine planting ratios. Common ratios of male to female plants are 2:8, 2:14, 4:8, 4:12, 4:18 or similar planting patterns (Fig. 11.12). Due to the indeterminate nature of flowering and the increasing temperature and day length triggers for flowering, nicking is not usually a problem for F1 hybrid spinach seed production. The male and female lines for F1 seed production are often "split planted" to assure there is adequate pollen during the pollination period. For example, half of the male line may be planted and then a week later, the female and the second half of the male may be planted. The timing of plantings of the males and female parents can vary for each hybrid especially when a faster male line is being crossed with a slower bolting female. Sometimes seed priming is used to promote earlier bolting and more uniformity in flowering to improve nicking for certain parent lines. After the pollination period is complete, the male rows are removed to prevent contamination of the male parent with the F1 hybrid seed.

Isolation

Distances between other seed fields depend on several factors and are based on leaf type which includes savoy, smooth-leaf, and Oriental types. In areas of concentrated seed production, separate growing areas may be designated for each leaf type. Since most cultivars produced are F1 hybrids, isolation distances are based on the male leaf type. Recommendations for separation distances vary among seed companies and types of seed produced. Commercial seed producers are most confident with isolation distances of at least 3.2 km (2 miles) when there is difference in type of leaves. Some seedsmen allow isolation distances of 1 km (0.6 miles) between seed fields of cultivars with similar characteristics. Others require only 0.5 km (0.3 miles) isolation if cultivars of the same leaf (savoy, smooth-leaf, or Oriental) or seed type are produced. Shorter distances can be acceptable if natural and artificial barriers are in place and strong winds do not occur.

Because of the heavy concentration of spinach seed production in some areas, a "pinning" system is utilized and effectively controlled. High resolution maps of a production area are placed in a public area and seed companies use color-coded pins to make their spinach production locations known. Marked production locations should also specify the type of spinach grown. It is wise for a local official to adjudicate conflicts caused by close field placements among competitors to make a pinning system affective.

Irrigation

Since spinach is a cool-season crop grown when transpiration is generally low, water requirements are less than for many other seed crops produced in summer or in desert areas. In areas with consistent rainfall, spring and overwinter crops can often be produced without irrigation. At least 254 mm (10 in.) of water is required to produce a spinach seed crop. Approximately 2.5 cm (1 in.) of water is needed each week during months of active growth in fall and spring. Less water is needed during winter when evapotranspiration is low. Irrigation can vary with cultivar, soil, and environment conditions. Because of its shallow root system, spinach plants are susceptible to drought stress during dry periods. Prolonged drought stress will reduce leaf size and seed yields. Overhead irrigation may be used in times of drought, although prolonged wetting of spinach foliage can increase foliar disease.

Rogueing

The first rogueing should be conducted when plants are vegetative in the rosette stage of development for both the female and male parents. Plants that fail to form a rosette, early bolters, extreme male plants, plants with incorrect leaf shape and size, diseased plants (especially seed-borne diseases such as CMV) and any other plants that are not genetically true to the breeder's cultivar description should be removed. The second rogueing stage should be conducted at flowering and may require more than one rogueing pass through the field. Field inspections should be conducted on a regular basis to determine when the timing is best for effective rogueing to manage labor costs. Remove extreme males, monoecious plants, diseased plants, plants with flowering anomalies, or incorrect phenotypes.

Genetics

The number of F1 hybrid cultivars has increased with time. The fact that spinach is a dioecious species is an advantage in hybrid spinach seed production. Sex expression of parental lines may be male, female, or monoecious. One method of F1 seed production to create a female parent is to repeatedly rogue out male plants from a dioecious line. However, this method performed on older hybrids requires rogueing up to seven times to ensure removal of all male plants. This technique was very labor intensive, not 100% effective, and costly. A missed male plant in the female parental line could pollinate many females creating many “sibs” in the hybrid seed lot.

Sex expression in spinach is genetically elastic and will change due to a wide range of factors. For example, high temperature, drought stress, or even plant competition may cause female plants to produce some male flowers. Another approach to F1 hybrid spinach production is to develop gynoecious lines by very diligently rogueing out males in a foundation seed/stock seed increase before any pollen is shed. If no male pollen is available in a field after rogueing, female plants usually produce male flowers in a process known as “sex reversal”. Pollen produced from sex reversal of female plants will pollinate other female plants in the population. The seed produced from only female plants through sex reversal in this way will produce female offspring. This is a method for producing an all-female line that can be used for F1 hybrid seed production. To obtain the high-quality female lines, seed production is often grown in greenhouses separate from the main seed production areas to prevent foreign pollen from contaminating the crop. Monoecious plants in the germplasm used as the female parent cause confusion during seed production when early sex reversal occurs in the plants. Very close and frequent monitoring is critical to identify and remove monoecious plants from a system if using sex reversion to produce the female breeding line. During the development of female lines for hybrids, breeders must perform meticulous selection on the female plants ensuring that they uniformly produce only female flowers for at least five to six weeks before sex reversing. Then, when used as female parents in hybrid seed production, these lines should have adequate time to receive all their pollen from the male parent to complete the hybrid cross.

Traditionally, the male parents for hybrid spinach seed production have been dioecious spinach lines producing equal numbers of male and female plants. Male parents in F1 crosses are typically dioecious to maintain their vigor and ability to produce ample pollen. Such lines are therefore referred to as “open-pollinated parents” or “OPs” among spinach seed producers. The male plants in these OPs tend to have the two distinct forms, “extreme males” and “vegetative males” (described above). As 50% of the plants in the male parent produce ample pollen to make the cross it is not an issue that the other 50% of the plants are female and do not contribute to hybrid production. Male plants flower earlier and usually for a shorter time than the female and often requires a split planting to assure good synchronization during the pollination period. The male rows are usually destroyed several weeks before the hybrid seed is harvested from the female seed-producing rows, to avoid any chance of seed mixing at harvest.

In recent years there has been a trend to develop male parental lines that are monoecious, having both male and female flowers on the same plant. These monoecious male parents are derived through inbreeding and tend to be more uniform in their reproductive characteristics than the traditional dioecious males. Monoecious male parents often start with producing female flowers, with male flowering beginning later. The timing of planting needs to be considered for ideal nicking. A good monoecious parent should have less than 30–40% female flowers. Also, the elimination of extreme male plants from these monoecious males means there is later, more uniform flowering. This trait is passed on resulting in later bolting in the hybrids produced using these lines. The main reservation with the use of monoecious males currently is that many of them do not produce as much pollen as traditional dioecious males.

Another method for making F1 parental lines is to use a highly female monoecious inbred line as the female parent and a highly male monoecious or a dioecious line as the male parent line (Thompson, 1955; Janick, 1998). Female parents are developed that almost exclusively produce pistillate flowers early in development and male flowers are produced later after female flowers have set (Fig. 11.13).

Female monoecious parents can show sex reversal ranging from two to five weeks before they begin flowering. Using parent lines that do not show sex reversal until after three weeks after flowering begins should be adequate for seed production provided there is good nicking. The male parent provides pollen at the same time the pistillate flowers are open on the female parent ensuring hybridization. The male parent is removed after flowering but before seed harvest. The elasticity of spinach expression can cause sex reversal under stressful conditions, so it is possible for sex reversion to increase staminate flowers on the female line particularly earlier in the development. Therefore, producing hybrids with a predominately female monoecious line has potential to produce a higher percentage of seeds from self-pollinated plants under stressful environmental conditions. Modern F1 hybrid spinach cultivars tend to have homogeneous sex expression because more uniform sex expression is selected for in the parents. Older open-pollinated heirloom cultivars tend to have more variable sex expression (Rubatzky and Yamaguchi, 1997).

To maintain a genetic population of spinach, a minimum of 120 plants should be included in a seed harvest to avoid inbreeding depression. Two hundred plants or more are recommended for seed production to maintain adequate genetic diversity (Organic Seed Alliance, 2007).

Seed Harvest

Spinach seed develops acropetally, so the lower seeds are older and younger seeds continue developing up the flower stalks. Due to this upward sequential maturation, new seeds are continuously setting throughout the season. However, only about 75% of seeds on any plant reach maturity by harvest (Organic Seed Alliance, 2007). Maturity can be visually assessed by determining the percentage of seed with mature tan-brown color (Fig. 11.14).

When 60 to 80% of seeds are brownish gray the crop is nearing harvest. This determination is

Fig. 11.13. Highly female monoecious spinach parent. (Courtesy of Tim Loeffler.)

Fig. 11.14. Maturing spinach seed field in Denmark with the male rows removed. (Courtesy of Tim Loeffler.)

not precise because genetic and environmental variation or disease may lighten color. For example, Stemphylium/Cladosporium leaf spot fungus complex may prematurely lighten plant and seed color (Organic Seed Alliance, 2007). Some spinach cultivars produce naturally greenish seed when fully mature. Maturity of the seed endosperm is the most reliable trait to monitor for maturity. The endosperm and embryo mature concurrently, so monitoring endosperm maturity suggests the state of seed maturation. Endosperm appearance will transition with maturity from translucent or milky early in development to what is described as "flinty" with a grayish, waxy appearance at the midpoint of development. When seeds in the middle of the stalk reach the "flinty" stage, irrigation, if being applied, may be stopped to speed maturation and desiccation of the crop. This is important if maturation may extend into wet conditions conducive to disease. At full maturity, the endosperm is starchy, white, and firm when observed by cutting through or cracking open a seed. In the seed-producing areas with relatively calm and dry weather, spinach crops may be combined directly when most plants have dried, and seeds are mature. In some production areas, the seed fields may be chemically desiccated to hasten dry down of the crop for direct combine harvest. Spinach seed is subject to shattering. Windy or wet weather may lead to considerable loss from shattering, disease, or birds. Alternatively, the crop can be cut and placed in windrows to final dry as soon as senescence begins, and the earliest seeds are mature. When most plants have at least 75% mature seeds, they can be cut near the base and windrowed. If weather is warm and dry, windrowed stalks will be ready to combine or thresh in 10–14 days after cutting. Turning stalks in the windrows facilitates uniform drying, especially if the crop has received rain after cutting. For small productions, stacks of stalks are sometimes placed on sheets to dry to avoid seed loss from shattering. Fully dried cut plants may be run through a combine or small drum threshing machine. To prevent damage during harvest, the seed should be inspected and the drum speed adjusted to the type of harvesting equipment used. Concaves should be set apart to minimize broken stalk debris. Threshing also helps break up seed clusters, producing individual seeds.

Post-harvest Seed Conditioning

After harvest, the seed crop is scalped to remove stem pieces, weed seeds, and other crop debris, then dried and then stored in bins until the seed is scheduled for conditioning. The conditioning process includes the debearder to break up any seed clusters and seeds with sticks attached, followed by the air screen machine and density gradient table to separate debris, undeveloped seed, and light seeds. Sometimes an indent cylinder is used to remove any remaining sticks or similar material. Most spinach seeds are sized into several size categories to facilitate precision seeding for vegetable growers.

Seed Yield

A good spinach seed yield is approximately 2000 kg/ha (1784 lbs/acre). Yields may vary dramatically with cultivar and environment. One hundred smooth and 90 prickly seeds each weigh approximately 1 g. Seed yields of open-pollinated and many F1 hybrid cultivars are comparable when a low male-to-high female plating ratio is used.

Pest Management

Weeds

Weed control issues are like those discussed for beet seed production above. Spinach seedlings grow slowly especially in cold soils and do not compete well with weeds, making early-season weed control critical. Weed seed can contaminate spinach seed lots, so keeping fields weed free is a priority. Lambsquarters (*Chenopodium album*) and goosefoot (*Chenopodium albescens*) are related to spinach (USDA ARS, 2023a). Selective herbicides cannot be used to control these weeds because they would damage spinach. Non-selective herbicides can only be used before planting or for spot treatments of weeds by hand, while shielding spinach plants, after emergence. Obtaining uniform stand establishment, frequent cultivation, and the stale bed technique can provide effective weed control. Weed pressure later in development makes rogueing more difficult, will reduce seed yields due to competition, and produces leggy plants that lodge easily (Organic Seed Alliance, 2007). Other weeds that are difficult to remove during seed conditioning are bedstraw (*Galium aparine*), wild buckwheat (*Polygonum convolvulus*), and field bindweed (*Convolvulus arvensis*) which is a noxious weed. It is important to avoid fields known to have a history of these weeds and to closely monitor fields during the growing season.

Pathogens

Downy mildew (*Peronospora farinosa* f. sp. *spinaciae*) is one of the most serious diseases of spinach. Downy mildew is difficult to control using fungicides particularly after symptoms appear. The fungus thrives under cool wet conditions. Initial symptoms of downy mildew are dull to bright yellow spots that can form anytime on cotyledons and leaves. These spots may enlarge becoming tan and dry. Purple growth of the fungus occurs on the underside of leaves. Downey mildew is very adaptive, overcoming genetic plant resistance and rapidly evolving into new races. Main controls are genetic resistance and rotating of systemic fungicides. Most resistance to four major physiological races is available in modern F1 hybrid cultivars (LeStrange and Koike, 2012). Many breeding programs are focused on developing new hybrids with genetic resistance as new races are continually being identifed. Using cultivars with resistance is important for organic production where non-organic fungicides are not allowed. There are now more than 15 races of downy mildew for spinach.

A complex of fungi is associated with leaf spot disease on spinach. Cladosporium (*Cladosporium variabile*) is apparently the primary pathogen under cool, 18°C to 23°C (65°F to 74°F) and moist conditions (du Toit and Hernandez-Perez, 2005). This leaf spot is characterized by round, tan lesions that rarely exceed 0.64 cm (0.25 in.) in diameter. Dark green spores and mycelium later develop in the centers of these spots. For control, disk or bury volunteer plants and residue underground before planting new spinach crops. Hot-water treatment of infested seed at 40°C (104°F) for 10 min effectively eliminated *Cladosporium* without affecting germination (du Toit and Hernandez-Perez, 2005). Stemphylium (*Stemphylium botryosum*) leaf spot is another seed-borne pathogen causing circular to oval, gray-green leaf small spots 0.33 to 0.64 cm (0.13 to 0.25 in.) in diameter. The optimum temperature for *Stemphylium* development is over a range of 15°C to 20°C (59°F to 68°F) with high humidity (du Toit and Hernandez-Perez, 2005). As the disease progresses, leaf spots enlarge, remain circular to oval, turning tan. Older spots coalesce, dry up, and become papery in texture. Overall, symptoms resemble circular spots caused by pesticide or fertilizer damage. These leaf spot diseases are spread by spores splashed from volunteer spinach or plant debris on the soil surface by rain or irrigation onto the leaves of new plants. Weeds or other reservoir hosts have not been identified.

Members of the virus genera *Hordeivirus, Ilarvirus, Nepovirus, Potyvirus, Cucumovirus, Tobamovirus,* and *Tobravirus* and viroids are seed-borne (Hull, 2014). Cucumber mosaic virus (CMV, *Cucumovirus*) is a seed-transmitted virus that affects spinach. Symptoms include slight chlorosis and narrow or puckered young leaves. Leaves may have margins that roll inward. In advanced stages, the plants appear stunted, and the crown may die completely. Chemical control of the vector aphids (*Aphis fabae* and *Myzus persicae*) reduces incidence of CMV (LeStrange and Koike, 2012).

Other common viral disease of spinach is beet western yellows virus (BWYV). This infection causes interveinal and leaf margin chlorosis on older leaves. Secondary fungi often invade older infected leaves, leading to necrosis that supports dark green to black fungal growth. BWYV is not seed-transmitted. Beet curly top virus infections stunt leaf development and cause chlorosis. Younger leaves in the center of the rosette are often very chlorotic, extremely curled, and rigid. Plants usually die a few weeks after symptoms appear. Impatiens necrotic spot and tomato spotted wilt are both insect-vectored viruses that cause ringspots, circular leaf spots, and necrotic spotting on foliage. Tobacco rattle virus transmitted by nematodes causes yellow and necrotic spotting, mottling, and leaf crinkling on spinach leaves (LeStrange and Koike, 2012).

Weeds may serve as virus reservoirs and should be removed from areas in and around seed fields. However, weed removal will not necessarily prevent virus infections. No virus-resistant spinach cultivars are available. Spraying for insect vectors may not totally effective (LeStrange and Koike, 2012).

Damping-off affects spinach production throughout the world and can be caused by multiple soil-borne fungi. *Pythium* (several species) and *Rhizoctonia solani* infect seedlings and reduce stands in seed fields. Above ground symptoms are similar to symptoms of root rot. Newly emergent seedlings are most susceptible to damping-off, which causes infected plants to collapse at the soil line. Seedling damping-off caused by *Pythium* spp. and *Rhizoctonia* spp. can be prevented by reducing soil moisture levels, seed dressing with fungicides, and planting fungicide treated seeds. Severe damping-off is associated with clay or poorly draining soils used repeatedly for spinach production (LeStrange and Koike, 2012).

Verticillium wilt (*Verticillium dahliae*) is a fungus that infects the vasculature of plants. Verticillium wilt is a problem in seed production in some areas but is less of an issue in commerce for fresh market or processing. The pathogen can be seed-borne as well as being present in the soils of seed production areas. The disease develops late in the growing season and can affect seed yields (Villarroel-Zeballos *et al.*, 2012). Soil grow-out tests can be conducted to select those fields where levels of the pathogen are low and acceptable for seed production.

One of the most important diseases affecting spinach seed production is Fusarium wilt (*Fusarium oxysporum* f. sp. *spinaciae*). Fusarium is controlled primarily by crop rotation and high pH. Due to high susceptibility of spinach to *Fusarium*, long rotations of typically 6 to 10+ years depending upon the levels of contamination must be followed. Biologicals (e.g. *Trichoderma* spp.) may be effective in some cases.

Insects

Lygus (western tarnished, *Lygus hesperus*) bugs damage spinach seed crops by puncturing plant tissues with their piercing mouthparts, including developing flowers and seed, and sucking sap. The lygus bug adult is about 6.4 mm (0.25 in.) long and about half as wide. Lygus are brownish with a prominent, V-shaped marking near the center of their bodies at the base of the wings. Eggs are bean-shaped, with blunt outer ends. Eggs are difficult to spot especially in flowers and on and around developing seeds. A single lygus bug per flower or seed head can cause total loss of seeds or development of embryo-less seed. A black sticky substance is often produced at the site of feeding injury. Control is difficult. Lygus bugs have developed resistance to certain organophosphate, carbamate, and pyrethroid insecticides in areas where these are heavily used against lygus. Manage lygus resistance by scouting populations, then spraying as few times as possible and rotate among insecticides with different modes of action. Lygus bugs have a wide host range of over 200 plants including many weeds. Russian thistle, black mustard, wild radish, and goosefoot are other lygus bug hosts. When weedy fields and orchards are located nearby, the lygus bug may migrate to more succulent spinach seed fields as surrounding weeds dry. Remove weed hosts before lygus populations become winged adults.

Leafminer (*Liriomyza trifolii*) larvae tunnel within a spinach leaf blade beneath the epidermis. Control treatments must be applied before leafminers enter the leaf where they will be protected by epidermal tissues. Some leafminer populations are resistant to organophosphates. Control measures include deep plowing in early spring to destroy infested weeds and plant material from the previous season, as well as removing weed hosts such as pigweed, lambsquarters, plantain, chickweed, and nightshade from in and around seed fields. Releasing parasitic wasps can also provide control.

Caterpillars (loopers, earworms, and armyworms) are some of the most problematic pests because they directly feed on leaves and can quickly destroy a crop. These caterpillars damage the crown, severely stunting or killing spinach plants. The potential for damage and contamination continues until harvest. Cutworms (*Peridroma saucia* and *Agrotis subterranean*) feed at or below ground level. Beet armyworms lay their eggs in distinctive cottony masses on leaf surfaces as described above. The western yellow-striped armyworm can also develop high populations, rapidly damaging spinach.

Several management techniques are effective for controlling spinach caterpillars. Natural insect predators that systematically attack spinach caterpillars can be very effective. Among the most common parasites are the wasps, *Hyposoter exiguae* and *Chelonus insularis*, and the tachinid fly, *Lespesia archippivora*. Naturally occurring viral diseases may also kill caterpillars in a balanced ecosystem. Cultural control methods may also be effective. For example, disking fields immediately following harvest to kill larvae and pupae as well as destroying weeds along field borders are effective cultural techniques (LeStrange *et al.*, 2012).

Cultural and biological controls mentioned above along with sprays of *Bacillus thuringiensis* or the Entrust® formulation of spinosad are approved for Certified Organic production in the USA and certain other countries (LeStrange *et al.*, 2012). Spinosad is a fast-acting, somewhat broad-spectrum material that acts on the insect primarily through ingestion, or by direct contact, causing loss of muscle control. Continuous activation of motor neurons causes insects to die of exhaustion within one or two days after ingestion. Foliar applications of spinosad are not highly systemic although some movement into leaf tissue occurs. The addition of a penetrating surfactant increases spinosad absorption by tissues and activity on pests that mine leaves (Larson, 1997).

Effective control of insect pests in spinach seed fields requires careful monitoring before seedlings emerge by checking for eggs and young larvae in surrounding weeds. If populations are high on weeds, larvae may also be present on field seedlings. Once seedlings emerge, they should be checked twice weekly for egg masses and young larvae. Insecticides, if used, are more effective against young larvae than eggs, so applications should be delayed until a majority of eggs have hatched before beginning treatment. For beet armyworm control, the best time to apply insecticide is at dawn during twilight hours when worms are active (LeStrange *et al.*, 2012).

A partial list of other spinach insect and nematode pests include: *Ditylenchus dipsaci* (nematodes), *Clavigralla tomentosicollis* (African pod bug), *Helicotylenchus dihystera* (common spiral nematode), *Heterodera schachtii* (beet cyst eelworm), *Liriomyza sativae* (vegetable leafminer), *Listroderes costirostris* (vegetable weevil), *Myzus persicae* (green peach aphid), *Nacobbus aberrans* (false root-knot nematode), *Tetranychus cinnabarinus* (carmine spider mite), *Trichoplusia ni* (cabbage looper), and *Helicoverpa zea* (Corn earworm) (LeStrange *et al.*, 2012).

References

Abegg, F.A. (1936) A genetic factor for the annual habit in beets and linkage relationships. *Journal of Agricultural Research* 53, 493–511.

Atherton, J.G. and Farooque, A.M. (1983a) High temperature and germination in spinach. I. The role of the pericarp. *Scientia Horticulturae* 19, 25–32.

Atherton, J.G. and Farooque, A.M. (1983b) High temperature and germination in spinach. II. Effects of osmotic priming. *Scientia Horticulturae* 19, 221–227.

CABI (2023) *Beta vulgaris* (beetroot). List of pests. In: *Crop Protection Compendium*. CAB International, Wallingford, UK. https://www.cabidigitallibrary.org/doi/10.1079/cabicompendium.8778 (accessed 11 September 2023).

Davis, J.G. and Westfall, D.G. (2015) Fertilizing Sugar Beets. Colorado State University Extension. https://extension.colostate.edu/topic-areas/agriculture/fertilizing-sugar-beets-0-542/ (accessed 20 September 2023).

Du Toit, L.J. and Hernandez-Perez, P. (2005) Efficacy of hot water and chlorine for eradication of *Cladosporium variabile*, *Stemphylium botryosum*, and *Verticillium dahliae* from spinach seed. *Plant Disease* 89,1305–1312.

George, R.A.T. (2009) *Vegetable Seed Production.* CAB International, Wallingford, UK.

Goldman, I.L. and Navazio, J.P. (2007) Table beet. In: Prohens-Tomás, J. and Nuez, F. (eds) *Handbook of Plant Breeding*, Vol. 1: *Vegetables: Asteraceae, Brassicaceae, Chenopodicaceae, and Cucurbitaceae.* Springer, New York, pp. 219–238.

Grubben, G.J.H. and Denton, O.A. (2004) *Plant Resources of Tropical Africa. 2. Vegetables*. PROTA Foundation, Wageningen, the Netherlands.

Hanelt, P., Büttner, R., Mansfeld, R. and Kilian, R. (2001) *Mansfeld's Encyclopedia of Agricultural and Horticultural Crops*. Springer, New York.

Harrington, J.F. and Minges, P.A. (1954) *Vegetable seed germination*. Mimeo leaflet. University of California Cooperative Extension, Davis, California.

Hembree, K.J. and Norris, R.F. (2010) *UC IPM Pest Management Guidelines: Sugarbeet*. UC ANR Publication 3469. UC Statewide IPM Program, University of California, Davis. www.ipm.ucdavis.edu/PMG/r735700111.html (accessed 11 September 2023).

Hill, G. and Langer, R.H.M. (1991) *Agricultural Plants,* 2nd edn. Cambridge University Press, Cambridge, UK.

Hoyt, G.D., Monks, D.W. and Monaco, T.J. (1994) Conservation tillage for vegetable production. *HortTechnology* 4, 129–136.

Hull, R. (2014) Chapter 7. Replication of plant viruses. In: *Plant Virology*, 5th edn. Academic Press, London, pp. 341–421. doi: 10.1016/B978-0-12-384871-0.00007-8. https://www.ncbi.nlm.nih.gov/pmc/articles/PMC7184227/ (accessed 11 September 2023).

Jarmin, M.L. and Thornton, R.E. (1985) *Table Beet and Swiss Chard Seed Production in the Pacific Northwest.* Pacific Northwest Cooperative Extension Publication 271. Oregon State University.

Janick, J. (1998) Hybrids in horticultural crops. In: Larnkey, K.R. and Staub, J.E. (eds) *Concepts and Breeding of Heterosis in Crop Plants.* ASA-CSSA-SSSA, Madison, Wisconsin, pp. 45–56.

Larson, L.L. (1997) Effects of adjuvants on the activity of Tracer™ 480SC on cotton in the laboratory. *Arthropod Management Tests* 22, 415–416.

LeStrange, M. and Koike, S.T. (2012) Downy mildew. In: *UC IPM Pest Management Guidelines: Spinach.* UC ANR Publication 3467. UC Statewide IPM Program, University of California, Davis. www.ipm.ucdavis.edu/PMG/r732100111.html#SYMPTOMS (accessed 11 September 2023).

LeStrange, M., Koike S.T. and Chaney, W. E. (2012) Caterpillars (Lepidopterous pests). In: *UC IPM Pest Management Guidelines: Spinach*. UC ANR Publication 3467. UC Statewide IPM Program, University of California, Davis. www.ipm.ucdavis.edu/PMG/r732300411.html (accessed 11 September 2023).

Natwick, E.T., Summers, C.G., Haviland, D.R. and Godfrey, L.D. (2010) Sugarbeets: insects and mites. In: *UC IPM Pest Management Guidelines: Sugarbeet*. UC ANR Publication 3469. UC Statewide IPM Program, University of California Davis. www.ipm.ucdavis.edu/PMG/r735300811.html (accessed 22 January 2013).

Organic Seed Alliance (2007) *Principles and Practices of Organic Spinach Seed Production in the Pacific Northwest*. Organic Seed Alliance, Port Townsend, Washington. https://projects.sare.org/wp-content/uploads/1626spinach_seed_manual.pdf (accessed 11 September 2023).

Organic Seed Alliance (2018) *Beet Seed Production*. Organic Seed Alliance, Port Townsend, Washington. https://seedalliance.org/publications/beet-seed-production-quick-reference/#:~:text=Beets%20should%20be%20direct%20seeded,sprout%20from%20a%20single%20seed (accessed 11 September 2023).

Owen, F.V. (1954) The significance of single gene reactions in sugar beets. *Proceedings of the American Society of Sugar Beet Technology* 8, 392–398.

Owen, F.V., Carsner, E. and Stout, M. (1940) Photothermal induction of flowering in sugar beet. *Journal of Agricultural Research* 61, 101–124.

Rubatzky, V.E. and Yamaguchi, M. (1997) *World Vegetables: Principles, Production and Nutritive Values*, 2nd edn. Chapman & Hall, New York.

Schrader, W. and Mayberry, K. (2002) *Beet and Swiss Chard Production in California*. ANR Publication 8096. University of California, Division of Agriculture and Natural Resources, Oakland, California.

Sharma, P., Kikkert, J. and Pethybridge, S. (2021) Bacterial Leaf Spot of Table beet. College of Agriculture and Life Sciences, Cornell University, New York. www.vegetables.cornell.edu/pest-management/disease-factsheets/bacterial-leaf-spot-of-table-beet/ (accessed 20 September 2023).

Smit, A.L. (1983) Influence of external factors on growth and development of sugar beet (*Beta vulgaris* L.). PhD dissertation. Landbouwhogeschool te Wageningen, Agricultural Research Reports 914. Centre for Agricultural Pub. and Documentation, Pudoc, Wageningen, the Netherlands.

Thompson, A.E. (1955) Methods of producing first-generation hybrid seed in spinach. *Cornell University Agricultural Experiment Station* 336, 1–48.

UC IPM (2023) Powdery mildew. In: *UC IPM Pest Management Guidelines: Sugarbeet*. UC ANR Publication 3469. UC Statewide IPM Program, University of California, Davis. https://ipm.ucanr.edu/agriculture/sugarbeet/powdery-mildew/ (accessed 11 September 2023).

USDA APHIS (2023) *Beta vulgaris* ssp. *vulgaris*. https://acir.aphis.usda.gov/s/cird-taxon/a0u3d000000FFGsAAO/beta-vulgaris-ssp-vulgaris (accessed 20 September 2023).

USDA ARS (2023a) *Chenopodium album* L. lambsquarters. National Resources Conservation Service. https://plants.usda.gov/home/plantProfile?symbol=CHAL7 (accessed 20 September 2023).

USDA ARS (2023b) *Beta vulgaris* L. common beet. National Resources Conservation Service. https://plants.usda.gov/home/plantProfile?symbol=BEVU2 (accessed 20 September 2023).

Villarroel-Zeballos, M.I., Feng, C., Iglesias, A., du Toit, L.J. and Correll, J.C. (2012) Screening for resistance to Verticillium wilt in spinach and isolation of *Verticillium dahliae* from seed of spinach accessions. *HortScience* 47, 1297–1303.

Welbaum, G.E. (2015) Chapter 17. Family Amaranthaceae subfamily Chenopodioideae. *Vegetable Production and Practices*. CAB International, Wallingford, UK, pp. 349–366.

Zohary, D. and Hopf, M. (2000) *Domestication of Plants in the Old World*. Oxford University Press, Oxford, UK.

12

Family Cucurbitaceae

Overview

Crops of the family Cucurbitaceae, often referred to as Cucurbits, are primarily herbaceous perennials grown as annuals in temperate regions. The family includes major vegetable crops: melon, cucumber, watermelon, squash, and pumpkin. Many are of tropical origin with different genera originating in diverse locations such as Africa, tropical America, and Southeast Asia. Some cultivars are adapted to humid conditions while others are better suited for arid regions with low humidity.

Most Cucurbit fruit are botanically classified as fleshy berries or hard-rind pepos. Cucurbit seed can be used for food or propagation. Seeds of watermelon are popular snack foods in Southeast Asia and the Middle East, while pumpkin seeds are consumed in the Americas. Tea and edible oil can also be produced from Cucurbit seeds. Cucurbit seeds both for food and propagation are grown commercially in warm, dry, irrigated desert regions with long growing seasons in the Middle East, North America, South America, Asia, and Africa to avoid disease problems. Melon seed production including human food use estimates are over 608,000 ha (1.5 million acres) with total yield of 576,000 metric tonnes (634,941 US tons) (FAO, 2023). This chapter is organized to cover basic characteristics of Cucurbits followed by unique properties of individual crops.

Taxonomy

The family Cucurbitaceae consists of about 120 genera and more than 800 species. Cucurbit food crops generally fall within the subfamily Cucurbitoideae and belong to the tribes Cucurbiteae and Sicyoideae. (Maynard and Maynard, 2000). The genera *Cucurbita*, *Cyclanthera*, and *Sechium* originated in the New World, while other genera originated from Africa or the Asian tropics. This chapter will consider seed production of cucumber (*Cucumis sativus* L.), netted and non-netted melons (*Cucumis melo* L.), watermelon (*Citrullus lanatus* Thunb.), and squash and pumpkin (*Cucurbita* spp.). The principles of seed production for these four species are applicable to other related Cucurbits. General information about growing these crops for food as well as basic agronomic and climatological requirements are available elsewhere (Whitaker and Davis, 1962; Robinson and Decker-Walters, 1997; Rubatzky and Yamaguchi, 1997; Welbaum, 2015).

Generally, successful sexual crosses among species are impossible. One exception is *Cucurbita* where plant breeders have successfully crossed certain species. Flower morphology is often similar within a genus. Among genera, flowers may vary considerably in size and color and traditionally have been used for taxonomy along with stems and seeds.

DOI: 10.1079/9781789243260.0012

Botany

Cucurbits commonly produce vines with anchoring tendrils at the nodes with a range of internode lengths. Some "bush" cultivars have very short internodes and are available commercially for some Cucurbits. Most commonly, Cucurbits are monoecious, although other forms of sex expression occur including hermaphrodite, andromonoecious, gynomonoecious, trimonoecious, gynecious, and dioecious. A usual flowering sequence for many cultivated Cucurbits is first production of staminate flowers, followed by pistillate and staminate flower production in a somewhat random order. This cycle of flower formation tends to be repeated on lateral branches. Cultivars differ as to when the first female flowers form, but within plants of a cultivar, female flower formation tends to be consistent. Development of the first female flower is an important determinant of early yields. The ratio of male to female flowers early in development is usually heavily skewed to favor maleness with ratios of 25 male to 1 female commonly occurring. Although sex expression is an inherited trait, environment also affects floral gender. Sex expression is associated with plant hormone levels in developing flower buds. The environment can affect hormone levels. For example, high temperatures and long days usually favor male flower production whereas low temperatures and short days tend to favor the formation of female flowers.

Seed morphology is a tool for identifying species, particularly within *Cucurbita*, where seed color, testa margin and texture, as well as chalaza shape define each species. Some cultivars of *Cucurbita* are grown for their edible naked seeds that do not develop seed coats because of a natural mutation. Watermelon seeds have anatomy similar to other Cucurbits, but seeds vary in size, number, and color from white through tan, dark brown, and black. Watermelon cultivars grown for their edible seeds are generally small fruit and contain larger seeds, while edible fruit cultivars have larger fruit with small seeds or no seeds.

Stand establishment

Establishments of seed crops can be by direct seeding on flat fields or raised beds. Seed crops can be established from plug transplants started in a greenhouse. Direct-seeded, open-field production is typically planted 2.5–5 cm (1 to 2 in.) deep into pre-irrigated beds. If seed is of high quality, planting can be to a final stand spacing or overseeded to achieve the target plant population. Cucurbit seeds germinate and emerge in as little as four days at a soil temperature of 25°C (77°F) and from six to 12 days at 20°C (68°F). Most Cucurbit seeds do not germinate well below 15°C (59°F).

In short season areas, for tetraploid lines, or to enhance the timing of pollination (e.g. nicking), Cucurbits are sometimes transplanted in the three- to five-leaf stage. Plug tray transplants are preferable since plants are sensitive to transplant shock and do not easily produce adventitious roots. Seedless (tetraploid) watermelon breeding lines are often grown from transplants because germination is slower and higher temperatures optimize germination.

Growth and Development

Most commercial Cucurbits are frost-intolerant, although some species are more tolerant of low temperature than others. Cucurbits are also chilling sensitive and are damaged by prolonged exposure to temperatures below 15°C (59°F) (Welbaum, 2015). Cucurbits are extensively cultivated from seed as annuals in temperate regions with long growing seasons. In tropical and even temperate regions where seeds can overwinter, Cucurbits can become weeds.

Cucurbits show variation in traits including rapid indeterminate growth for long internode types, variations in fruit color, and variable sex expression as examples. Genetic diversity for these traits, and others, allows plant breeders to combine genes in modern cultivars to improve horticultural characteristics compared to their wild progenitors. For example, through domestication and in more recent times plant breeding, the size of both fruits and seeds have increased, disease resistance has improved, dry mass has increased, bitterness has been reduced, and seed dormancy has decreased. Specific traits selected by the breeder that are characteristic of each cultivar and are often referred to as true-to-type. It is the job of seedsmen to ensure these traits are passed on from generation to generation without genetic drift, genetic changes that are not characteristic of a cultivar. Sometimes traits selected by breeders are not advantageous to seed production. Breeding Cucurbits that produce hard-rind fruit to resist cracking and dehiscence has improved shipping characteristics and fruit storage life. An unintended consequence of increased shelf life for fruit production is that mature seeds are sometimes "trapped" in mature fruit and exposed to harsh osmotic and extreme temperature conditions that lead to accelerated aging and damage in seed production fields if harvest is delayed (Welbaum, 1993).

Cucumber and melon have similar seed anatomy with a transparent mucilage endocarp that adheres tightly to the outside of the seed coat and is the outermost layer of the seed. Inside the endocarp is the testa with successive layers of epidermis, subepidermal layer, sclerenchyma, parenchyma, and inner epidermis nearest the embryo (Welbaum and Bradford, 1990; Ramakrishna and Amritphale, 2005).

The endosperm tissue surrounding the embryo in melon and cucumber is unique (Welbaum and Bradford, 1990; Ramakrishna and Amritphale, 2005). In melon, mature endosperm is a single layer of dense cytoplasm tissue enclosing the embryo that deposits a layer of callose on the cell wall adjacent to the testa (Yim and Bradford, 1998). Microscopy and differential staining show the callose and suberin layer forms 30 to 35 days after anthesis. This layer was incorrected identified as perisperm tissue in earlier studies (Yim and Bradford, 1998). The callose and suberin layer makes the endosperm envelope semipermeable, preventing the diffusion of molecules as small as potassium ions but allowing water to pass freely (Welbaum and Bradford, 1990). This semipermeability traps osmotically active solutes that leak from embryo tissue when seeds are damaged resulting in osmotic swelling, a condition known as "fish mouth" in the seed industry. Fish mouth disorder occurs after seeds have aged in the field under adverse conditions prior to harvest, when damaged during seed extraction, fermentation, or processing, or in storage. Seeds exhibiting the fish mouth characteristic are damaged, not of high quality, and have lower germination. Apparently, endosperm tissue has evolved from a storage to protective role that limits electrolyte leakage from the embryo, slows fungi and bacteria attacks, and protects embryos from potentially harmful fruit decomposition products (Welbaum, 1993). Endosperm with similar characteristics surrounds embryos of other Cucurbits, including cucumber, although not all Cucurbits have been examined in detail (Ramakrishna and Amritphale, 2005).

High nitrogen promotes vegetative growth and male flowers and may even delay or even inhibit formation of female flowers. Frequent harvesting tends to extend flower production. Parthenocarpic fruit development occurs in cucumber but is not common in other species. Parthenocarpic fruit apparently have little influence on subsequent plant growth or flowering.

F1 hybrids dominate the market for most commercial Cucurbit crops, particularly in developed countries where growers can afford their higher cost and can afford to buy new seed each year. Many growers prefer F1 hybrid cultivars because of their greater vigor, higher yield, disease resistance, and greater uniformity compared to open-pollinated cultivars. F1 hybrid seeds offer the advantages of combining desirable characteristics from diverse parental backgrounds and protect the investment of seed companies, since the parent lines are proprietary, and seed cannot be saved by growers for a second generation. For F1 hybrid cucumber and squash production, ethylene-releasing compounds suppress male flower production, so plants produce mostly female flowers. For some hybrid cucumbers, the female parent has been genetically manipulated to produce lines of predominately female flowers (gynoecious plants) that can be harvested for hybrid seed. Adjacent rows of the male parent lines are not treated, supplying a source of pollen for the cross. In both the genetic and chemical methods of altering sex expression, only the female rows are harvested as hybrid seed. For hybrid melons (honeydew, cantaloupe, Crenshaw, Persian, casaba) or watermelon, because sex expression is not easily altered genetically or chemically, extensive hand labor is required to emasculate flowers and make crosses. Consequently, hybrid seed production often occurs in countries where labor is abundant and less inexpensive.

Cucurbit embryos develop an ability to germinate progressively well before the accumulation of maximum dry mass if removed from the fruit and incubated in water (Fig. 12.1).

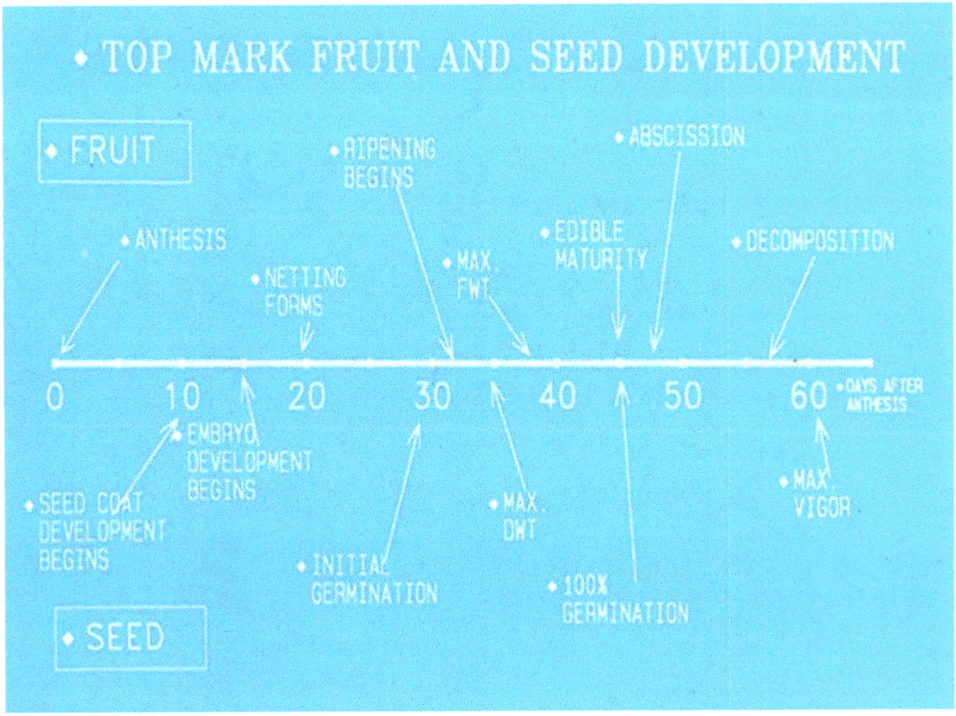

Fig. 12.1. Top Mark fruit and seed development comparison. FWT: fresh weight; DWT: dry weight.

Studies on *Cucumis melo* cv. Top Mark illustrate that from 15 to 18 days after anthesis (DAA), the cotyledons of isolated immature melon embryos expand and produce chlorophyll when incubated on blotter paper saturated with water at 25°C (77°F) (Welbaum and Bradford, 1990). At approximately 20 DAA, hypocotyls elongate, and cotyledons expand, but root and shoot meristems did not grow when isolated embryos were incubated under favorable conditions. At 25 to 30 DAA, roughly 40% of epicotyls and roots from isolated embryos grew, while the remaining 60% showed only cotyledon and hypocotyl expansion (Welbaum, 1999). By 35 DAA, the time of mass maturity (maximum dry weight), all embryonic tissues grew, and normal seedlings developed if the testa and endosperm were removed. Intact seeds were not fully viable until 45 DAA, suggesting the endosperm and testa maintain developmental control (Fig. 12.1). The highest concentrations of abscisic acid (ABA) in melon seeds occurred 25 DAA when embryo tissues only grew if all surrounding maternal tissues were removed (Welbaum *et al.*, 2000). As germination percentages increased, ABA concentrations fell to very low levels in all tissues after 40 DAA, so the development of germinability is well correlated with a decline in endogenous ABA concentrations (Welbaum *et al.*, 2000). Also, immature melon seeds were the most sensitive to exogenous ABA early in development (Welbaum and Bradford, 1990). Based on germination rate, low temperature germination, and resistance to controlled deterioration, occurred 50 to 55 DAA or about 10 days past edible maturity but before the onset of severe fruit decomposition (Fig. 12.1). Physiological maturation in Cucurbit seeds inside fruit increased vigor and continued until peaking about 20 days after seed mass maturity and full viability were attained. Physiological maturation proceeds quickly inside maturing and senescing fruits because of the high seed moisture content and high temperatures, similar to after-ripening in dry seeds (Fig. 12.1). This environment also favors rapid aging if seeds are not harvested and dried once full physiological maturity is attained (Welbaum, 1993). In the field, the transition from high quality to aged or even dead seeds occurs swiftly over a few days particularly at high temperatures. For example, the additional ten days within the fruit caused a fraction of 65 DAA muskmelon seeds to age beyond the point of peak physiological maturity, reducing seed quality (Fig. 12.1).

Although most Cucurbit seeds display some degree of embryo dormancy early in development, the degree varies widely among species and even cultivars. Another factor that helps maintain quiescence in developing seeds is the low osmotic potential in developing fruit, particularly melon and watermelon. This phenomenon has been called "osmotic inhibition" of germination because, even in the absence of phytohormones, germination is prevented by the low fruit water potential (Welbaum and Bradford, 1988). For example, precocious germination occurs more frequently in Armenian cucumber, a type of *Cucumis melo*, which accumulates less sugar and therefore has a higher fruit osmotic potential throughout development (Welbaum and Bradford, 1988).

To better understand factors contributing to seed vigor, the control of radicle emergence has been extensively studied in muskmelon. When seeds are fully hydrated, they reach zero water potential, but embryos are prevented from fully hydrating by surrounding tissues because the testa and endosperm exert opposing pressure to water uptake. When the endosperm and testa are broken the external pressure is relieved and water potential is reduced, allowing axis water content to increase by almost 50%. Weakening and rupture of the endosperm tissue is required for radicle growth and thus germination to occur. Enzymatic degradation of the endosperm surrounding the radicle and not an increase in radicle turgor is the mechanism that allows growth to initiate. In imbibed seeds, glucanase, endo-β-mannanase, and chitinase activity all increase in hydrated endosperm tissue adjacent to the radicle prior to protrusion and are correlated with endosperm weakening as measured by Instron analysis (Welbaum *et al.*, 1995; Whitmer *et al.*, 2003).

Open-pollinated non-hybrid Cucurbit seed crops are often harvested destructively by machine or by hand for smaller fields when most fruit is mature. For example, the times from plant emergence to harvest for irrigated desert regions of California, USA varies with the crop from about: 78–90 days for netted melon; 120 to 150 days for honeydew and other late-maturing melons; 85–100 days for watermelons; 100–120 days for summer squashes; 120–130 days for cucumbers; to over 130 days for winter squashes. Optimal seed quality coincides with fruit maturity. Seed quality can decline rapidly if harvest is delayed, particularly if temperatures are high and the fruit are exposed to direct sunlight which heats the partially hydrated seeds inside causing rapid aging as described for melon above. Sunburned fruit is an indication of high temperature exposure and the potential for damaged seed. Combining seeds from different stages of physiological maturity in a "once-over" production system may adversely affect seed quality if there is variation in fruit development at harvest. F1 hybrid seed crops produced from hand-pollinated fruits are often harvested by hand as individual fruits ripen, reducing variation in seed quality. Hand-harvested fruits are selected based on ripeness and fed into seed extraction equipment. The optimum stage of development for best seed quality is after full physiological maturity but before the onset of severe fruit decomposition which can expose seeds to aging conditions (Fig. 12.1).

Pollination

Honeybees (*A. mellifera*) are the most reliable and cost-effective way to achieve pollination of Cucurbits although alternative species, including bumble bees (*Bombus* spp.) and other wild species, may contribute in some locations (Kremen *et al.*, 2002). For commercial seed production, a common recommendation is for the introduction of one to two honeybee hives per acre when 5 to 10% of flowers open. If hives are placed near other flowering crops or in weedy locations, the number of hives per acre should be increased. After the percentage of open flowers increases, additional hives should be placed up to a total of three to four hives per acre. Clusters of hives should be placed around the periphery of fields with additional hives placed inside larger fields. It is important to monitor the strength and activity of the hives throughout the pollination season and add additional hives when necessary. Multiple bee visits are required to transfer the 300 to 500 pollen grains from anthers to the stigma for successful pollination. Pollen grains are sticky, preventing direct anther to stigma transfer even on perfect flowers without insect or human assistance. For many species, flowers are only receptive to pollination for a short period of time, e.g. morning and afternoon of a single day. Flowers of some species remain open during the night or for longer periods. Inadequate pollination causes aborted flowers and may result in misshapen or underdeveloped fruit. For successful fertilization, pollen uniformly distributed on all lobes must germinate and pollen tubes must grow through the style into the ovary to fertilize individual ovules.

Genetics

New cultivar development of Cucurbits occurs mainly through traditional plant breeding. The use of molecular markers has been a key advancement to assist in more rapid development of new inbred lines and cultivars. Recent advanced technologies include the use of genetic engineering and gene editing; however, these methods are not accepted in all countries. Currently, the only Cucurbit species where genetic engineered hybrids are used is squash. Genetically engineered squash with enhanced virus disease resistance has been approved for commercial production in the USA and certain other countries. However, to date transgenic cultivars of zucchini and summer squash have had little market impact in North America and are not widely grown because of public concerns about genetically engineered crops.

Seed Yield

Melon fruits produce from 350 to 500 seeds each, arising from unique individual fertilization events. Fruits with less than 300 ovule fertilizations often abort before maturity. Non-parthenocarpic cucumber fruit contain approximately 250 seeds for small-fruited cultivars to 350 for large fruit cultivars. The average seed yield for melon and cucumber is about 300 kg/ha (268 lbs/acre), although the best melon seed production areas can achieve up to 600 kg/ha (535 lbs/acre). There are approximately 1100 cucumber seeds per ounce (30 to 35 per gram) (Lorenz and Maynard, 1988). Others estimate a range from 25 to 55 seeds/g for both melon and cucumber depending on the cultivar (George, 2009).

Cucurbita spp. fruits produce from 350 to 600 seeds depending on cultivar. The average seed yield for *Cucurbita* spp. is about 500 kg/ha (446 lbs/acre), but under optimum pollination and cultural conditions up to 1000 kg/ha (892 lbs/acre) can be obtained. For *Cucurbita* spp. there are approximately 4 to 10 seeds/g and size may vary with cultivar (George, 2009).

Average seed yield of most open-pollinated watermelon cultivars under good conditions is 400 kg/ha (357 lbs/acre). An open-pollinated cultivar such as Charleston Gray is a relatively low seed yielder by comparison and averages only 250 kg/ha (223 lbs/acre). Under the relatively poor cultural conditions found in some tropical countries, seed yields of the better yielding cultivars are as low as 100 kg/ha (89 lbs/acre). A relatively short growing season also affects seed yield adversely. A lower yield is achieved when fruit are left on the plant for a once-over destructive harvest at the end of the season. Highest yields and best seed quality are obtained when fruits are promptly harvested and extracted after reaching edible maturity. On average there are approximately 8 to 20 seeds/g for watermelon as seed size varies dramatically with cultivar (George, 2009). Cultivars bred for food have larger seeds than those developed for fruit consumption and this accounts for the differences in size.

After-Ripening and Storage

Cucurbit seeds after-ripen during dry storage. After-ripening is a period of dry storage, the duration of which varies with seed moisture content and temperature. During after-ripening the vigor and sometimes the germinability of seeds increase above what they were at harvest (Bewley, 1997). Some seed companies routinely store seeds at controlled higher temperatures for a certain period or use priming to sell seeds shortly after harvest without after-ripening. Often Cucurbit seeds are stored for several months to a year at room temperature so seeds after-ripen before sale (Oluoch and Bradford, 1996; Bewley, 1997; Finch-Savage and Leubner-Metzger, 2006). After-ripened Cucurbit seeds germinate faster and over a wider range of temperatures and moisture stress compared with when they were newly harvested (Welbaum and Bradford, 1991; Finch-Savage and Leubner-Metzger, 2006).

Harvesting

Large-scale Cucurbit seed production is often mechanically harvested with specialized equipment. In advance of harvest, vines are cut and fruit windrowed on top of the bed with a tractor implement. In a destructive harvest, fruit are collected into the harvester on a chain conveyer and crushed to release seeds. The seed and pulp travel through an inclined rotating screened drum. Seeds fall through the screen while fruit rind, pulp, and trash move out the rear of the harvester back into the field. Screens of various sizes are used for different Cucurbit crops. The seeds with juice and pulp are then augured into bins for transport to a washer-dryer system located in a nearby processing area. To prevent contamination or mixing of seed lots, the harvester must be thoroughly cleaned between fields of different cultivars or species.

Fruits may also be hand-harvested direct from the original rows. The fruits from small fields, plots, net houses, or greenhouses are hand-harvested and brought to a portable or stationary seed extractor.

Although fermentation for one to two days was once a standard post-harvest practice, mechanical washing of newly harvested seeds is more commonplace because seed temperature and hydration can be more closely controlled. Seed must be washed and dried promptly after harvest because microbial fermentation begins immediately in the slurry of seeds, juice, and fruit tissue and can expose seeds to excessively high temperatures. Seeds are spray washed as they move through inclined rotating drum screens and in some cases the seeds are also passed through a water flume to remove fruit and other debris. It is important to wash and dry seeds quickly, so germination does not begin, particularly in seeds that have no deep dormancy at harvest.

The washed seed can be dried in a rotating drum, precision, or table dryers through which ambient or heated air is forced from below. Final seed moisture should be in the range of 5 to 8%. Seeds with higher moisture do not store well, while excessively dry seeds may be damaged. It is critical that the moisture and temperature be monitored frequently to prevent under- or over-drying. If controlled-temperature drying is not available, seeds of lower value may be sun dried in warm regions with low humidity if spread flat and mixed frequently to prevent overheating. Cucurbit seeds dry rapidly since they have no significant barriers to water movement in the seed coats of most commercial species. Excessively rapid drying using heated air or sun drying may damage seeds and reduce seed vigor. In humid regions, air-dried seeds may not achieve a low enough water content for long-term storage. Rechargeable drying beads may be used to final dry seeds under these conditions (Timsina *et al.*, 2014). Seed temperature should not exceed 35°C (95°F) during drying.

Once seeds have been washed and dried, they are cleaned, sized, upgraded, and tested for physiological, physical, genetic quality, and seed health. The presence of weed seed, variability in seed sizes, and excess foreign material can require additional seed cleaning or conditioning at a seed-processing facility. Seed must be precleaned to remove crop trash, weeds, and other material to meet phytosanitary requirements prior to shipping to the final seed-cleaning facility in another country.

Seed Storage

Cucurbit seeds are long lived and if sealed in airtight containers at 5 to 8% moisture content they can be successfully stored at 3 to 5°C (37°F to 41°F) for 20 or more years (Fig. 12.2). Cucurbit seeds may be successfully frozen at −12°C if their moisture content is sufficiently low (5 to 6%). Cryopreservation, the storage of Cucurbit seeds is at −196°C, is possible for ultra-long-term storage of germplasm. Typically, seed companies produce enough seeds to maintain sufficient

Fig. 12.2. Dried cucumber seed ready for cleaning. (Courtesy of Tim Loeffler.)

inventory of a given cultivar for two to three years. Older seeds of a popular cultivar can be sold if quality is acceptable and meets market requirements. Commercial seeds are usually stored under controlled temperatures ranging from 5°C to 15°C (41°F to 59°F) with a relative humidity between 30 to 50%.

The stage of development at harvest is a major determinant of quality. Maximum viability and vigor are obtained for a relatively short period of development determined by cultivar and environmental conditions.

Many Cucurbit seeds are dormant at harvest and after-ripening (a period of dry storage) is required before Cucurbit seeds to reach maximum viability and vigor as measured by germination rate and tolerance of low water potential. Vigor continues to increase during after-ripening even after seeds are fully viable. The depth of dormancy varies with species, some species showing deep dormancy and others showing little or no dormancy. Because of this dormancy, seeds can be heated treated under controlled temperature or primed if they must be sold soon after harvest. Sometimes the seeds are routinely stored for one year prior to sale to ensure after-ripening is complete. Embryo ABA concentrations were positively correlated with the weak dormancy displayed by developing muskmelon seeds (Welbaum *et al.*, 2000). Some Cucurbit seeds exhibit dormancy when placed in pure water but germinate rapidly under a slight water stress (–0.2 MPa) or low concentrations of ABA. Dormancy in some Cucurbit seeds apparently involves phytochrome because seeds harvested and maintained in the dark fail to germinate. Exposure to red light stimulates germination while exposure to far red light inhibits it (Bewley, 1997).

Post-harvest Treatment

Most Cucurbit seeds are comparatively large, so pelleting is generally not required to change the shape for ease of handling. Seed coating, increasingly film coating in a range of colors, is used to deliver pesticides, particularly fungicide and insecticides. Biological agents are added to Cucurbit seeds using biopriming or as inoculants to improve seed performance and protect against disease. Such treatments remain largely experimental and are used on a limited basis commercially. Some evidence suggests that systemic acquired resistance (SAR), the expression of natural resistance genes in plants, can be induced through pregermination seed treatment with salicylic acid derivatives. Other evidence suggests that SAR is only a post-germination plant response that helps protects seedlings from certain pathogens. Chitinase, an enzyme produced as part of systemic acquired resistance response, was found in both developing and germinating muskmelon seeds (Whitmer *et al.*, 2003).

Seed priming, controlled hydration followed by redrying, can improve Cucurbit germination performance. In Cucurbit seeds, priming advances physiological maturity by substituting for after-ripening, thus making immature Cucurbit seeds perform as if they were more mature (Welbaum and Bradford, 1991). The water potential inside developing muskmelon fruits is approximately −1.3 MPa, similar to an osmotic priming solution. Apparently, muskmelon seeds are primed *in situ* during the later stages of development after seed mass maturity. Therefore, seed lots comprising immature seeds that are not physiologically mature tend to show a greater response to priming than seed lots comprising older seeds. Priming also initiates the process of endosperm envelope weakening, a key event that helps determine the timing of radicle emergence. Priming may also repair damage incurred during seed storage although studies with muskmelon seeds show that priming will not restore the viability of seeds lots subjected to preharvest aging due to delayed harvest (Welbaum *et al.*, 1998).

CUCUMBER

Botany

Commercial cucumber cultivars are warm-season, frost-sensitive, and grown as annuals in temperate regions. Plants commonly have a trailing or climbing growth habit, although cultivars with shortened internodes form bushes and are available for most types. Root systems are extensive but shallow. Stems are square with stiff abrasive bristle hairs and generally range in length from 0.4 to 3.0 m (1.3 to 10 ft). Unbranched tendrils at each node help anchor plants and allow climbing on supports. Cucumber petioles are 3 to 15 cm (1.2 to 5.9 in.) long. Rough leaf blades have a triangular ovate shape from 5 to 25 cm (2.0 to 10 in.) wide with three- to five-angled regions, or shallow-lobed sinuses, and a pointed apex (Rubatzky and Yamaguchi, 1997).

The sex expression of cucumber plants may be monoecious (separate male and female flowers on a plant), andromonoecious (separate perfect and male flowers on the same plant), or gynoecious (all female). In nature, the flowers are cross-pollinated by bees even in types when flowers are perfect because the pollen is sticky and must be physically moved from anthers to stigmas. The development of monoecious or gynoecious parental lines is preferred for F1 seed production. Ideally, a strong gynoecious female parent only producing pistillate flowers and a monoecious male parent are used for F1 seed production but this depends on the characteristics of the desired hybrid to be produced. Sex expression is genetically controlled but modified by the environment or plant growth regulators. Application of auxin or ethephon favors both earlier and a higher percentage of female flowers, while gibberellin delays female flower formation and percentages and stimulates the formation of male flowers. Ethephon, which breaks down to produce ethylene, is most used to increase female flower formation. Multiple applications are applied during early development to promote only pistillate flowers on the seed parent. Ethephon also causes shorter internodes and later applications may stunt plants and reduce seed yield. It is important to monitor the development of male flowers during pollination as the effectiveness of the ethephon decreases with time. Inbreeding may occur if staminate flowers develop on the female parent. Beehives should be removed if male flowers form on the female parent to avoid inbreeding. However, late developing fruit are usually too immature to provide viable seed at harvest.

In monoecious cultivars, staminate flowers appear first and are more abundant than pistillate (Fig. 12.3). Flowers occur at the nodes, staminate in clusters or singly with only one open flower on a cluster at a time. Pistillate flowers are borne singly on the main stem and lateral branches in monoecious types and singly or in clusters on the main stem and lateral branches on gynoecious types. The large inferior ovary resembles a miniature fruit on pistillate flowers and is easy to identify (Fig. 12.3). Both staminate and pistillate flowers are 1 to 3 cm (0.4 to 1.2 in.) in diameter with a yellow, showy five-lobed corolla (Fig. 12.3). Flowers are open for a single day and, if not adequately pollinated, rapidly abort (Rubatzky and Yamaguchi, 1997).

Flowering is influenced by environmental factors as well. Photoperiod affects both the number and sex of flowers. Short days generally

Fig. 12.3. Monoecious cucumber plant. (Courtesy of Tim Loeffler.)

cause earlier and more frequent pistillate flowering. Low temperatures cause pistillate flowering as well. Conversely high temperature and long days promote male flowers (Rubatzky and Yamaguchi, 1997). High nitrogen promotes vegetative growth and inhibits flowering in general. Environmental stress (water, nutrient, shading, etc.) tends to promote female flowering.

Cucumber fruits may vary dramatically in size among cultivars and may be spherical, blocky, oblong, or elongated (Fig. 12.4). Fruit surfaces vary in the number and size of spiny warts, which are usually more apparent on young fruit. Fruit colors vary from dark to light green and the amount of white epidermal tissue varies with cultivars and maturity (Fig. 12.4). However, for seed production, fruit should reach full physiological maturity before seed extraction. Mature fruit have a much different appearance compared to edible fruit sold commercially. Cultivars are divided into the general categories of fresh market or processing to describe their intended uses.

Types and Cultivars

Pickling cucumbers

Pickling cucumbers are cultivars developed for processing, although when small may also be sold for consumption in some fresh markets. Immature processing cultivars are processed with vinegar, spices, and herbs (such as dill seed) to make pickles and other products. Pickling cucumber cultivars often have smaller vines, fruit with a length-to-diameter ratio of about 2.5:1, bicolored pale and darker green immature color, thin skins, blocky shape, and more prominent warts. The warts bear white or black loosely attached spines (spicules) in their centers. Fruit with white spines turn from green to white or yellow at full physiological maturity when seeds have fully developed. Black-spine fruit change from green to orange at maturity. Black-spine fruit have thinner skin and lighter fruit color compared to white-spine fruit and for many years were preferred for processing because they have a more attractive appearance after pickling. However, black-spine cucumbers tend to prematurely develop their mature orange color at the stem and blossoms ends under environmental stressors or when over mature, so more white-spine cultivars are used for processing (Welbaum, 2015).

Seed production of pickling cucumbers occurs primarily in open-field environments due the large volume of seed required for the market and lower production costs. The first flowers developing on a gynoecious cultivar are female, so every flower has the potential to produce fruit resulting in a concentrated early set.

Parthenocarpic small-fruited pickling cultivars are sometimes grown for processing in vinegar and are popular in parts of Europe. For the processing of larger whole fruit, parthenocarpic cultivars are less desirable than seeded types because they lack seed development, which contributes to texture and flavor associated with

Fig. 12.4. Maturing and immature slicing cucumber fruit with angular leaf spot foliar disease symptoms. (Courtesy of Tim Loeffler.)

quality pickles. Parthenocarpic pickling cucumbers often fail to set fruit in the field and have lower yield potential. However, they still have a place for early-season production because they don't need a pollinizer cultivar or require bees.

For F1 hybrid seed production, parthenocarpic pickling cucumbers can set seed if crossed with another inbred that has staminate flowers. Usually, one gynoecious inbred is pollinated by another gynoecious inbred designated as the male parent that has been treated an ethylene inhibitor. Currently, only aminoethoxyvinylglycine hydrochloride (AVG) is labeled for use in the USA. Products containing silver nitrate may be used in some other countries. These compounds are less expensive than gibberellic acid, which also induces male flower formation on gynoecious plants but is less consistent. The gynoecious character may not always be stable, so some plants may be monoecious in a gynoecious cultivar particularly under stressful conditions. Crowding may increase the percentage of monoecious plants in a gynoecious field depending on the genotype. Usually, gynoecious fruit produce fewer seeds. Hybrid cucumber seed is more expensive than for open-pollinated cultivars. If chemicals are not used to alter sex expression for organic seed production or for breeder seed production, hand-crosses are often made in the greenhouse or isolated field, and seeds extracted one fruit at a time when physiologically mature. Parthenocarpic cucumber seeds are also more expensive than other types because of the extra labor and lower yields involved. Pickling cucumbers for seed crops mature 75 to 95 days from seeding depending on the cultivar.

Slicing cucumbers

Slicing cultivars are often sold fresh for use in salads. The fruit shape is nearly cylindrical with little or no white color at blunt ends and fruits appear circular in cross-section. Cultivars grown for fresh market have comparatively solid dark-green skin and a length-to-width ratio of 5:1 or greater, 4 to 6 cm (1.5 to 2.5 in.) at edible maturity, 8 to 10 cm (3 to 4 in.) in diameter at seed maturity, and 20 to 25 cm (8 to 10 in.) long or more (Fig. 12.4). Fruit with these characteristics is sometimes called American slicing cucumbers. The pericarp (fruit) walls are thicker, and the seed cavity greatly reduced although larger than processing cultivars (Welbaum, 2015). Many "slicing" cultivars have warts with spines (spicules) in the center that are white. Modern cultivars tend to have fewer warts and spines compared to older heirloom cultivars (Rubatzky and Yamaguchi, 1997). Many modern cultivars are resistant to Fusarium and Verticillium wilts, virus diseases, and mildews.

Beit Alpha cucumbers were developed in Israel but are popular in many parts of the world for fresh market use. Beit Alpha parthenocarpic slicing cucumber fruit have unique characteristics with a length-to-width ratio of about 6:1. Beit Alpha, also known as Beta Alpha, is a type of parthenocarpic, all female, multi-fruited, dark green hybrid that does not require pollination and offers high yield potential. The seedless fruit is 14 to 19 cm (5.5 to 7.5 in.) long, with a thin smooth skin with few warts that does not need to be peeled before eating. Fruits can be harvested when 3 to 4 cm (1.2 to 1.6 in.) in diameter or for specialty markets with flowers still attached.

Greenhouse slicing is another distinct parthenocarpic type also known as Long Dutch or English cucumbers. They have a length-to-width ratio of 8 or 10 to 1 or greater, diameter of 4 to 6 cm (1.5 to 2.5 in.) at edible maturity. F1 hybrid cultivars of both Beit Alpha and greenhouse parthenocarpic cultivars are made by hand crossing inbred lines. Parthenocarpic seed production takes place by hand pollination within protected environments such as glasshouses or tunnels. The male parent is typically treated with products containing silver nitrate or gibberellic acid to obtain staminate flowers for pollination. The skin of greenhouse parthenocarpy types is very thin and fruit can be easily damaged (Rubatzky and Yamaguchi, 1997). Parthenocarpic slicing cucumbers (greenhouse or Beit Alpha) are harvested at physiological maturity, which occurs at approximately 60 to 70 days after pollination. The fruits are hand-harvested in multiple pickings and stored for five to seven days prior to seed extraction, washing, and drying.

Asian types

Asian cultivars produce fruit that are usually long and slender with lengths of 20 or 25 cm (8 or 10 in.) or greater and diameters of 4 to 5 cm (1.5 to 2.0 in.) at edible maturity. Maturing fruits contain seeds, have thin dark green skins, and considerable warts and spines. They are popular in Asian markets and are grown to a limited extent in other parts of the world. Asian cultivars are often divided into two groups: the day-neutral North China group and the short-day South China group that is grown primarily for winter production. The South China group produces primarily pistillate flowers under short-day conditions and many staminate flowers under long-day conditions.

The Sikkim cucumber, also called the Concombre Apple, is popular in India and is unique because the fruit has reddish-brown skins. Most cultivars are open-pollinated, and plants freely intercross to make seeds although isolation is needed to maintain cultivars. The Sikkim cucumber is a fat, large fruit reaching 38 cm (15 in.) long and 15 cm (6 in) wide. The ripe fruit is eaten cooked, fermented, or raw and has a mild flavor (Tamang *et al.*, 1988).

Armenian

The Armenian cucumber, also called the serpent or snake cucumber, is a type of melon (*C. melo*) that is eaten fresh like a slicing cucumber when immature. The sex expression is monoecious with small and pale-yellow flowers with five petals similar to other melons. Armenian cucumber is long and slender with pale to dark green skin that is wrinkled longitudinally without spines. The slender fruit have their best flavor when 25 to 35 cm (10 to 15 in.) in length but if allowed to grow will reach a length of about 0.9 m (3 ft) and up to 15 cm (6 in.) in diameter. Fruit is rarely straight and often bent and twisted. The fruit color changes to yellow when ripe with a mature odor similar to cantaloupe (Rubatzky and Yamaguchi, 1997). Seed crops should be harvested at full physiological fruit maturity for best seed quality. Most cultivars are open-pollinated and should be isolated from other *C. melo* during seed production.

Growth and Development

Cucumbers need warm soil temperatures for germination and reliable stand establishment. The minimum, optimum, and maximum soil temperatures for germination are 15.5°C (60°F), 35°C (95°F), and 40.5°C (105°F), respectively. The optimum range of soil temperatures for stand establishment ranges between 15.5°C and

35°C (60°F to 95°F) (Texas Vegetable Growers Handbook, 2023) with faster emergence at higher temperatures over this range. Little vine growth occurs when temperatures fall below 15.5°C (60°F). The growth rate increases steadily as temperatures rise above 21.1°C (70°F) (Delahaut and Newenhouse, 2011).

Cucumbers should not be planted in a rotation after related Cucurbit species like melons, squash, or watermelon. Buildup of various insects and soil-borne diseases, which survive season-to-season in crop residue, is common among related species. Cucumbers are very sensitive to some herbicides, so carryover from previous crops must be considered when selecting fields for conventional production.

Soil

Cucumbers can be grown on most fertile, well-drained soils. Loam-textured soils with organic matter are well suited for cucumber production. Soils with more than 70% organic matter should be avoided because they may encourage luxuriant vegetative rather than reproductive growth. Light, sandy soils are acceptable provided they are adequately fertilized and irrigated. A pH of 6.0 to 7.0 is suggested for cucumbers for optimum growth. Acid soils with pH below 5.5 should be avoided (Lorenz and Maynard, 1988).

Fertilization

Cucumbers respond favorably to fertilizer, so proper fertility management is a key to maximizing productivity. Fertilizer requirements vary, depending on the soil type, the environment, and cultivar. There is no single recommendation to fit all production situations. Fertilization is most efficient and effective if based on preplant soil test recommendations and foliar analyses throughout the season. Most soil testing labs will rate essential minerals as being high, medium, or low and recommend how much fertilizer should be applied to successfully produce cucumbers, although such recommendations are often for edible food and not seed crops. It is important to determine how many nutrients a cucumber crop is likely to remove during a season and apply these during the growth cycle as needed. The nutrients removed from a field by a typical cucumber food crop would be 175 (196), 25 (28), 175 (196) lbs/acre (kg/ha) for N, P, K, respectively (Texas Vegetable Growers Handbook, 2011). Soil test recommendation for N and K often are less than these amounts because of the assumption that some nutrients will be provided from organic matter decomposition and the nitrogen cycle. Recommendations for P additions often exceed the amount removed by the crop because some P fertilizer will be fixed or immobilized in soil. Micronutrients may also be needed to avoid deficiencies.

Nitrogen is a very important nutrient for cucumber seed crop production. Early in the season, adequate nitrogen promotes vegetative growth and canopy formation that is essential for maximizing photosynthesis to support later fruit and seed development. Lower available nitrogen after vining begins promotes flower formation and fruit set, while excessive nitrogen delays reproductive growth.

Petiole or plant tissue analysis can assess crop nutrient status during the growing season. Later in the season, side dressing or fertigation with nitrogen, and possibly other nutrients, may correct deficiencies and maximize yields particularly on sandy soils. Supplemental nitrogen applications are often needed following heavy, leaching rains on light, sandy soils.

Production

Large-scale field production of cucumbers is usually without mulch because it interferes with harvest equipment. F1 hybrid and small plot production may utilize mulch and drip irrigation when fruit are hand-harvested. Seed production is often on raised beds, as recommended to improve drainage and to reduce fruit belly rot. Mineral nutrients should be incorporated into the bed prior to planting. The soil should be tilled so it is friable and free of clods to provide good seed to soil contact for direct-seeded crops for uniform emergence and growth.

Precision planting of cucumber seeds to the right spacing and depth will encourage uniform seedling emergence. A uniform stand will lead to concentrated fruit set for best crossing of inbred lines and uniform crop development for efficient hand or mechanical harvesting. Precision seed spacing produces the desired stand of plants, saves expensive parent seed, and eliminates the need for thinning after establishment. Ideally it is important to manage the timing, field conditions, and use high-quality parent seed to promote the uniform development of the crop for timely pollination and seed set. Crops that are not uniform tend to produce lower seed quality and yields. For F1 seed production, typically female-to-male planting ratios are 3:1 and 2:1 (Fig. 12.5). This is particularly important for once-over destructive machine harvests. Cucumbers are normally direct-seeded to a depth of 1.3 to 5 cm (0.5 to 2 in.), depending on soil type, time of season, and soil moisture availability.

Spacing is dependent on cultivar, environments, and cultural practices. In-row spacing for slicing vining cucumber cultivar seed crops ranges from 23 to 31 cm (9 to 12 in.) with row spacing of 0.9–1.8 m (36 to 72 in.). This results in populations of 7300 to 21,780 plants per acre (18,000 to 29,600 plants per ha).

Pickling cucumbers usually have smaller vines and are grown at high populations. Depending on environmental conditions, maximum yields for hand-harvest may be obtained with in-row spacing of 15 to 31 cm (6 to 12 in.) with 0.9 to 1.8 m (36 to 72 in.) between rows. This spacing results in populations of 7300 to 29,040 plants per acre (18,000 to 71,700 plants per ha).

Pollination

Bees are extremely important for field production of cucumber. In most areas, native bee populations are insufficient pollinators. Typically, three to four hives are set per acre at peak flowering for open-field F1 seed production. Additional hives may be required if bee activity is low. Most cucumber flowers are open for only one day and a successful pollination event requires multiple bee visits. Cool wet weather significantly reduces bee activity and pollination, resulting in low yields and/or a high percentage of misshapen fruit. Pesticides that harm bees

Fig. 12.5. F1 hybrid cucumber seed production field on raised beds with three female rows per single male pollinator row. (Courtesy of Tim Loeffler.)

should not be used and when necessary, spray applications should be scheduled for late afternoon or early evening. For F1 seed production, it is important to monitor the presence of staminate flowers within the female parent and remove beehives toward the end of the pollination period. The minimum isolation distance for cucumber seed crops ranges from 1600 m (1.0 mile) to 3200 m (2.0 miles) depending on type of cucumbers (George, 2009).

Irrigation

Cucumber seed crops require large amounts of water. Fruits contain about 95% water. Since cucumber seed crops are often grown in warm desert regions with seasonal rainfall to minimize disease, irrigation is necessary to maximize both fruit and seed yields. The rule of thumb is for at least 2.5 to 3.8 cm (1 to 1.5 in.) of water per week as a minimum for cucumber seed production, depending on soil type, temperature, and humidity. In areas where humidity is low and temperatures high, 5 cm (2 in.) per week may be required particularly on lighter soils. Wilting and drought stress will increase seed abortions depending on the severity. The most critical need for water is during the fruiting when drought may cause "nubbing" (small fruit shriveled behind the blossom). In seed crops destined for once-over mechanical harvest, the last irrigation should be timed so the soil will dry sufficiently to allow harvest equipment into the field.

Keeping the foliage dry reduces foliar disease. Furrow and drip irrigation wet soil while keeping foliage dry. Furrow irrigation is used in some areas where topography, water availability, and infrastructure permit. However, furrow irrigation doesn't use water efficiently. Some growers have converted to subsurface drip lines in areas where fields are mechanically harvested. Drip lines above the soil can be used when crops are hand-harvested. Prolonged periods of wet soil and leaves promote infection by mildew, *Alternaria*, angular leaf spot, and other fruit-rotting and foliage diseases.

Rogueing

Rogueing cucumber seed fields occurs over multiple stages of crop development to identify and remove off-types as well as diseased plants. Plants should be checked before flowering during the five- to seven-true-leaf stage of vegetative growth before flowering begins. Fields should be monitored weekly and rogued as needed for trueness-to-type. An inspection is advised during advanced fruit development when color and shape are discernable. Finally, fruit should be checked at fruit harvest for correct cultivar characteristics.

Seed Harvest

For best seed quality, cucumber fruits must reach full physiological maturity which for white-spined cultivars is yellow-white skin color and for black-spined cultivars is an orange-brown fruit skin color. Fruits should be cut open and examined to verify seed maturity (Fig. 12.6). Maturity is also signaled by a vinegar-acidic taste of the interior of the fruit or vine senescence.

Mature fruit should not remain in the field for very long after vine decline because solar heating of fruit can age seeds leading to development of osmotically distended seeds, aging of seeds with a loss of vigor, and precocious germination in some cases. Seeds that separate easily from the flesh of a sliced cucumber fruit are ready for harvest. Fruits grown on small plots and F1 hybrid fruit may be harvested and extracted by machinery in the field or at a processing facility (Fig. 12.7). Seed production that was hand-pollinated is usually harvested in multiple picks based on a certain number of days after pollination. These fruits are typically stored under ambient conditions to allow after-ripening in the fruit for five to seven days prior to seed extraction. For large-scale production of commercial open-pollinated seed, vines are cut, allowed to sun-dry and wither in the field before fruits are windrowed prior to mechanical harvesting.

Prior to washing and drying, the seeds are naturally fermented, treated with pectinase enzyme, or receive acid treatments. For natural fermentation, fruits are first crushed and macerated, and the surrounding pulp, juice, and seeds fermented for about 24 to 48 h in large bins. Adding water to the seed slurry can begin the germination process and damage seeds. Seeds can be similarly damaged by over-fermentation or prolonged hydration during cleaning which has increased popularity of acid and enzyme

Fig. 12.6. Inspecting cucumber fruit for seed maturity. (Courtesy of Tim Loeffler.)

Fig. 12.7. Hybrid cucumber destructive mechanical harvest. (Courtesy of Tim Loeffler.)

treatments. Pectinase enzyme treatment can effectively remove the mucilage from seed coats. Before pectinase treatment, seeds should be free of gross fruit debris. Another method for removing mucilage is hydrochloric acid 90 ml (3 fluid oz) or sulfuric acid 30 ml (1 fluid oz) added to and stirred with 11.3 kg (25 lbs) of the sliced, macerated fruit tissue, but this process is more typical for small seed volumes or protected production. After 15 to 30 min, water is added to the mixture and the mature seeds sink to the bottom while the digested pulp floats to the top. Whether fermentation, enzymes, or acid treatments are used, seeds must be immediately washed and promptly dried to preserve quality (Fig. 12.8). Washing is accomplished with pressurized water and screens to collect cleaned seeds. Cucumber seeds are frequently run through water plenum troughs with riffles that catch seeds while debris and chemical residues float off.

MELON

Botany

Melon is a general term for fruit produced by *C. melo*. In some areas, the term is applied to both netted and non-netted fruits but not watermelon. Naudin classified different domesticated and wild forms of *C. melo* into ten separate botanical varieties in 1859, including *C. melo* var. *flexuous*, *inodorus*, and *cantaloupensis* (Stepansky *et al.*, 1999). Although Naudin's classification was not recognized taxonomically, it was useful for many years to delineate obvious horticultural differences among fruit types within *C. melo*. Botanical designations have been classified into up to 18 separate botanical varieties or horticultural groups under *C. melo* (Pitrat *et al.*, 2000). The genome of *C. melo* has been sequenced (Garcia-Mas *et al.*, 2012).

Melon can describe netted fruit but in some markets is used for cultivars of *C. melo* that lack netting and are also called "mixed melon" or sometimes "winter melon". For the North American market there are two important Groups (formerly botanical varieties) of *C. melo*. The Group Inodorus contains the so-called "winter melons" like casaba, Persian, canary, honeydew, Crenshaw, and Santa Claus that do

Fig. 12.8. Cucurbit seed washery. (Courtesy of Tim Loeffler.)

not abscise from the vine at full physiological maturity, are not climacteric but many ripen in response to exogenous ethylene, lack heavy netting, are generally less aromatic, are susceptible to sunburn, are late maturing, and have longer storage life than the Group Cantaloupensis (formerly Reticulatus), the netted melon group. One exception is the Crenshaw melon which has a short shelf life and is odoriferous.

The terms cantaloupe and/or muskmelon are used to describe netted melons in some areas. The Group Cantaloupensis contains both western and eastern shipping melon cultivars popular in North America. Charentais and Galia types are popular in Europe. The name cantaloupe is more commonly used in the commercial vegetable trade to refer to smaller shipping cultivars developed for low humidity regions of the western USA and northern Mexico such as PMR 45, Topscore, Top Mark, or Easy Rider. Muskmelon often describes larger-fruited ribbed Bender-type cultivars that are better adapted to humid conditions. Iroquois, Delicious, or Burpee Hybrid are classic muskmelon cultivars maturing in 75 to 95 days, are climacteric, heavily needed with a large wet seed cavity, abscise from the vine at maturity, and change from green to orange-yellow when mature. Cantaloupensis and Inodorus Groups can be crossed to create intermediate cultivars like Honeyloupe.

Since cucumber and melon are both *Cucumis* species, some of the agronomic information on cucumber applies to melon, although they do not sexually cross. Commercial melons are warm-season (frost-sensitive) annuals that are cross-pollinated by bees. Melon sex expression is andromonoecious. Commercial cultivars are day-neutral. Plant habit is vining with varied internode length. Short internode "bush" cultivars are available. Most cultivars have tendrils and flowers have five yellow petals that are open for a single day like cucumber. However, leaves differ from cucumber in being circular, oval, or kidney shaped with multiple rounded lobes. Truly wild forms are found only in eastern tropical Africa (south of Sahara). Related wild forms of melon are reported in India, and are likely undomesticated escapees derived from local cultivars. Care should be taken to prevent unintended crossing with wild plants that may exist in some seed production areas.

Within *C. melo*, fruit size and characteristics vary widely among groups and even cultivars. These characteristics should be considered when rogueing off-types. Fruit shape varies from spherical to oblong. Some fruits are covered with corky reticulate netting that varies in density and depth with cultivar but is also affected by vine health and environmental conditions. The vein tracts or sutures are longitudinal indentations on the fruit surface associated with vascular bundles, which may not be visible in round fruits with heavy netting. The edible melon flesh is fruit pericarp and varies in thickness, color, and texture. Pericarp colors may be white, green, pink, or orange. Aroma varies from strong to odorless and depends on the quantity and quality of volatile compounds produced, usually associated with the ethylene burst characteristic of climacteric fruit ripening. Some netted melons are climacteric and ripen in response to ethylene produced in the seed cavity. Fruit softening, color change, and volatile compounds production are associated with climacteric ripening. Some melon cultivars like honeydew are not climacteric but respond to exogenous ethylene which accelerates ripening.

Strongest contributions to commercial melon development in North America came from Colorado in the form of Rocky Ford strains of cantaloupe developed from Netted Gem, a cultivar introduced from France by the Burpee seed company in the late 1880s. The use of F1 cultivars, especially for commercial production, has accelerated over the past 30 years because of their greater uniformity and heterosis compared to open-pollinated. Heirloom cultivars are popular in some areas with hobbyists, organic producers, and direct marketers.

Soil

Melons require soils that are fertile and well drained. Maximum seed yields are achieved on medium-textured soils with high water-holding capacity and good internal drainage. Melon seed crops can be grown on a wide range of soils, although peat or heavy clay soils are not recommended. Clay soils generally have poor aeration and restricted drainage, which reduces root growth and thus fruit and seed yield.

Fertilization

Melons develop extensive root systems that efficiently explore the soil for water and available nutrients. For this reason, fertilizer requirements are considered moderate compared to other crops. Soils should always be tested prior to planting a seed crop to obtain an accurate determination of crop nutrient needs. Fertilization should initially be tailored to soil tests and supplemental additions during the season based on foliar test results. The goal is to grow balanced plants with a large healthy canopy of leaves early in development to support the photosynthesis capacity for a high yield of healthy fruits and seeds. This can be accomplished by maintaining adequate levels of N, P, K, S, Ca, Mg, B, Cu, Fe, Mn, Mo, and Zn. Melons are sensitive to acid conditions but will grow well on slightly acidic (pH 6.8) to moderately alkaline (pH 8.0), soils. They are also quite sensitive to salinity.

Nitrogen is the most required fertilizer, although phosphorus promotes seedling vigor, maximum productivity, and high fruit quality, especially in alkaline soils. Most mineral soils contain adequate potassium, but it may be deficient in areas with light soils and certain cropping histories. Nitrogen is commonly applied in two side dressings, the first at the two- to four-leaf stage and the second when vines start to run. Care must be taken to not over-fertilize the crop, especially with nitrogen. Like cucumber, excessive nitrogen favors vegetative growth over reproductive growth and can delay or inhibit fruit set. The N level, especially ammoniacal N, must be low enough at flowering that the plant will form fewer new leaves after fruit growth begins. This allows more photoassimilates to accumulate in fruit, rather than developing leaves and vines which are competing sinks for carbon compounds.

Plant tissue analysis is the most effective means of diagnosing the crop nutrient status during crop development. Petiole analysis involves expressing small quantities of sap and analyzing it for nutrient content in real time using portable nutrient meters. Plant nutrient status should be monitored and optimally maintained throughout crop development. Common nutrient needs for melon seed crops grown on a loam soil are 112 to 135 N kg per ha (100 to 120 lbs N per acre) and 90 to 112 kg P_2O_5 per ha (80 to 100 lbs P_2O_5 per acre) as a general fertility program recommendation based on nutrients removed by the crop. Greater amounts may be needed in sandy soils. Alkaline mineral soils chemically tie up phosphorus, making it unavailable to plants. Thus, precision placement by banding phosphorus alone or as a complete analysis fertilizer near the seed is preferred to field-wide broadcast application, which is wasteful.

Pollination

Melon pollen is very viscous and does not move naturally the short distance from anther to stigmas on perfect flowers. Since melons are andromonoecious, bees and sometimes other insects move pollen from staminate to perfect flowers or from the anthers to the stigma on perfect flowers for open-pollinated seed production. Honeybees are the most reliable and cost-effective way to achieve pollination. One to two hives per acre are introduced when 5 to 10% of flowers have opened. If hives are placed in weedy areas or close to other flowering crops, more hives are needed. Hives should be placed in clusters around the periphery of fields, with additional hives placed inside of larger fields. The squash bee (*Peponapis pruinosa*) is a wild, native bee that pollinates melon seed crops in some areas including California, USA (Kremen *et al.*, 2002).

The ideal minimum isolation distance for commercial seed production is 1000 m (0.6 miles) although sometimes distances of only 500 m (0.3 miles) may be used especially if barriers such as buildings or woodlands are present. It is important that adequate isolation is maintained from melon seed fields, Cucurbit food crop fields, and home gardens that contain melons and wild relatives of *C. melo*.

Using plant hormones to manipulate melon sex expression is currently not possible. Hybrid crosses are made by emasculating bisexual flowers and hand pollination. Hybridization of these crops requires extensive hand labor. Typically, the pistillate flower is emasculated and the flower is covered with a small paper bag or plastic cover to prevent foreign pollen until it is hand-pollinated. After hand pollination is complete, a different color bag or plastic cover marks that pollination is complete. As a result, hybrid seeds are produced where labor is relatively inexpensive and abundant. South America and Asia are areas where seed-borne diseases are not an issue and labor is available.

Production

A crop rotation cycle of several years between planting other Cucurbitaceae crops (cucumber, pumpkin, watermelon, etc.) is needed if pathogen populations are very high and fumigation is not used. Grasses, corn, legumes, or sorghum make good rotation crops. Herbicide carryover (e.g. atrazine) may inhibit melon growth.

With the high cost of producing parent lines, the seeds for F1 seed production are started in a nursery and are transplanted at the three- to five-leaf stage to a production field, net houses, cages, or greenhouse (Fig. 12.9). Advanced production systems utilize substrates rather than soil within the greenhouse or glasshouse.

Starting the seeds in a nursery also allows for proper scheduling of each parent for timely nicking and harvest. Sometimes the plants are grafted on to disease-resistant root stalks particularly in areas where soil-borne diseases are problematic. Melons are either grown on trellis systems or flat raised beds (Fig. 12.9).

Melons for open-field seed production are often grown on beds raised 15 to 20 cm (6 to 8 in.) to ensure good drainage. Beds are shaped to 203 cm (80 in.) or half width 102 cm (40 in.) across before planting. Other widths are possible. For half-width beds, every other bed is seeded at planting. When plants begin to vine, unplanted beds are broken open and the field is reworked to position plants in the middle of the final 203 cm (80 in.) bed. Plastic mulches can be used to control weeds and conserve moisture in F1 hybrid production fields where fruits for seed extraction are hand-harvested.

Irrigation

Preplant or post-plant irrigations should be applied to ensure uniform stand establishment. Irrigation needs to be closely monitored throughout the growing season. The frequency of application and the amount applied depend on soil type, rainfall, evapotranspiration, and parental lines or cultivar. Excessive or lack of water should be avoided. Melon plantings are easily damaged by excessive water. Common problems caused by excessive water are: (1) poor root system that may lead to vine collapse as the crop reaches maturity, (2) poor fruit development, and (3) excessive fruit rot and fungal disease.

Fig. 12.9. Hybrid melon developing in a greenhouse after emasculation and hand cross-pollination. (Courtesy of Tim Loeffler.)

After establishment further irrigation may be delayed until the pre-vining stage to avoid cooling the soil and encouraging damping-off diseases, and to promote deep-rooting. Surface and subsurface drip irrigation is more efficient and has replaced furrow irrigation in many areas. Furrow irrigation is the least efficient but requires minimum equipment, is still used where fields are flat and mass flow water delivery infrastructure exists, where drip irrigation conversion to deemed too costly, or where drip is incompatible with harvesting machinery. Care must be taken with furrow irrigation to ensure that water gets to the seed or root zone, and wet beds don't encourage fruit rot. In warm desert production areas, furrow irrigations are needed every seven to ten days. The last irrigation is typically applied ten to 20 days before harvest begins, but may vary depending on soil type, air temperature, humidity, etc.

Drip irrigation may be used for F1 hybrid production where fruits are hand-harvested and drip lines will not interfere with mechanical harvesting equipment. Subsurface drip is better suited for mechanical harvesting. Drip irrigation provides the most uniform application and efficient use of water and enables efficient fertigation. Sprinkler irrigation is frequently used during stand establishment, but this is not a good system for melon once the vine canopy has developed, because it enhances the potential for vine and fruit diseases.

Rogueing

Melon rogueing must occur at multiple stages of crop development to identify and remove off-types and diseased plants. Rogueing should be conducted during the following stages: vegetative growth prior to flowering, after flowering commences and early fruit development begins, during advanced fruit development when netting and shape are discernable, and at harvest when off-types fruits should be discarded.

Seed Harvest

Fruit maturity at harvest plays an important role in determining seed quality (Welbaum, 1999). Seeds extracted from immature fruit may be underdeveloped and have low germinability and vigor. Even after seeds have reached maximum dry weight physiological maturation must occur for maximum vigor (Welbaum and Bradford, 1991). Generally, highest seed vigor coincides with fruit development just past edible maturity at the onset of senescence (Welbaum, 1999), although the exact stage of fruit maturity for best seed quality varies with melon type. Netted cantaloupes naturally separate from the vine during climacteric ripening at full physiological maturity. Separation from the vine is classified as "full slip" and an indication fruit are ready for seed harvest.

Hand-pollinated fruit for F1 hybrid production can be selected over several days at full maturity to ensure high vigor seeds. The fruit types may be held in post-harvest storage for up to five days and extracted together. Seeds may be extracted by hand or with mechanical seed extraction equipment (Fig. 12.10).

Open-pollinated melon fruits for seed crops are often harvested when most of the field is at or near full slip. If harvest is delayed and ripe fruit begin to decompose in the field, particularly when unprotected fruit are exposed to intense prolonged direct sunlight, accelerated aging results in rapid loss of both viability and vigor (Fig. 12.11). In addition, delaying harvest will increase the percentage of osmotically distended or "fish mouth" seeds (Welbaum, 1993).

For once-over, destructive mechanical harvest, vines are cut and fruit are windrowed so they can be collected, broken, and seeds extracted during a single pass through a field (Fig. 12.11). Alternatively, fruit may be selected by hand and feed into a seed extractor positioned in the field. Hand-selecting fruit ensures they are fully mature and contain true-to-type, high-quality seed. Some cultivars in *C. melo* Group Inodorus do not abscise at maturity. For these cultivars, color change, days from anthesis, heat unit index, and/or fruit sampling should be used to judge full physiological maturity for harvest.

Cleaning

Melon seeds, like cucumber, have a mucilage layer associated with the seed coat that must be removed during post-harvest processing. Traditionally seeds have been fermented in fruit

Fig. 12.10. Melon stationary seed extractor. For this system, fruits are hand-selected for proper maturity and shipped in bins to a central processing facility for extraction, conditioning, and drying. (Courtesy of Tim Loeffler.)

Fig. 12.11. Melon fruit in open field ready for mechanical harvest and field extraction. Once-over, destructive mechanical harvesting and extraction of melon seed crops often combines fruit from a broad range of maturities at harvest leading to variation in seed quality.

juices, so the mucilage is microbially digested. Adding excessive water to the fruit/seed slurry may initiate germination and damage seeds, as will over-fermentation. During fermentation, less dense aborted and poorly developed seeds float to the top of the slurry and are removed. Seeds with greater density sink to the bottom of the fermentation container. After fermentation, seeds are cleaned by washing and brushing or passed through a water plenum where clean seed is retained on the bottom of the trough but debris floats off the top.

The fermentation for F1 seed or parent lines is usually overnight. Washing should occur as soon as possible after fermentation. Seed is dried either on table or in precision dryers to 7–8% seed moisture at temperatures not exceeding 32°C (Fig. 12.12).

Increasingly, pectinase enzyme treatment is applied after seed harvest instead of fermentation. Enzyme digestion of the mucilage causes potentially less damage to seeds and is easier to manage compared to fermentation. After pectinase treatment, seeds are washed and dried and can be graded on a density separation table to remove undeveloped or misshapen seeds.

WATERMELON

Botany

Watermelon (*Citrullus lanatus*) is a warm-season, frost-sensitive plant that is grown as an annual in temperature regions. Watermelon fruit has smooth skin with striped or solid colors that mature in 75 to 120 days after planting. Although there are short internode bush cultivars, most commercial cultivars produce vines 1.5 m or more (15 ft) long with some cultivars producing vines as long as 5 m (16 ft). Tendrils and yellow flowers that are roughly 2.5 cm (1 in.) across develop at the nodes. Watermelons are monoecious and require bees to move the sticky pollen from the female to male flowers. Watermelon has distinctive deeply lobed pubescent leaves. Roots are extensive but usually shallow and lack a taproot.

Commercial fruit range in size from 2.3 to 54.4 kg (5 to 120 lbs) depending on the cultivar and environmental conditions. Cultivars grown only for edible seed produce smaller fruit with thin rinds and many large seeds compared to fruit of cultivars grown for fruit pulp consumption. Seeds vary in size, number, and color from white through tan, dark brown, or black. Seed is imbedded in pericarp tissue unless fruit are parthenocarpic. Except for the rind, the fruit is edible pericarp. Globular to oblong-shaped fruit have mature pericarp colors that may be white,

Fig. 12.12. Precision seed driers. (Courtesy of Tim Loeffler.)

yellow, pink, orange, or red. Fruit wall is firm and usually thick with a transition zone of 1–4 cm of white inner flesh between hard outer rind and epidermis and the juicy edible pericarp.

Also of interest is the citron melon (*Citrullus caffer*), also called *Citrullus lanatus* var. *citroides* and *Citrullus amarus*, a relative of the watermelon. Other common names include fodder melon, preserving melon, red-seeded citron, jam melon, stock melon, Kalahari melon, or tsamma melon (Nesom, 2011; USDA ARS, 2023). Citron melons have hard white flesh that is rarely eaten fresh. More commonly the rind is pickled or used to make fruit preserves and as cattle feed. The citron melon has become a feral species in the southern plain states of the USA and parts of Mexico. In these areas, steps must be taken to prevent citron from contaminating seed fields or wild plants cross-pollinating seed crops (Laghetti and Hammer, 2007).

Production

Watermelons prefer fertile well-drained soil. Heavy clay soils are not recommended. Clay soils generally have poor aeration and restricted drainage, which can be detrimental to root growth and increase fruit rot. Watermelons grow well on slightly acid (pH 6.0–6.8) to moderately acid soils. A crop rotation of several years without growing Cucurbits reduces pathogen populations and fumigation is not used. Grasses, legumes, corn, or sorghum make good rotation crops. Herbicide carryover may be a concern if watermelon is rotated with some agronomic crops where chemical weed control is used. For lighter soils, irrigation is recommended.

An important factor required for the selection of watermelon seed production areas is control and detection of bacterial fruit blotch (*Acidovorax avenae* subsp. *citrulli*). Areas with a persistent presence of this disease are avoided for seed production. More recently, specialized sanitation procedures are required for field entry of personnel and field equipment to prevent disease contamination from outside sources.

Plant populations for seed production range from 16,000 to 40,000 plants/ha (6300 to 15,750 plants/acre) depending on the cropping system. Watermelon for seed can also be produced on a trellis or on stakes in some areas, allowing higher plant populations. Seed crops are planted on raised beds and the width between beds varies within and across growing areas. Often two rows can be planted on the same bed.

Plastic mulch may be used for F1 hybrid seed production to enhance soil temperature, maintain soil moisture, or reduce weed pressure. For F1 production, often the parent lines are started in plug trays in a nursery and later transplanted to a production field or protected environment. Benefits of transplanting include better timing of development for proper nicking of female and male parents during pollination, improved stand establishment, and more efficient use of costly parent seed. Normally the male parent line is planted seven to ten days earlier than the female parent. In some seed production areas, grafting of the either or both parental lines is employed to promote better seed yields.

Different categories of watermelon seed have unique production requirements. There are several different types of watermelon seed produced: open-pollinated, F1 hybrid, triploid, inbred lines, or hybrid male pollinizer cultivars planted with triploid hybrids. Each type has its unique spacing and production considerations.

Pollination

Open-pollinated cultivars are isolated from surrounding watermelon plants to prevent outcrossing by bees. The minimum recommended isolation distance of watermelons for seed production is 1 km (0.6 miles). Recommended distances for stock seed are at least 1.5 km (0.9 miles) (George, 2009). It is important to ensure that seed crops are also isolated from watermelons grown for edible use. Honeybees are the most reliable and cost-effective way pollinate watermelon. Three to six hives per hectare (one or two per acre) are introduced when 5 to 10% of plants have open flowers. The numbers of hives should be increased in weedy areas or close to other flowering crops. Hives should be clustered around the periphery of fields. Additional hives can be placed inside larger fields.

Hormones or growth regulators do not reliably change sex expression of watermelon and are not used for F1 hybrid watermelon seed production. Gametocide use has not been perfected for watermelon. Therefore, F1 hybrid watermelon seed production requires extensive emasculation and hand pollination using a female-to-male ratio of 4:1 or 5:1. Typically, the female plants are pruned to have two vines with the goal of producing one or two fruit per plant. The perfect flowers are emasculated, covered, and the peduncle tagged until hand pollination occurs the following day. Crews select staminate flowers from male parent and then hand-pollinate pistillate flowers by ensuring the anthers contact the stigmatic surface. After hand pollination, flowers are capped to prevent foreign pollen from entering. Developing fruit confirmed to result from hand-crossing are marked for harvest (Fig. 12.13). The goal is for emasculation and hand pollination to be completed in less than ten days, so only pistillate flowers pollinated during this period produce hybrid seed.

Some hybrids are produced using genetic male-sterile (GMS) female lines and can utilize bee pollination. The GMS lines are normally 50% sterile and 50% fertile flowering plants. The spacing for the female parent can be closer in the row since any male-fertile plants will be removed. Fertile plants bloom earlier and are removed prior to flowering of the sterile plants that will produce hybrid fruit (Dong *et al.*, 2021).

Seedless watermelons are produced by crossing a tetraploid inbred used as female parent and a diploid inbred line used as the male (Welbaum, 2015). Tetraploid lines are created by plant breeders using colchicine, a naturally occurring alkaloid. Colchicine increases ploidy by causing DNA replication without cell division. Tetraploid and diploid lines are grown in alternating rows, and the tetraploid female line is hand-pollinated with staminate flowers from the diploid parent by contacting the anthers to the stigma. Petals from the flowers to be crossed are often removed to improve contact. Bees may also be used for seedless watermelon production. The male flowers and any open female flowers in the female (tetraploid) parent are removed prior to introducing bees. Once the female parent is cleaned, bees are introduced in a uniform pattern throughout the field. It is critical to monitor and remove any male flowers in the female parent during the pollination period. Tetraploid parents for triploid production may germinate poorly in the field and are sometimes transplanted from plugs to ensure a uniform stand. The hybrid triploid seeds are often cleaned with advanced sorting or imaging equipment to eliminate any tetraploid seeds (self-pollinations), and to separate lighter undeveloped triploid seeds. Ploidy can also be verified by destructively counting chromosomes from isolated cells using flow cytometry.

Watermelon pollinizers can either be open-pollinated cultivars or F1 hybrid. Since triploid hybrid watermelons are sterile, pollinizers are planted as a source of pollen to produce seedless

Fig. 12.13. Hybrid watermelon fruit marked for harvest. (Courtesy of Tim Loeffler.)

watermelons for edible consumption. Pollinizers are usually selected for long stems, multiple branches, and high staminate flowering production over a long period but may produce female flowers later in development. Pollinizer fruit characteristics must also be easily distinguished from the commercial hybrid to avoid confusion at harvest. Certain open-pollinated cultivars such as Mickeylee may be pollinizers for seedless fruit production as well.

Seed increases are similar to other open-pollinated cultivars with isolation distances of at least 1.7 km (1 mile) from other watermelons. A common source of pollinizer germplasm is the Fusarium wilt-resistant citron melon line, PI 296341. F1 hybrid pollinizers are produced by hand-crossing two inbred lines from alternating rows as described above. Since female flower production occurs later, the female inbred line is often planted earlier than the male pollen parent.

Irrigation

Most watermelon seed crops are grown in warm dry areas with irrigation and low disease pressure. Watermelon requires a consistent supply of water particularly during fruit development. With advancing technology, it is easier to monitor soil moisture to apply irrigation as needed throughout the season. Watermelon crops require roughly 51 to 64 cm (20 to 25 in.) of water consistently applied during the growing season to maximize yields depending on soil type, environmental conditions, and cultivar. Since this amount of water is usually not consistently supplied by rainfall, irrigation is needed. Sprinkler irrigation may be used during stand establishment and is not recommended during the growing season. Increasingly, drip irrigation is used especially for F1 and subsurface systems are installed in areas where the crop is mechanically harvested. Furrow irrigation, although decreasing, is still used in parts of the USA, Mexico, Chile, and other regions where ground is level and irrigation infrastructure exists. However, furrow irrigation does not use water efficiently. Preplant or post-plant irrigations should be applied to ensure seed germination, emergence, and stand establishment. Irrigation is then delayed until the pre-vining stage to avoid cooling the soil and encouraging damping-off diseases, and to promote deep-rooting. Irrigation is terminated two or three weeks prior harvest to allow the soil to dry prior to harvest.

Rogueing

Rogueing is important for quality watermelon seed production to remove off-type plants. It is recommended that watermelon seed fields be rogued a minimum of four times. Before flowering, vegetative characters should be checked against the breeder's description and off-types removed. At early flowering, trueness-to-type of developing fruit should be checked. During fruit development, plants producing off-type fruits should be removed. As fruits reach marketable size fruit, characteristics should again be checked and off-types eliminated. Some fruit should be opened to examine internal characteristics. In triploid seed production, the female parent is tetraploid and often "tetraploid reversions" which are diploid need to be removed early in growing season. These are easily distinguished as they produce long internodes and thin stems.

Seed Harvest

For best seed quality, fruit harvest and seed extraction should not begin until at least seven to 14 days after edible fruit maturity. In other words, when fruit are overripe for eating but before severe decay and decomposition begin. Mixing seeds from mature and immature stages of fruit development is a common cause of reduced seed quality. The harvesting stage for seed production can be confirmed when the tendrils have withered on the same node as the fruit. Another sign of fruit maturity is the color change from a green/white to pale yellow on the underside of the fruit (i.e. the ground spot).

The best method of collecting fruit for seed extraction depends on the scale of operation. Where watermelon seed fields are large, at least 10–20 ha (25–50 acres), a once-over destructive mechanical harvest is often used (Fig. 12.14). Vines are cut, dried, and fruit mechanically moved into windrows in preparation for harvest. Once fruit have been collected, harvest should occur. Uncovered fruit are exposed to direct

Fig. 12.14. Mechanical hybrid watermelon seed harvest. (Courtesy of Tim Loeffler.)

sunlight and excessive heating which can cause seed aging. Self-propelled or pull-type vine seed harvesters drive through the field, collect fruit, crush them, separating seed from pulp which is deposited back in the field (Fig. 12.14).

When harvesting F1 hybrid seed, only marked fruits are collected. Hand-harvested fruits can be after-ripened up to five days before seed extraction. The fruit of parent lines or smaller hybrid production sometimes are cut open and evaluated for trueness-to-type and seed quality prior to extraction. In smaller fields or where hand labor is abundant, fruit may be hand-harvested when fully mature and fed into a mobile seed extractor placed in the field or centralized location.

Cleaning

Watermelon seeds are distributed throughout the fruit pulp and not contained in a central fruit cavity like cucumber or melon. Therefore, hand extraction methods are reliant on fruit maceration. The macerated pulp containing seeds is fed through an angled revolving cylindrical screen, so seeds fall through and are collected in a bin, but the larger fruit pieces are conveyed out the back end and returned to the field (Fig. 12.14). The better the initial maceration of pulp, the more efficient is the screen separation (Fig. 12.15). Watermelon seeds do not contain a persistent mucilage layer like melon and cucumber seeds. Typically, F1 diploid seeds are fermented for 24 to 36 h prior to washing and drying. Sometimes watermelon seeds are brush washed without fermentation. Enzyme treatment is usually unnecessary.

Hybrid triploid seed and tetraploid inbred production must be washed and dried immediately after seed extraction to avoid decline in seed quality. Sometimes hydrochloric acid is added during washing for control of seed diseases and to help remove coat debris.

Fig. 12.15. Watermelon seed and pulp slurry after mechanical field harvest ready for washing. Washing should effectively separate white empty seed coats from fruit pulp and the "good" black seeds before final cleaning. (Courtesy of Tim Loeffler.)

The drying temperature of watermelon seed should not exceed 32°C (90°F). Seed cleaning and drying is similar as described above for Cucurbits.

SQUASH

Botany

Squash comprises many related crops in the family Cucurbitaceae. Squashes can be members of several different species. One major delineation of squash is summer and winter. However, this designation has to do with fruit maturity for edible consumption and is not based on taxonomic differences. In North America, pumpkins are squash with orange skin at maturity that are used for ornamental purposes most often associated with fall or the Halloween holiday although they can also be eaten as a vegetable. Differences between pumpkin and squash have never been clearly or satisfactorily resolved botanically. In some competitions, only *C. pepo* are considered as true pumpkins while *C. maxima* are considered squash, but this is by no means universal. In other parts of the world, pumpkin generically describes winter squash regardless of species.

Pumpkin seeds are eaten as human food in some areas. Special cultivars are produced for the edible seed market. Some cultivars of edible-seed pumpkin lack fully developed seed coats because of genetic mutation. Such cultivars are called "naked" seed types. One pumpkin cultivar Triple Treat produces naked seeds, edible flesh, and orange skin for ornamental display.

Squashes are warm-season, frost-sensitive annuals including the following: *Cucurbita pepo* L. (pumpkin, winter squash, summer squash, gourds); *Cucurbita moschata* Duch. (pumpkin, winter squash); *Cucurbita maxima* Duch. (pumpkin, winter squash); *Cucurbita argyrosperma* Pang. (pumpkin, gourds); *Cucurbita ficifolia* Bouche (malabar or figleaf gourd); and the perennial *Cucurbita foetidissima* HBK (buffalo gourd). Some wild species are cross-fertile with the cultivated species listed above, and care should be taken that wild populations do not cross-pollinate with seed fields.

Wild species capable of contaminating seed fields (regions where they grow) are: *C. lundelliana* Bailey (Central America); *C. martinezi* Bailey (east coast Mexico); *C. andreana* Millan, cross-fertile with *C. maxima* (South America); and *C. texana* Gr. Millan, cross-fertile with *C. pepo* (southern USA and northern Mexico). *Cucurbita* species are similar but there are differences in

Table 12.1. *Cucurbita* seed identification based on appearance. (Modified from Whitaker and Bohn, 1950.)

Species	Funicular attachment of seed	Seed margin
C. pepo	Obtuse, symmetrical	Smooth, obtuse
C. moschata	Obtuse, slightly asymmetrical	Scalloped, obtuse
C. argyrosperma	Obtuse, slightly asymmetrical	Barely scalloped, obtuse
C. maxima	Acute, asymmetrical	Smooth, obtuse
C. ficifolia	Obtuse, slightly asymmetrical	Smooth, obtuse

seed anatomy that can be used for identification (Table 12.1; Whitaker and Bohn, 1950).

Soil

High yields are achieved on medium-textured soils with high water-holding capacity and good internal drainage. Squash is grown on a wide range of soils, although peat and heavy clay soils are not recommended. Clay soils generally have poor aeration and restricted drainage, which can be detrimental to root growth and increase fruit rot. Pumpkins and squash are sensitive to acid soils but tolerate slightly acidic (pH 6.8) to moderately alkaline (pH 8.0) soils and are sensitive to salinity. Fertility requirements should be based on soil testing prior to planting. Fertilization amounts vary but general recommendations for N-P-K applications are approximately 84 to 168 kg/ha N (75 to 150 lbs/acre N) applied before planting and after fruit set, 54 to 84 kg/acre P (50 to 75 lbs/acre P), and 112 to 196 kg/ha K (100 to 175 lbs/acre K). Nitrogen fertilization stimulates vegetative growth over reproductive growth but is needed in adequate amounts to ensure good canopy development for maximum photosynthetic capacity which contributes to high fruit and seed yield. A general recommendation is for moderate N until after crown fruit set, additional N can be applied via side dressing or fertigation to satisfy the needs of fruit and seed development.

A crop rotation should be practiced when growing Cucurbits in the same field, particularly if pathogen populations are high and solarization or fumigation is not used. Grasses, corn, or sorghum are good rotation crops. However, care must be taken that no carryover herbicide residues remain before planting squash or pumpkin seed crops.

Establishment

Squash seed production can be direct-seeded or transplanted. Seed production occurs in both open-field and protected environments usually in cages, net houses, or tunnels. Squash is often grown in rows on 15 to 20 cm (6 to 8 in.) high raised beds. Bed in-spacing is generally 203 cm (80 in) center-to-center. Large-fruited squash for open-pollinated seed production is planted at much closer row spacing than edible fruit since overall seed yield and not fruit size is the objective. Closer spacings generally produce smaller fruit and wider spacings result in larger fruit. Seed crop spacing as close as 0.6 to 1.8 m (2 to 6 ft) between rows is common. In-row spacings of 7.5 to 15.0 cm (3 to 6 in.) are the goal. Plants can be thinned to a final spacing at the two- to four-leaf stage during rogueing. Bush plantings are direct-seeded with a target in-row spacing of 20 to 30 cm (8 to 12 in.) between plants. Two rows spaced 1.5 to 1.8 m (60 to 70 in) apart or three rows spaced 0.6 m (24 in.) apart per bed may be used depending on plant size. Seeds are planted 1.3 to 2.5 cm (0.5 to 1.0 in.) deep.

The minimum, optimum, and maximum soil temperatures for seed germination are 18°, 35°, and 38°C (65°, 95° and 100°F), respectively. Germination is slow and erratic when soil temperature is below 20°C (68°F). Fruit maturing when daily mean air temperatures are below 20°C (68°F) have poorer quality. Squashes do well in warm weather, but very high temperatures 43°C to 46°C (110°F to 115°F) can cause temporary vine wilting, sunburned fruit, and soft fruit at harvest with reduced shelf life. Squash and pumpkin fruit are chilling sensitive and should not be stored below 13°C (55°F) for extended periods for maximum storage life (Welbaum, 2015).

In-bred lines for hybrid production are less affected by close spacing than open-pollinated cultivars because they often lack heterosis due to of inbreeding depression. In hybrid seed production, the planting is by direct seeding or occasionally plug transplants if inbred lines for F1 hybrid production mature at different times. Squash is sensitive to transplant shock and should be transplanted with a root ball and irrigated until established. Transplanting on hot sunny days should be avoided and plants should be "hardened" before transplanting by gradual exposure field conditions.

Successful hybrid production must also maintain field isolation distances and bees must be managed to ensure only F1 seeds are produced. For hybrid squash production, the female parent is naturally monoecious. Ethylene-releasing compounds like ethephon or tri-iodobenzoic acid are sprayed on the female rows to ensure all flowers are pistillate, so only cross-pollination with male rows can occur (Fig. 12.16).

Often multiple spray applications at specific growth stages are needed to totally suppress staminate flower production. Close monitoring for complete female flower conversion is needed. Any male flowers must be removed prior to opening of the female flowers. The ratio of male to female rows varies depending on the pollen productivity of the male line. Ratios ranging from 1 male to 2, 3, or 4 female rows are common (Fig. 12.17).

Adjacent rows of the male parent lines are not sprayed. Hand pollination is used in some areas, particularly in protected environments where bees may not be readily available nor easily managed. Only female rows are harvested for hybrid seed. Genetic male-sterile genes have been discovered in squash and can be used for hybrid seed production. However, because of the easy and effectiveness of chemical treatments to change sex expression, genetic male sterility is currently not widely used.

Fig. 12.16. Squash female rows receiving plant growth regulator treatment to suppress staminate flower production. (Courtesy of Tim Loeffler.)

Fig. 12.17. One male to two female rows in a squash F1 hybrid seed field. (Courtesy of Tim Loeffler.)

Pollination

The recommended isolation distance between for open-field squash seed crops varies among and between *C. pepo*, *C. moschata*, *C. argyosperma*, and *C. maxima* species from 0.1 km (0.2 miles) to 5.0 km (2.0 miles). It is also important to avoid areas where there is domestic fruit production or wild species grow. Coordination among seed companies is important in vine seed production areas. For example, in the USA, the California Crop Improvement Associate offers on-line pinning services to help companies isolate seed fields.

Squash fruit set requires pollen movement from staminate to pistillate flowers. Honeybees are the most reliable and cost-effective way to achieve pollination. Bumble bees (*Bombus* spp.), and other wild species, may contribute to pollination of seed crops in some locations but native populations should not be relied upon solely for seed production (Kremen *et al.*, 2002). One to two honeybee hives per acre are introduced when 5 to 10% of the plants have open flowers and up to four hives per acre during the peak pollination period. If hives are placed in weedy areas or close to other flowering crops, the number of hives per acre should be increased. Hives should be placed in clusters around the periphery of fields, with additional hives placed within larger fields.

In hybrid production when plant growth regulators are used, the goal is to retain the earlier fruit set near the crown of the plant for the highest genetic and physiological quality. The beehives are removed, and the plants' apical buds removed in a process called "tipping" before male flowers start to form on female plants. Sometimes the fruit are marked to ensure only hybrid seed is harvested. Later developing fruit is usually lower quality and has a higher percentage of inbreds.

Irrigation

Squash has extensive, moderately deep, root systems that efficiently explore the soil for water. Although irrigation is highly recommended for consistent water supply and is a necessity in desert production areas, in areas with adequate rainfall, squash seed crops may be grown without irrigation. Pre plant or post-plant irrigations ensure adequate soil moisture for seed germination, emergence, and stand establishment. After establishment squash should receive 2.5 to 3.8 cm (1 to 1.5 in.) of water per week at a minimum. Drip irrigation saves water and provides uniform applications. If mechanical harvesting is used, subsurface drip lines can avoid interference with harvesting equipment. Sprinkler irrigation is applied during stand establishment. Overhead irrigation is not recommended once vine canopy is established, because significant vine and fruit disease may result especially in humid climates. Furrow irrigation is declining because it is the least efficient and least uniform method. Furrow has the advantage of being simple and is still used in areas where water is abundant, land is flat, and infrastructure permits.

Rogueing

The following rogueing stages are recommended for squash crops:

- Early vegetative stage: Check that vegetative characters (e.g. bush or trailing type), foliage, disease resistance, and vigor are in accordance with cultivar description.
- Before first flowers open: Check all the above along with undeveloped fruit on female flower buds to ensure they are true-to-type.
- First female flowers setting: Check that developing fruits are true-to-type, and as above.
- Fruit developing: Fruit characters are true-to-type, and resistance to specific pathogens according to the cultivar description.
- Fruit removal: Any late developed or unmarked fruit on the plant is removed prior to harvest to ensure high-quality seed.

Seed Harvest

Most squash types take approximately 16 weeks from anthesis to seed maturity. After flowering and fruit set, male pollinator rows are removed from F1 hybrid seed fields prior to harvest to avoid mixing of hybrid and non-hybrid fruit (Fig. 12.18).

After 16 weeks, depending on the cultivar, the rind has hardened and cannot be penetrated by a fingernail. Physiologically mature fruits change to their mature color. The green types change to yellow-orange and the yellow-gold-colored types change to a darker straw at full maturity. Mature seeds have a white, shiny, enamel appearance, which is an indication the fruits are ready for harvest.

For large-scale production, fruit may be placed in windrows and collected by a vine seed harvester or manually fed into a fruit extractor designed for squash and pumpkin seed extraction. In some areas, the dried plants with fruit still attached are collected and extracted by a mechanical harvester that crushes fruit and extracts seeds in one pass through the field. Pull-type or self-propelled vine seed harvesters have a revolving mesh cylinder for seed separation as described previously (Fig. 12.19).

The seeds and small debris are captured in a bin underneath the revolving cylinder and taken to a seed-processing facility for cleaning, sorting, and drying. The fruits are relatively dry at harvest, so water is often introduced from a tank pulled behind the harvester into the separator to clean the screens and promote the movement of seeds and fruit debris through the harvester (Fig. 12.19). It is important to closely monitor the mechanical harvester to adjust and avoid mechanical damage or loss of good seed.

Cleaning

Squash seeds do not have a persistent mucilage layer like cucumber and melon seeds but are still more challenging to clean than watermelon seeds. It is important the seeds are washed as soon as possible after harvest and the time lag should not exceed six hours. The wet squash seeds are taken to a separate location and are subjected to pressurized washing and mechanical brushing to clean the seed coats. At some locations, seeds are also run through a water plenum to separate remaining debris. Sometimes seed lots that have substantial debris and may be treated with pectinase or dilute acid. Light seeds can be removed after seeds are dried using an air screen machine or density gradient table.

Fig. 12.18. Hybrid squash field with male pollinator rows removed. (Courtesy of Tim Loeffler.)

Drying

After final washing, Cucurbit seeds are immediately dried. Prolong hydration can lead to fungal growth, insect attack, seed aging, or germination. Cucubit seeds dry rather quickly because they have no significant barriers to water movement. The initial drying temperature should not exceed 35°C (95°F), and the seed should be dried to a moisture content of 6 to 8 % on a wet weight basis. Depending on volume, the seeds are usually dried in large rotating dryers, table dryers, or smaller precision dryers (Fig. 12.20). In areas with low relative humidity, sun drying is used with frequent stirring and turning. However, sun drying is not recommended for high-value seeds. Air drying seeds in hot humid climates may not reduce seed moisture content sufficiently. A recyclable clay bead drying system has been developed for humid conditions where air drying has been used historically (Timsina *et al.*, 2014).

Once seeds are dried to the desired low moisture content, they are hermetically sealed in containers with a barrier to moisture. This prevents seeds from reabsorbing moisture from the environment under humid conditions. Refrigerated dry storage prolongs seed storage life if available.

Pathogens and Pest Management

Blossom end rot is a disorder in developing Cucurbit fruit related to calcium deficiency. When fruit growth is rapid, insufficient calcium is delivered to the blossom end, regardless of the calcium available in the soil. The problem, more common in cultivars with elongated fruit shapes, may be minimized by maintaining a consistent water supply during fruit growth to aid calcium translocation (discussed in detail above). "Fish

Fig. 12.19. Squash harvester with water tank in tow. Water from the tank is sprayed into the separator to keep screens clean and promote the movement of seeds through the harvester. (Courtesy of Tim Loeffler.)

mouthed" or "osmotically distended" (OD) is a non-pathogenic disorder. Damaged embryo cells leak solutes that become trapped inside the endosperm envelope. These solutes cause osmotic swelling which splits seed coats, showing the swollen embryo and endosperm inside (see Growth and Development above).

Common Cucurbit viruses include cucumber green mottle mosaic virus (CGMMV), watermelon mosaic (WMV II), papaya ringspot (PRSV-W), zucchini yellow mosaic (ZYMV), squash mosaic (SqMV), cucumber mosaic (CMV), and Cucurbit aphid borne yellows (CABYV). Additional Cucurbit virus diseases include cucumber vein yellowing, Cucurbit chlorotic yellows, squash leaf curl, tomato spotted wilt, watermelon chlorotic stunt, watermelon silver mottle, and Cucurbit yellow stunting disorder. Squash mosaic virus (SqMV) is transmitted by cucumber beetles, and the others by several aphid species. Although specific symptoms may vary, viruses cause leaf mottling and stunted plant growth. Both direct and indirect enzyme-linked immunosorbent assay (ELISA) and polymerase chain reaction (PCR) tests can detect plant viruses.

Chemical control of virus spread is difficult because infection occurs immediately after insect feeding, and the insects migrate freely among fields. Yellow adhesive tapes/cards can provide affective control of aphids particularly in enclosed areas. A good control strategy is to maintain healthy and vigorous plants, monitor the fields often to detect unusual growth, correctly identify the disorder, and rogue defective plants from production fields.

Foliar diseases may also cause localized problems. Angular leaf spot (*Pseudomonas syringae* pv. *lachrymans*) on cucumbers is a seed-borne bacterium that may be spread through sprinkler irrigation. Bacterial fruit blotch (*Acidovorax avenae* subsp. *citrulli*) is a serious problem in some watermelon seed production regions. There are several symptoms including seedling blight, leaf lesions, and fruit spots that may cover the entire surface in severe infections. The disease is transferred to new generations by infected seed, transplants, or infected weeds (Lessl *et al.*, 2007).

Fig. 12.20. Squash seed on a table drier. (Courtesy of Tim Loeffler.)

Since the disease can be seed-borne, close monitoring of production fields, transplant houses, and testing for contamination are always required. Bacterial fruit blotch disease can infect both embryos and the seed coat. There is no effective treatment for embryo infection. Peroxyacetic acid is reported as a treatment for coat infection, but its effectiveness as a commercial practice is questionable (Hopkins *et al.*, 2003). Since this is a quarantined disease, seed companies destroy lots infected with bacterial fruit blotch to eliminate financial risk. Another bacterium that may affect watermelon is *Xanthomonas Cucurbitae*, which causes pumpkin spot.

Fungal diseases affect Cucurbits as well, especially in humid areas. Damping-off results in seedling collapse at the soil line and is most frequent in cold wet soils. It is caused by *Phythium* spp., *Rhizoctonia* spp., or *Fusarium* spp. and controlled by fungicide seed treatment, managing soil moisture, or *Trichoderma* spp. biological treatment. Phytophthora root and fruit rot (*Phytophthora capsici*) affects maturing fruit (Noh *et al.*, 2007). Gummy stem blight (*Didymella bryoniae*) is a fungal disease that causes several diverse symptoms including damping-off, crown rot, leaf spot, stem canker, and fruit rot (Gusmini *et al.*, 2005). Gummy stem blight causes characteristic round or irregular, brown lesions on the leaves and may also attack the stem causing elongated water-soaked areas that turn gray. Gum may ooze from the stem cracks and plant dieback is an identifying feature.

Anthracnose (*Glomerella cingulata* var. *orbiculare*) is a fungus that attacks watermelon foliage and fruit. In severe cases, leaves die back leaving only bare stems. Anthracnose is most common under warm rainy conditions. Powdery mildew (*Erysiphe* or *Sphaerotheca* spp.) can be a late-season problem in Cucurbits, but there is little evidence that it reduces seed yield unless symptoms occur early in the season. Downy mildew (*Pseudoperonospora cubensis*) occurs on Cucurbits worldwide and attacks the foliage causing yellow spots that coalesce into brown areas, leaf curl, and a scorched foliage. Other

Cucurbit fungal diseases include black root rot (*Chalara elegans*), fruit rot (*Choanephora Cucurbitarum*), Alternaria leaf spot (*Alternaria alternata*), and cucumber blight (*Alternaria cucumerina*).

Insects

Several species of cucumber beetle attack Cucurbits, including *Diabrotica undecimpunctata*. Both the beetles and larvae can damage stems and foliage, while larvae feed on roots. Cucumber beetles are vectors of bacterial wilt (*Erwinia tracheiphila*) disease. This disease doesn't usually affect watermelon and *Cucurbita* spp. but can devastate cucumber and melon plants, producing symptoms similar to drought stress. In India, the red pumpkin beetle (*Aulacophora foveicollis*) is a significant pest, which feeds on the immature leaves and flowers. Epilachna beetles (*Epilachna* spp.) are common pests of watermelon in Africa. The larvae and adults feed on the leaf and chew holes in the stems and fruits.

The melon aphid, *Aphis gossypii*, and the green peach aphid, *Myzus persicae*, may infest Cucurbits. Heavy aphid populations can seriously weaken plants causing yellowing, wilting, and stunted growth. Aphids also vector several viruses, which stunt growth. Several species of thrips may also infect watermelon and other Cucurbits.

Melon fruit fly (*Bactrocera Cucurbitae*) is a pest of watermelon in some seed-producing countries. Adult flies lay eggs that hatch into larvae within a week and feed on the fruit, causing damage and allowing fruit rot pathogens to enter. Watermelon crops can be grown under cover to protect against fruit flies. Mites (*Tetranychus urticae* and related species) may be a pest of watermelon particularly during warm, dry conditions. Mites distort growth, create chlorotic spotting, and produce distinctive fine webs on foliage. Armyworms, sometimes referred to as melonworm or rindworms, feed on developing fruit. There are several types of armyworms that feed on watermelon fruit including: southern (*Spodoptera eridania*), beet (*Spodoptera exigua*), and fall (*Spodoptera frugiperda*).

Seed corn maggots (*Delia platura*) are favored by early planting, heavy residues, and cool, wet weather. Seed corn maggots attack many horticultural crops including beans, peas, cucumber, melon, onion, corn, pepper, potato, and watermelon. Maggots feed primarily on decaying organic matter but will feed on seeds and seedlings of Cucurbits and especially watermelon. Similarly, cutworms (*Agrotis segetum* and *A. ipsilon*) feed on seedlings reducing watermelon stands (Welbaum, 2015).

References

Bewley, J.D. (1997) Seed germination and dormancy. *The Plant Cell* 9, 1055–1066.

Delahaut, K.A. and Newenhouse, A.C. (2011) Growing Pumpkins and Other Vine Crops in Wisconsin. Wisconsin Horticulture, University of Wisconsin-Madison. https://hort.extension.wisc.edu/articles/growing-pumpkins-and-other-vine-crops-wisconsin/ (accessed 20 September 2023).

Dong, W., Wu, D., Yan, C. and Wu, D. (2021) Mapping and analysis of a novel genic male sterility gene in watermelon (*Citrullus lanatus*). *Frontiers in Plant Science* 12, 639431. doi.org/10.3389/fpls.2021.639431 (accessed 12 September 2023).

FAO (2023) FAOSTAT (Food and Agriculture Organization of the United Nations statistics). Crops and livestock products. Melon seed. https://www.fao.org/faostat/en/#search/melon%20seed (accessed 20 September 2023).

Finch-Savage, W.E. and Leubner-Metzger, G. (2006) Seed dormancy and the control of germination. *New Phytologist* 171, 501–523.

Garcia-Mas, J., Benjak, A., Sanseverino, W., Bourgeois, M., Mir, G., *et al.* (2012) The genome of melon (*Cucumis melo* L.). *Proceedings of the National Academy of Sciences USA* 109, 11872–11877. doi.org/10.1073/pnas.1205415109 (accessed 12 September 2023).

George, R.A.T. (2009) *Vegetable Seed Production,* 3rd edn. CAB International, Wallingford, UK.

Gusmini, G., Song, R. and Wehner, T.C. (2005) New sources of resistance to gummy stem blight in watermelon. *Crop Science* 45, 582–588.

Hopkins, D.L., Thompson, C.M., Hilgren, J. and Lovic, B. (2003) Wet seed treatment with peroxyacetic acid for the control of bacterial fruit blotch and other seedborne diseases of watermelon. *Plant Disease* 87, 1495–1499.

Kremen, C., Bugg, R.L., Nicola, N., Smith, S.A., Thorp, R.W. and Williams, N.M. (2002) Native bees, native plants and crop pollination in California. *Fremontia* 30, 41–49.

Lessl, J.T., Fessehaie, A. and Walcott, R.R. (2007) Colonization of female watermelon blossoms by *Acidovorax avenae* spp. *citrulli* and the relationship between blossom inoculum dosage and seed infestation. *Journal of Phytopathology* 155, 114–121.

Laghetti, G. and Hammer, K. (2007) The Corsican citron melon (*Citrullus lanatus* (Thunb.) Matsum. et Nakai subsp. *lanatus* var. *citroides* (Bailey) Mansf. ex Greb.) a traditional and neglected crop. *Genetic Resources and Crop Evolution* 54, 913–916. doi.org/10.1007/s10722-007-9220-y (accessed 12 September 2023).

Lorenz, O.A. and Maynard, D.N. (1988) *Knott's Handbook for Vegetable Growers*, 3rd edn. Wiley, Hoboken, New Jersey, pp. 128–156.

Maynard, D. and Maynard, D.N. (2000) II. Chapter 6. Cucumbers, melons, and watermelons In: Kiple, K.F. and Ornelas, K.C. (eds) *The Cambridge World History of Food*. Cambridge University Press, Cambridge, UK, pp. 298–313.

Nesom, G.L. (2011) Toward consistency of taxonomic rank in wild/domesticated Cucurbitaceae. *Phytoneuron* 13, 1–33.

Noh, J., Kim, W., Lee, K., So, S., Ko, B. and Kim, D. (2007) Effect of furrow mulching with PE black film and dripping of phosphorous acid on control of Phytophthora root and fruit rot (*Phytophthora capsici*) occurred in field-grown watermelon. *Korean Journal of Horticultural Science and Technology* 25, 24–28.

Oluoch, M.O. and Welbaum, G.E. (1996) Effect of postharvest washing and post-storage priming on viability and vigour of six-year-old muskmelon (*Cucumis melo* L.) seeds from eight stages of development. *Seed Science and Technology* 24,195–205.

Pitrat, M., Hanelt, P. and Hammer, K. (2000) Some comments on infraspecific classification of cultivars of melon. In: Katzir, N. and Paris, H.S. (eds) Proceedings of Cucurbitaceae 2000: the 7th Eucarpia Meeting on Cucurbit Genetics and Breeding. *Acta Horticulturae* 510, 29–36.

Ramakrishna, P. and Amritphale, D. (2005) The perisperm-endosperm envelope in *Cucumis*: structure, proton diffusion and cell wall hydrolysing activity. *Annals of Botany* 96, 769–778.

Robinson, R.W. and Decker-Walters, D.S. (1997) *Cucurbits.* CAB International, Wallingford, UK.

Rubatzky, V.E. and Yamaguchi, M. (1997) *World Vegetables: Principals, Production, and Nutritive Values*, 2nd edn. Chapman & Hall, New York.

Stepansky, A., Kovalski, I. and Perl-Treves, R. (1999) Intraspecific classification of melons (*Cucumis melo* L.) in view of their phenotypic and molecular variation. *Plant Systematics and Evolution* 217, 313–332.

Tamang, J.P., Sarkar, P.K. and Hesseltine, C.W. (1988) Traditional fermented foods and beverages of Darjeeling and Sikkim – a review. *Journal of the Science of Food and Agriculture* 44, 375–385.

Texas Vegetable Growers Handbook (2023) original eds: Masabni, J., Dainello, F. and Cotner, S. Texas A&M AgriLife Extension Service. aggie-horticulture.tamu.edu/vegetable/guides/texas-vegetable-growers-handbook (accessed 12 September 2023).

Timsina, K., Dahal, P., Bradford, K.J., Kunusoth, K., Van Asbrouck, J., *et al.* (2014) Impacts of a new post-harvest drying technology on the horticultural seed value chain in Nepal. In: *Proceedings of the 29th International Horticultural Congress*, Queensland, Australia, Springer Nature, London, pp. 17–22.

USDA ARS (2023) *Citrullus amarus*. Germplasm Resources Information Network (GRIN). US Dept of Agriculture, Agricultural Research Service. https://npgsweb.ars-grin.gov/gringlobal/search?q=Citrullus%20amarus (accessed 20 September 2023).

Welbaum, G.E. (1993) Water relations of seed development and germination in muskmelon (*Cucumis melo* L.). VIII. Development of osmotically distended seeds. *Journal of Experimental Botany* 44, 1245–1252.

Welbaum, G.E. (1999) Cucurbit seed development and production. *HortTechnology* 9, 341–348.

Welbaum, G.E. (2015) Chapter 10. Family Cucurbitaceae. *Vegetable Production and Practices*. CAB International, Wallingford, UK, pp. 136–165.

Welbaum, G.E. and Bradford. K.J. (1988) Water relations of seed development and germination in muskmelon (*Cucumis melo* L.). I. Water relations of seed and fruit development. *Plant Physiology* 86, 406–411.

Welbaum, G.E. and Bradford, K.J. (1990) Water relations of seed development and germination in muskmelon (*Cucumis melo* L.). V. Water relations of imbibition and germination. *Plant Physiology* 92, 1046–1052.

Welbaum, G.E. and Bradford, K.J. (1991) Water relations of seed development and germination in muskmelon (*Cucumis melo* L.). VII. Influence of after-ripening and ageing on germination responses to temperature and water potential. *Journal of Experimental Botany* 42, 1137–1145.

Welbaum, G.E., Hill, D.R., Pavel, E., Grayson, R.L. and Gunatilaka, M. (2000) Chapter 8. Compartmentation of abscisic acid in developing muskmelon (*Cucumis melo* L.) seeds. In: Black, M., Bradford, K.J. and Vazquez-Ramos, J. (eds). *Seed Biology: Advances and Applications*. CAB International, Wallingford, UK, pp. 85–100.

Welbaum, G.E., Muthui, W.J., Wilson, J.H., Grayson, R.L. and Fell, R.D. (1995) Weakening of muskmelon perisperm envelope tissue. *Journal of Experimental Botany* 46, 391–400.

Welbaum, G.E., Shen, Z., Oluoch, M.O. and Jett, L.W. (1998) The evolution and effects of priming vegetable seeds. *Seed Technology* 20, 209–235.

Whitaker, T.W. and Bohn, G.W. (1950) The taxonomy, genetics, production, and uses of the cultivated species of *Cucurbita*. *Economic Botany* 4, 52–81.

Whitaker, T.W. and Davis, G.N. (1962) *Cucurbits: Botany, Cultivation, and Utilization*. Interscience, New York.

Witmer, X., Nonogaki, H., Beers, E.P., Bradford, K.J. and Welbaum, G.E. (2003) Characterization of chitinase activity and gene expression in muskmelon seeds. *Seed Science Research* 13, 167–178.

Yim, K.O. and Bradford, K.J. (1998) Callose deposition is responsible for apoplastic semipermeability of the endosperm envelope of muskmelon seeds. *Plant Physiology* 118, 83–90.

13

Family Fabaceae

Overview

Fabaceae is the third-largest plant family, behind only Orchidaceae and Asteraceae, with 730 genera and over 19,400 species (Stevens, 2023). Crops from this family are commonly called "legumes", a term derived from an archaic family name. Fabaceae includes a diversity of economically important agricultural plants such as forages, cover crops, green manure, vegetables, and grains. Worldwide, volume and calories contributed by grain legumes, often called pulses, significantly outweigh vegetable legumes, which include fresh pods and seeds, foliage, shoots, and storage roots. This chapter will focus on vegetable legumes with limited mention of grains. Examples of major types of vegetable legumes include the following.

Common bean

The common bean, *Phaseolus vulgaris*, is an herbaceous, annual, warm-season, frost-intolerant plant grown worldwide. Common beans are versatile with many uses: dry seeds harvested at full maturity, snap beans (immature fruit harvested before significant seed development), and shelled beans harvested for "green" mature seeds. Bean leaves are occasionally used as a vegetable in some cultures.

Soybeans

Vegetable soybean (*Glycine max*) seeds, often called edamame, are eaten in an immature green stage after cooking. They are an important source of protein in Southeast Asia (Fig. 13.1; Li *et al.*, 2022). Agronomic or dry grain soybeans are a more important crop that dominate world production. Agronomic soybeans are harvested as dried grain and processed into animal feed, cooking and industrial oils, tofu (bean curd), soy sauce, soymilk, and many other food and industrial products.

Peas

Peas (*Pisum sativum*) are among the oldest cultivated crops (Fig. 13.2). The earliest archaeological use of peas dates from the Neolithic era in what is today Syria, Turkey, and Jordan. Segregation of pulse, fodder, and vegetable green peas occurred after domestication. Edible-podded peas (sugar snap and snow peas) are more recent. Sugar snap peas resulted from a spontaneous mutation that were collected and preserved.

Lima bean

Large-seeded Lima beans (*Phaseolus lunatus*) found in archeological sites date back several

DOI: 10.1079/9781789243260.0013

Fig. 13.1. Edamame vegetable soybean plant with seeds and pods at edible maturity.

Fig. 13.2. Determinate "bush" pea plant at pod set.

thousand years. Evidence suggests small-seeded Lima beans were grown in Central America 2500 years ago. Lima bean was a novelty for the Spanish who introduced them to Europe, and cultivation spread to Asia and Africa. Since Lima beans can tolerate humid tropical weather, they have become an important crop in parts of South America as well as parts of Indonesia and Southeast Asia.

Botany

Vegetable legumes are herbaceous dicot annuals or perennials. Seed production of annual beans is common in many temperate areas. Legumes display a broad range of temperature tolerance. Many are warm-season and frost-intolerant (snap beans and soybeans) while others are cool-season and frost-tolerant (pea or fava bean; Table 13.1).

Leaves are generally alternate and mostly compound, and pinnate, trifoliate, or shaped like an open hand. Flowers are perfect and characteristically butterfly shaped, consisting of an upright dorsal petal, two lateral petals or wings, and two lower petals often more or less united in a keel-like structure (Fig. 13.3). Enclosed within the petals are the stamens and pistil (Rubatzky and Yamaguchi, 1997). Legumes are predominantly self-pollinated although some are cross-pollinated by bees.

Legumes produce simple, dry, dehiscent fruit consisting of a mature ovary wall or pericarp containing few to many seeds (Fig. 13.3). The pod is a single-carpel fruit that splits along two sutures at maturity releasing the seeds (Rubatzky and Yamaguchi, 1997). The pod itself may be green, purple, or yellow or solid color with speckles of a second color. Cultivars of edible pod bean have thicker fruit walls and develop seed at a slower rate than agronomic beans harvested as a grain (pulses) (Fig. 13.3).

Legume seed consists of two large cotyledons surrounding a tiny but well-developed embryonic plant. The cotyledons contain carbohydrates, fats, and protein in varying proportions depending on the species. The whole seed is enclosed in a tough testa (coat). Some legumes have "hard" seed coats that resist rapid water uptake. The seeds vary widely in their size, shape, and color. Some legumes, such as pea, have hypogeal germination where the cotyledons and seed coat remain beneath the soil during germination because of limited hypocotyl elongation. The epicotyl is fully differentiated prior to germination and pushes through the

Table 13.1. Common legumes grown from seed.

Common name	Genus and species
Adzuki bean	*Vigna angularis*
Broad bean	*Vicia faba*
Catjang cowpea	*Vigna unguiculata* Group *cylindrica*
Chickpea	*Cicer arietinum*
Cluster bean	*Cyamopsis tetragonolobus*
Cowpea	*Vigna unguiculata* Group *unguiculata*
Fenugreek	*Trigonella foenum-graecum*
Garden and field pea	*Pisum sativum*
Lentil	*Lens culinaris*
Lima bean	*Phaseolus lunatus*
Lupines	*Lupinus* spp.
Mat bean	*Vigna aconitifolia*
Mung bean	*Vigna radiata*
Rice bean	*Vigna umbellata*
Scarlet runner bean	*Phaseolus coccineus*
Common and snap bean	*Phaseolus vulgaris*
Soybean	*Glycine max*
Tepary bean	*Phaseolus acutifolius*

Fig. 13.3. Flowering bush edible pod bean plant at fruit set and early development.

soil. Epigeal germination occurs in the common bean where the cotyledons emerge above the soil surface because of rapid hypocotyl elongation.

Bean and soybean

The root system of common bean and soybean is generally shallow and spreading. However, some roots may extend 1 m (3.3 ft) deep or more in friable soils under optimal conditions. Plant architecture is classified as indeterminate or climbing and non-climbing or bush types. Half-runner beans have an intermediate growth habit between bush and indeterminate types. Determinate "bush" cultivars differ from indeterminate "climbing" types because they are self-topping with a terminal flower and have less apical dominance. The stem length of indeterminate climbing types can grow 3 m (10 ft) or more depending on the growing season and have 25 or more flowering nodes. However, indeterminate vines tangle in seed fields when plants are unsupported. Determinate bush types are shorter, approximately 45 to 60 cm (1.5 to 2 ft) tall, with fewer nodes compared to indeterminate types, and the inflorescence are terminal (Rubatzky and Yamaguchi, 1997). Many modern cultivars of common bean and soybean are bush types whose concentrated pod set and maturity are best adapted to labor-saving, once-over, destructive mechanical harvesting.

Most commercial cultivars of both determinate and indeterminate common bean, pea, and Lima bean are day-neutral. There are short-day cultivars in each group but most are heirlooms. Soybeans are short-day plants. Exposure to a several successive nights longer than the critical length induces flowering. Different cultivars are developed for different day lengths or latitudes, so it is essential that soybean seed crops be produced under the right day lengths to induce flowering when the plants are large enough to support seed production. The plant pigment phytochrome is responsible for photoperiod detection in soybean, is converted into an active form by sunlight, and returns to an inactive form in the dark.

Common bean flowers are medium to large and petals may be white, pink, or purple with ten stamens, nine of which are united in a tube enclosing the long ovary while one upper stamen is free of the rest. Flowers are self-pollinating with little out-crossing. Foliage is pinnately trifoliate. Many modern bush cultivars have smaller leaves to improve light penetration into the canopy for high-density plantings grown for once-over, destructive mechanical harvesting. Smaller leaves tend to increase yield, but are also genetically linked to smaller pod size (Rubatzky and Yamaguchi, 1997).

Common bean pods for edible consumption are longer than wide with lengths ranging from 8 to 20 cm (3 to 8 in.) or more, with widths of less than 1 cm (0.4 in.) to several centimeters. Depending on the cultivar, pod tips may be pointed or blunt. Cross-sectional pod shapes may be round, elongated flat, or heart shape (Welbaum, 2015). Pods of most modern cultivars are relatively straight, although some are naturally curved. The fruits of most cultivars are light to dark bluish-green but novelty cultivars may be yellow, purple, or multicolored. Cultivars with yellow pods are often called “wax” beans. The amount and rate of pod fiber development also vary. The term “string bean” describes edible pod cultivars with long, thread-like fibers that run along the dorsal and ventral pod sutures (Rubatzky and Yamaguchi, 1997). Such fibers have been removed by plant breeding and do not exist in modern edible pod cultivars. Stringless-type cultivars also contain less fruit wall fiber. Presently, only heirloom and specialty cultivars possess string-like suture fibers that must be removed before eating. Plant breeders develop new cultivars with varying “sieve size” to meet the demands of the processing industry. Sieve size is related to the diameter of the pod for non-flat cultivars. There are six sieve size categories ranging from 4.74 to 5.76 mm for sieve “1” to greater than 10.72 mm for sieve size “6”. Most cultivars utilized in the processing industry fall into the “3” or “4” sieve size range.

The number of seeds per pod varies among bean types. Most edible pod bean cultivars contain three to five seeds, while cultivars for dry bean production have more. Mature seed sizes exhibit variation in size and weight, ranging between 5 and 20 mm (0.2 and 0.8 in.) in length and individual seed weights vary from 0.15 to more than 0.80 g (0.005 to 0.03 oz). Seed shapes are variable and may be round, orbicular, ovoid, oblong, or kidney. Seed coat colors vary and are cultivar specific. In different countries of Latin America, certain seed coat colors are preferred: black seed coat in Brazil, El Salvador, Mexico, and Venezuela; red in Colombia and Honduras; yellow in Peru; white in Chile (Rubatzky and Yamaguchi, 1997). Edible pod cultivars grown for canning usually have white or light-colored seeds so leachate from dark seed pigments doesn't cloud liquid in cans.

Pea

Pea is a frost-tolerant, cool-season, herbaceous, annual dicot, having alternate leaves with the tip of the compound leaf modified as a tendril. Growth habits range from indeterminate vine to determinate bush or dwarf forms (Fig. 13.2). Foliage surfaces have a waxy cuticle and leaf colors ranging from yellowish green to deep blue green. Pea stems are singular, slender, angular, and hollow, but solid at the base. Some cultivars have an upright habit but are generally not self-supporting and branching is usually limited (Makasheva, 1983).

Taproots may reach 80 cm (31 in.) deep, but root systems are generally not extensive. Pea flowers develop at leaf axils and are perfect containing both male and female parts. Flowers are self-pollinating, often before fully opening, and cross-pollination is unusual. Flowers are usually white, but may be pink, purple, or blended colors. The number of non-flowering nodes is a cultivar characteristic. Very-early-maturing cultivars produce flowers in as few as five or six nodes. Some late cultivars produce 15 or more vegetative nodes before flowering (Rubatzky and Yamaguchi, 1997). Pod number per node is a genetically controlled trait that is also influenced by the environment. For example, environmental stressors like high temperatures or drought may reduce pod numbers. Early-flowering cultivars average one or perhaps two pods per node, while late-flowering cultivars usually average more than two per node. Some cultivars develop four or more pods per node. Indeterminate cultivars produce a main stem that sets pods over an extended period under favorable conditions. Determinate cultivars tend to group flowers in a terminal inflorescence instead of among the nodes that ends further vegetative growth (Makasheva, 1983). Determinate cultivars are preferred by the commercial canning and freezing industries due to their more concentrated flowering and fruit set and suitability for mechanical harvest.

Peas are also classified by leaf type. Pea leaf types have variation in the number of leaflets and tendril formation. Traditional wild-type peas have pinnate leaves with four to six opposed leaflets with tendril formation at the end of each leaf. Leaves of other types have only leaflets with no tendrils. The *afila* mutation results in formation of only branched tendrils without leaflets. Another mutation, *afila-tendril-less*, forms many small leaflets where tendrils would form in *afila* mutant types. Sexual crosses among these types yield intermediates (semi-*afila*) that have some leaflets and greater tendril development compared to wild-type peas (Hofer *et al.*, 2009). In general, the more tendrils a plant has the less lodging occurs (Wang *et al.*, 2003). Tendrils are advantageous because they support adjacent plants which develop upright to facilitate improved destructive mechanical harvesting. Normal leaf types tend to lodge and lie flat on the ground making mechanical harvest difficult. *Afila* and semi-*afila* cultivars are gaining popularity among processors due to their open plant structure which improves aeration, are less shaded, and have fewer diseases. *Afila* types also yield similar to normal leaf cultivars.

A pea pod is a dehiscent fruit formed from flower carpels. Seed are attached alternately to either side of two fused carpels. Pod walls of garden and field peas have a hard, lignified-parchment layer. The pod walls of snow and sugar snap edible pod peas lack lignification. Pod size and seed number vary among cultivars and usually range between five and 11 peas per pod. The pea seed consists of two large cotyledons surrounding the embryo and enclosed by either a pigmented or colorless seed coat that may be either smooth or wrinkled. A single gene controls the presence of or absence of the green pigment in the seed coat. Presence of the pigment contributes to the expression of dark green colored peas that are used primarily for freezing, and pigment absence results in light-colored fresh peas used primarily for canning. There are cultivars that are acceptable for both freezing and canning and are known as freezer-canner types. Mature smooth-seeded types have higher starch content, a mealy texture, better cold tolerance, are often processed by canning, and are becoming popular as an ingredient in plant-based meatless and other food products. Peas with wrinkled seed coats are sweeter when young. The sugars do not convert entirely to starch so the higher sugar and lower starch contents are maintained during processing.

Lima bean

The species name *P. lunatus* comes from the moon-shaped seed. Lima beans are often divided into three groups within *P. lunatus*. Small-seeded Lima beans are Sieva types and are sometimes referred to as “baby Limas”. Large-seeded green Limas are a second type sometimes referred to as Butterbean and Madagascar bean. The third type is “potato” Lima sometimes called Fordhook (Fofana *et al.*, 1997). Fordhook cultivars have larger seeds that are pale green.

Limas are considered short-lived perennials but are grown commercially as annuals. Both indeterminate and bush cultivars are available. Indeterminate cultivars may climb to 3 to 4 m (10 to 13 ft), whereas bush types only reach 50 to 90 cm (20 to 35 in.). Limas have highly branched roots reaching more than 1 m (39 in.) deep if soil conditions allow. Like other legumes, roots develop symbiosis with *Rhizobium* bacteria.

Flowers are self-fertile although a small percentage of out-crossing may occur. The slightly curved oblong pods range from 5 to 15 cm (2 to 6 in.) in length to 2 to 3 cm (0.8 to 1.2 in.) wide.

Pods of most cultivars contain two to six seeds. Pods of some cultivars are thick, and others are relatively thin. The large, flat, oblong seed of large-seeded Limas are as long as 3.5 cm (1.4 in.). Sieva seed types are also flat but more rounded and about 1 cm (0.4 in.) long. Fordhook seeds have an intermediate shape.

Lima commercial cultivars in North America tend to have light green or nearly white seed coats. In other areas, seed coats may be red, purple, brown, or black. Two large cotyledons account for most of the seed volume. Seeds of wild Limas found in South and Central America have a high cyanogenic glucosides content and must be leached before or during cooking. Lima beans should not be consumed raw (Rubatzky and Yamaguchi, 1997). Modern cultivars, especially those with light-colored seeds, contain little or no glucosides.

Bacterial symbiosis

Members of Fabaceae form a unique symbiosis with the soil *Rhizobium* spp. bacteria. *Rhizobium* is a non-spore forming, rod-shaped, motile, aerobic, gram-negative soil bacterium. *Rhizobium* bacteria in soil are attracted to potential legume hosts by flavonoids secreted by roots. *Rhizobium* attached to roots secrete *nod* factors signaling a change in root hair morphology to allow symbiosis. The initial step of symbiosis occurs by intracellular infection when *Rhizobium* enters legume roots through deformed root hairs. A second mechanism called "crack entry" occurs when the bacteria enter between cells through cracks caused by lateral root emergence (Jones *et al.*, 2007). Eventually, the bacteria are internalized into cells through tubular cell wall invaginations called infection threads. The infection triggers cell division in the cortex of the root where a nodule develops. Nodules are specially adapted organs that develop on the legume roots to accommodate *Rhizobium*. While in the nodule, the *Rhizobium* differentiate morphologically into bacteroids and fix gaseous atmospheric nitrogen (N_2) into ammonium ($NH_3 + H^+ \rightarrow NH_4^+$), that can be used by the enzyme nitrogenase in plants (Jones *et al.*, 2007). The reaction occurring in all nitrogen-fixing bacteria is:

$$N_2 + 8H^+ + 8e^- \rightarrow 2NH_3 + H_2$$

The legume–*Rhizobium* symbiosis is a classic example of mutualism because *Rhizobium* supplies ammonia or amino acids to the plant that in return provides the bacterium with organic acids, carbohydrates, and proteins.

Since legumes get most of their nitrogen from the atmosphere through *Rhizobium* N_2 fixation, they typically need less nitrogen fertilization. However, legumes do respond to nitrogen additions in the seedling stage before *Rhizobium* symbiosis is established, especially when soil nitrogen is low. Fertilizing legumes later in the season can be counterproductive because nitrogen fixation declines rapidly when soil nitrogen is high. Additions of phosphorus and/or potassium may increase nodule numbers, fresh weight, and nitrogen fixation. An important micronutrient for nitrogen fixation is molybdenum. Soils should be tested for molybdenum prior to growing legumes. A pH below 6.0 may reduce molybdenum availability. Molybdenum fertilizer may be applied as a seed treatment since only small amounts are required. Some *Rhizobium* inoculants have molybdenum incorporated (Erker and Brick, 2006).

Legume inoculation is the process of introducing commercially prepared sources of *Rhizobium* directly to seeds or soil to promote nitrogen fixation. When there is doubt about the *Rhizobium* population in field soils, inoculants should be used (Erker and Brick, 2006). If the legume crop was grown in the field previously, there is a good chance that the soil already contains sufficient *Rhizobium* for nodulation, since the bacteria can survive without a host. However, *Rhizobium* survival is poor in acid soils. Commercial inoculants are *Rhizobium* spp. selected for their nitrogen fixation potential. Healthy nodules that are actively fixing nitrogen are pink to reddish inside (Erker and Brick, 2006).

Vegetable legume seeds may be inoculated before planting either as a seed pretreatment or on farm. Inoculates can be purchased separately from seeds and added by the grower before planting. The three forms of commercial inoculation are solid, liquid, and freeze-dried (Erker and Brick, 2006). The most common is peat-based powder that is applied to the seed either wet or dry. The dry seed application has the disadvantage of uneven distribution and poor adhesion. A slurry is prepared by mixing the inoculum with water for better seed coverage and applied to dry seeds just prior to planting (Erker and Brick, 2006). Liquid inoculants are sold in culture broth or as a frozen concentrate. For soil application, inoculants are mixed with water and sprayed into the seed furrow at planting. Because liquid inoculants must be kept frozen or refrigerated, they have limited availability through normal distribution channels (Erker and Brick, 2006).

Growth and Development

Common bean

Common bean types are often grouped according to how they are utilized. French beans, which are simply called green beans in some markets, have stringless fleshy edible pods, and develop seeds later in fruit development (Fig. 13.3). Other names for this type included snap bean and Italian green bean. Another edible pod type is the string bean. Although similar to the French beans, the string or lignified vascular bundle is removed before cooking and eating. String beans have faster seed development and less fleshy pods. Haricot filet or vert is a third type of edible pod bean with a stronger bean flavor and thin, small, pods. Fruits may be either string and stringless, and the pod is less fleshy than French beans with earlier seed enlargement.

Common beans can be characterized based on plant growth habit as determinate, semi-determinate or indeterminate. Determinate bush bean cultivars are compact and self-supporting (Fig. 13.4). With a heavy fruit load, bush types may lodge during wet conditions particularly in shallow soils when anchor roots fail to develop. Bush cultivars are better adapted to high-density plantings and their concentrated fruit set makes mechanical destructive seed harvest efficient (Fig. 13.4).

Pole beans for edible consumption are often grown on vertical supports as the name implies. They have a vining habit and spread 1.8 m (6 ft) or more if allowed to sprawl. They require wider spacing than bush types. Pole bean seed crops

Fig. 13.4. Bush bean seed production field Salinas Valley, California, USA. (Courtesy of Tim Loeffler.)

are grown unsupported, but vines often intertwine making mechanical seed harvest difficult. Half-runner beans have an intermediate growth habit between pole and bush types. Commercial seed crops of half-runner cultivars are grown without support and plants can reach 40 cm (3.3 ft) across.

Most modern edible pod bean cultivars are not sensitive to photoperiod. A shift from short-day photoperiod response to day-length neutrality occurred with many biotypes (Rubatzky and Yamaguchi, 1997). Nevertheless, there are a few cultivars that flower under short days. Beans grow best in full sun and yields are reduced when plants are shaded. Bean leaves face and track the sun across the sky during the day, which is an adaptation to improve photosynthetic efficiency. However, during periods of excessive heat and low soil moisture, leaves will turn parallel to the sun's rays to decrease leaf temperature (Rubatzky and Yamaguchi, 1997).

Initiation and flower development is delayed by suboptimal temperatures. If temperatures are < 10°C (< 50°F), fertilization is poor and pods are small and misshapen. Blossom drop and ovule abortion also occur at high temperatures > 35°C (> 95°F). Lateral branching is beneficial because it provides more flowering nodes for fruit and seed development. Pole bean cultivars generally have more lateral branches, making them better adapted to stress. Determinate cultivars are more susceptible to stress that interferes with pod set, reducing seed yields. However, under favorable conditions, determinate plants have the advantage of concentrated fruit set and uniform pod development (Fig. 13.4).

Bush snap beans (*Phaseolus vulgaris*) cultivars generally produce a mature seed crop from 100 to 125 days after seeding under stress-free conditions. Pole cultivars require about 20 to 30 days longer and half-runner beans are generally intermediate between pole and bush cultivars. Dry beans such as kidney, navy, cranberry, etc. require from 100 to 130 days from seeding to field dry seed harvest.

The word pulse may refer to only the seed of beans harvested as dry grains or the entire plant (Table 13.2). Pulses are used as human and animal food but exclude vegetable, oil, and forage legumes. Pulses are destructively harvested at

Table 13.2. The Food and Agriculture Organization of the United Nations (FAO) recognizes the following 11 pulse crops.

Type or genus and species	Example
Phaseolus spp. beans including several species now classified as *Vigna* spp.	Kidney bean, haricot bean, pinto bean, navy bean (*Phaseolus vulgaris*)
	Lima bean, butter bean (*Phaseolus lunatus*)
	Adzuki bean (*Vigna angularis*)
	Mung bean, golden gram, green gram (*Vigna radiata*)
	Black gram, urad (*Vigna mungo*)
	Scarlet runner bean (*Phaseolus coccineus*)
	Rice bean (*Vigna umbellata*)
	Moth Mat or moth bean (*Vigna aconitifolia*)
	Tepary bean (*Phaseolus acutifolius*)
Vicia faba	Horse bean (*Vicia faba* subsp. *equina*)
	Broad bean and field bean (*Vicia faba*)
Pisum spp. (dry peas)	Garden pea (*Pisum sativum* subsp. *sativum*)
	Protein pea (*Pisum sativum* subsp. *arvense*)
Cicer arietinum	Garbanzo, chickpea, and Bengal gram
Vigna unguiculata	Dry cowpea, black-eyed pea
Cajanus cajan	Pigeon pea, Arhar/Toor, cajan pea, Congo bean
Lens culinaris	Lentil
Vigna subterranean	Bambara groundnut, earth pea
Vicia sativa	Common vetch
Lupinus spp.	Lupins
Minor pulses	Lablab, hyacinth bean (*Lablab purpureus*)
	Jack bean (*Canavalia ensiformis*), sword bean (*Canavalia gladiata*)
	Winged bean (*Psophocarpus teragonolobus*)
	Velvet bean, cowitch (*Mucuna pruriens* subsp. *utilis*)
	Yam bean (*Pachyrrizus erosus*)

full maturity before pods dehisce by combining, or sometimes hand-harvested for small plot production. The word gram also describes some legume plants harvested for seed. Gram bean is a general term for any one of several leguminous food crops grown in India, such as *Phaseolus mungo* (black gram or urd) and *P. aureus* (green gram) (Rubatzky and Yamaguchi, 1997). Gram beans are harvested when seeds are fully mature and dry (Table 13.2).

Bean seed crops are grown in areas with low relative humidity, low rainfall, and favorable conditions for harvest. A three- to four-year crop rotation is preferred to control diseases and avoid mixtures from volunteer plants from a previous crop. Major considerations for high-quality seed production include high germination, seed vigor, and absence of critical seed-borne diseases discussed in a later section. Bean seed exports must meet the phytosanitary requirements of the importing country.

Peas

Peas grow best during prolonged cool weather and can tolerate light frosts before flowering. Although peas are a cool-season vegetable, subfreezing temperatures may damage plants. Young plants are more tolerant of low temperatures than mature ones. Blossoms and pods are more sensitive to subfreezing temperatures than leaves and stems. For seed production, the planting window is relatively short. The emphasis is to plant as early as possible but avoid hard freezes that could damage the developing crop. Optimum mean temperatures for vegetative growth of peas are between 13°C and 18°C (55°F and 64°F) and growth is very slow above 29°C (84°F).

Growing degree days (GDD) are sometimes used to predict development of both beans and peas. While largely used to separate planting dates for edible vegetable production and to determine harvests, GDD have utility in tracking development of seed crops (Delahaut and Newenhouse, 1997). For peas, a base temperature of 4°C (39°F) with an upper limit of 29°C (84°F) is usually the range for calculating GDD. Outside this range, growth is negligible and few heat units are accumulated. Cultivars are often identified by their heat unit requirements. Early-producing cultivars require as few as 1200 GDD to achieve seed harvest maturity. Some late-maturing cultivars require more than 1800 GDD (Delahaut and Newenhouse, 1997). Generally early-flowering cultivars are day-neutral, but flowering is accelerated by long days for some late-maturing cultivars. Moderate temperatures also decrease the time to flowering, but temperatures above 30°C (86°F) can cause flower or ovule abortion. Moderate diurnal temperature fluctuations improve plant growth (Rubatzky and Yamaguchi, 1997).

Pea seed crops are often produced in cool desert areas (Chapter 4) with different spacings for determinate and indeterminate types. Major seed production areas include Columbia Basin of Washington state and the Treasure and Magic Valleys of Idaho in the USA, Hungary, England, France, Chile, and New Zealand as well as other countries. In the tropics and subtropics, pea seed crops are grown at high elevations when temperatures are relatively cool.

Three distinct types of peas exist:

1. cultivars producing well-developed, succulent but immature seed for edible consumption;
2. cultivars grown for their immature, succulent, edible pods and seeds that are eaten together; and
3. plants producing fully developed, mature, dry seed for cooking in soups and other dishes.

English, garden, vining, or green peas are synonymous terms describing the first type. Frozen garden pea cultivars are distinct from those utilized for canning. Garden pea cultivars developed for freezing have distinctive dark green color, wrinkled thick seed coats, and higher sugar content. Canning cultivars have smooth, thin pale green seed coats, and higher starch content. Pea pods are difficult to harvest by hand because there is no natural abscission zone between the plant and fruit. For seed production, all three types fully mature and are harvested by combines before they senesce.

Lima bean

Many characteristics described for common bean cultivars also apply to *P. lunatus*. Lima beans often require a longer growing season and do not pollinate well at high temperatures which can be a problem in some desert production areas. Lima bean seed crops are mechanically harvested when fully dry from 95 to 140 days after planting depending on type and cultivar. Lima bean cultivars are differentiated by seed size, shape, color, maturity, and plant architecture, e.g. indeterminate, determinate, or semi-determinate. Within the USA, Lima bean seed production takes place solely in coastal and central California and requires specialized harvesting equipment to minimize mechanical damage.

Soil

Well-drained, non-crusting, medium-textured, loam soils with organic matter are suited for common bean and Lima bean seed crops. A favorable soil temperature range is 18°C to 30°C (68°F to 77°F) (Rubatzky and Yamaguchi, 1997). Most beans, particularly bush cultivars, have comparatively small root systems, so the absorption of water and nutrients occurs near the soil surface. The optimum pH range is between 6.0 and 6.5. Fertilizer damage to seedlings is greater on acid soils when the pH is below 5.5. Crop rotations help reduce root rot diseases. Bean and pea seed crops should only be grown in the same field once every four years to control diseases and avoid seed mixtures from volunteer plants from a previous crop.

Growth is strongly reduced by soil compaction and poor drainage. Heavy clay soils tend to form a surface crust as they dry. Soil crusts inhibit epigeal emergence because cotyledon protrusion is inhibited by the dense, hard, dry, clay surface layer. Crusted soils can also cause damage or loss of cotyledons and slow seedling growth and development. Irrigation or shallow tillage using a rotary hoe can reduce soil crusting and improve emergence.

Peas can be grown in many types of soil from light sandy loams to medium clays. Peas are less susceptible to soil crusting because their cotyledons remain beneath the surface. The pH range for pea is from 6.0 to 7.5. Poor soil drainage and compaction reduce productivity and increase susceptibility to root disease. Crop rotations help reduce root rot diseases, and peas should be grown in the same soil only once every four years.

Fertilization

Fertilizer should be placed in the root zone for most efficient uptake and utilization. Beans are sensitive to high salts. If fertilizer is applied at planting it should be placed a minimum of 5 cm (2 in.) to the side and below the seed (Davis and Brick, 2013).

Many believe that legumes do not need N fertilization because of their symbiotic relationship with *Rhizobium*. This may only be true if beans immediately follow crops that were heavily fertilized or leave residue that will release N upon decomposition (Davis and Brick, 2013). Since symbiosis does not occur immediately after germination, particularly for determinate cultivars, early growth may be N limited (Rubatzky and Yamaguchi, 1997). Therefore, N fertilization may aid establishment depending on soil type and cropping history. Fifty-six to 90 kg/ha (50 to 80 lbs N/acre) are often applied depending on residual soil N, organic matter, and soil texture. Excessive N fertilization causes excessive vining, delayed maturity, reduced pod set, and inhibits nitrogen fixation by *Rhizobium*.

Phosphorus is especially important for early bean growth and crops respond to applied phosphorus on soils that test low or medium. Fertilizer should be banded in the root zone because P is immobile in soil. Phosphate fertilizers may also be surface applied and tilled into the soil prior to planting, so seeds and seedlings are less likely to be damaged by salt in dry soil. Monoammonium phosphate (11-52-0), diammonium phosphate (18-46-0), or ammonium polyphosphate (10-34-0) are equally effective. The best choice depends on availability and cost per unit of P (Davis and Brick, 2013).

Preplant soil testing results should determine K fertilization recommendations. No application is recommended when soil tests exceed 75 ppm K as tested by M1 extraction. Sandy or highly eroded soils often test low in extractable K. Incorporating K before planting or row banding is recommended. On heavier soils, a single application of K prior to or at planting is often sufficient. A split-K application is sometimes used for sandy soils with overhead-irrigation to decrease the potential for fertilizer burn and leaching. Preplant application of 25 to 50% of K fertilizer with the remainder applied as side dressing at flowering, similar to N, is often recommended on sandy soils (Hochmuth and Hanlon, 2010).

Combined N and potash (K_2O) applications should not exceed 112 kg/ha (100 lbs/acre) when between-row spacings are 0.9 m (36 in.) or 123 kg/ha (110 lbs/acre) for 0.75 m (30 in.) spacing to reduce seedling injury from high salts. Beans are highly sensitive to excessive soil boron. Zinc may inhibit growth particularly at high pH. Bean crops remove a minimum of 56-67.5-50 kg/ha (50-60-45 lbs/acre) of N-P-K, respectively, per season (Lorenz and Maynard, 1988). Nutrient applications should provide these amounts as a minimum and more for seed crops since the recommendations are for edible vegetables.

Pollination

Beans, peas, soybeans, and Lima beans are self-compatible and predominantly self-pollinated (Fig. 13.3). Beans and peas are cleistogamous, which means that many ovules are already self-pollinated when flower buds open. Therefore, cross-pollination is very rare under most circumstances but can sometimes occur. It has been proposed that out-crossing is greater in the tropics where there is more insect activity (George, 2009). Many seed companies stipulate a minimum distance of 2 m (6.6 ft) between seed crops of *Phaseolus vulgaris* (Fig. 13.5) and use another crop such as one or two rows of corn or safflower as a barrier to separate different cultivars. Barrier crop rows are common when increasing parent seed of multiple genotypes within the same large field. However, some prefer isolation distances of 50 m (164 ft) for legume seed crops in the final multiplication stage and 150 m (492 ft) for breeder, foundation, or basic seed (George, 2009).

Pea flowers are also cleistogamous and almost totally self-pollinated. Pollination often occurs in late bud stage before flowers are fully

Fig. 13.5. Bush bean pure-line stock seed field with safflower border between cultivars. (Courtesy of Tim Loeffler.)

open (Fig. 13.2). Recommended isolation distances for peas are relatively short and are mainly to prevent mixing during harvest. It is important that adjacent cultivars be separated by at least 2 m (6.6 ft). The distance is increased by some producers to those similar to beans described above for basic seed production (George, 2009).

Irrigation

Moisture management for legume seed crops is important for successful stand establishment, disease control, pod set, and pod quality. To avoid soil crusting and promote uniform emergence, irrigation should be applied prior to planting in areas with insufficient rainfall (OSU, 2010b). Irrigating should supply adequate moisture for pod development while minimizing wetting the canopy to avoid foliar disease. Sprinkler irrigation is sometimes used to ensure uniform emergence. After full bloom, early day irrigation is recommended to allow canopy drying as quickly as possible (OSU, 2010b). Surface irrigation is preferred over sprinkler for legume seed production to minimize seed-borne diseases. Furrow irrigation is used on flat soils where water is plentiful (Fig. 13.6). Some companies require drip irrigation for parent seed production. Drip irrigation is less common for legume seed production but conserves water in desert production areas. Irrigation water quality should be monitored, especially in desert soils prone to salt accumulation where leaching rainfall does not occur. Since legumes are sensitive to salinity, irrigation water should have an electrical conductivity <100 mS/m.

Moisture stress during legume flowering and pod fill may reduce seed quality and yield. Peak water demand for bean and pea seed crops ranges from 0.5 and 0.4 cm (0.20 and 0.16 in.) per day during pod fill. In areas without frequent rain, weekly irrigation during pod formation is adequate; however, on sandy and sandy loam soils, irrigation may be required as frequently as

Fig. 13.6. Furrow irrigation of bean seed crop to reduce foliar disease. (Courtesy of Tim Loeffler.)

every three to four days. Ideally, 406–483 mm (16–18 in.) of moisture evenly distributed throughout growth season should be sufficient for bean and pea seed production.

Rogueing

The amount of rogueing varies with the level of seed production. Breeder, basic, or foundation seed production requires more rogueing than stock seed. Initial rogueing is conducted before flowering to confirm plant habit, vigor, and height. At the onset of flowering, foliage, flower color, and plant vigor should be checked for the correct characteristics. Any plants showing symptoms of seed-borne pathogens at seed set and pod formation should be immediately sampled for confirmation. If tests confirm widespread infection, a decision should be made on whether to continue or reject the production. If the infection is isolated, infected and surrounding plants should all be removed from the field and destroyed to prevent any possibility of spread. Late-flowering off-types should also be eradicated. Plants with irregular, off-color, or poorly formed seed pods should be eliminated. Seeds from any genetic off-type or poor performing plants should be removed from the field and properly disposed.

For pea, the first rogueing stage occurs after emergence when seedlings are approximately 15 cm (6 in.) tall. Remove any off-type or diseased plants. Check foliage, leaf shape, color, plant height, tendril development, etc., for trueness-to-type. Eliminate early- and late-flowering plants, and any plants with off-type flowers. Verify flower number per node. Check pod shape, size, and color to ensure they are typical of the cultivar. For parent seed, remove low yielding plants before seed harvest, if practical.

Genetics

Cleistogamy, along with few seeds per fruit, makes production of F1 hybrid beans, peas, soybeans, and Lima beans difficult and economically impractical on a commercial sale. Therefore, bean, soybean, pea and Lima bean are pure-line cultivars. Pure-line cultivars are developed from an initial cross and taken through as many as ten generations of self-pollination to ensure that they are pure breeding before release as new cultivars (Kelly, 2010). Bean and pea cultivars grow true-to-type, so seed can be saved by growers for replanting.

While some cultivars of agronomic grain soybeans have been genetically engineered for herbicide and insect resistance, this technology has not been widely applied to vegetable (edamame) soybeans. If questions about seed genetic identity arise because of labeling or possible contamination from unwanted transgenic agronomic soybeans, polymerase chain reaction (PCR) assays can differentiate transgenic from non-transgenic seeds. Commercial common bean, pea, and Lima bean cultivars are also non-transgenic and new cultivars are currently produced by sexual crossing and selection of desired phenotypes.

Seed Harvest

Beans

Bean seed is sown in late spring after the frost-free date after soils have warmed. In warm, moist soils, uniform emergence can be expected within six to ten days of planting. Planting in cold soil slows germination and increases seed decay. Seed coat cracking and leakage historically are greater in some white-seeded cultivars of *P. vulgaris*, reducing emergence in cold soils and increasing susceptibility to pathogenic attack compared to black-seeded cultivars (Mohamed-Yasseen *et al.*, 1994). Poor germination is not characteristic of all white-seeded cultivars. Plant breeding and selection have improved crack resistance and low temperature performance in white-seeded beans (Dickson, 1971). Fungicide treatments are standard to protect bean seed planted in cold soils.

Sowing depths range from about 2.5 to 5.0 cm (1.0 to 2.0 in.) depending on seed size. At planting, dry bean seeds are less dense than soil and tend to rise to the surface after heavy rain or irrigation if planted too shallow. If seeds are close to the surface, water uptake may be insufficient for germination, seeds may dry out faster before germinating, or they may be eaten by predators.

Beans are sometimes planted on raised beds especially on heavier soils to improve drainage and aeration or when a field will be furrow irrigated (Fig. 13.7). Between-row and in-row spacing varies greatly depending on plant architecture and is cultivar dependent. Seed production research is the best way to determine optimum spacing for each cultivar. Increasing plant populations reduces the number of pods per plant but increases overall seed yield on an area basis. Higher plant populations also tend to concentrate maturity. At close spacing, plants are more erect, pods form closer to the stem, and higher in the canopy. Bush cultivars are often sown in rows 45 to 90 cm (18 to 35 in.) apart, according to cultivar and stage of multiplication. Climbing cultivars are spaced in rows 90 to 120 cm (35 to 47 in.) apart (Fig. 13.7). Wider spacing is usually needed for seed crops than food crops to accommodate cultivation, spraying, crop inspection, or harvesting equipment. A general recommendation is approximately 100 kg seed/ha (89 lbs seed/acre) for determinate "bush" cultivars and 50 kg seed/ha (45 lbs seed/acre) for running and indeterminate cultivars for seed crop production (George, 2009).

Harvesting of bean seed is determined by the cultivar, stage of pod and bean development, and harvest method. Some shatter-resistant types or cultivars such as edamame soybean can be direct harvested in a single process. Common bean seed harvest usually involves a two-step process. The crop is first cut, windrowed, and cured for several days until mechanically harvested at the target seed moisture content (Fig. 13.8). Bush types are cut when 10% of pods are dry and parchment like, but have not dehisced, and the remaining pods are changing color.

Plants may be cut for final drying when fruit pod pericarp viewed in cross-section has lost its clear wet appearance and mature seeds are loose in the pod. For colored cultivars, it is important to observe some color in the "eye" of the seed before cutting. Harvesting can be delayed to allow further field drying if the cultivar resists shattering. Seed moisture content for

Fig. 13.7. Edible pod bean seed field establishment. (Courtesy of Tim Loeffler.)

mechanical harvest is normally between 10 and 14% for most cultivars (Fig. 13.9). Some cultivars are particularly sensitive to mechanical injury and should be threshed with a target seed moisture of 15–16%. Seed harvested above 15% must be closely monitored and carefully dried to the 12–14% range for safe seed storage. Further drying after harvest with supplemental heat must be done at low temperatures not exceeding 27°C (85°F). The preference is for field drying and timely harvest at optimal moisture (Fig. 13.9). This avoids supplemental drying of large seed lots which are easily over-dried making them very sensitive to mechanical damage during handling and transport. Bean flowers tend to set fruit over a relatively long period and sometimes this leads to a "split set". This leads to shattering of the earliest maturing pods if harvesting is delayed or the inclusion of immature seed if harvest of early set is prioritized. Large dry legume seeds are especially susceptible to mechanical damage at moisture contents < 10% and additional precautions should be taken to adjust combine settings to minimize seed damage. Rubberized harvesting and handling equipment softens impacts and reduces physical damage during legume seed handling.

Indeterminate cultivars mature over an even longer period than bush types, complicating harvest determination. If labor is available, early set pods can be hand-harvested at three or four successive intervals as they mature for small seed lots. Hand-harvest is best suited to small seed lots of breeder, foundation, or basic seed. Small quantities of seed can be threshed as they mature. Wilson and McDonald (1992) concluded that where feasible, manual threshing is superior to mechanical harvesting methods for small seed lots. Manual threshing minimizes cotyledon damage and seed cracking. When labor is not available, indeterminate and semi-determinate types can be planted at closer spacing. Closer spacing concentrates fruit set for combining after windrowing when most fruit have matured. Bean seed is harvested using rotary or conventional combines and the settings

Fig. 13.8. Bean plants cut and windrowed for final drying. (Courtesy of Tim Loeffler.)

and speed of the threshing mechanism vary depending on the type of combine (Fig. 13.9). The rotary combine has a threshing rotor that runs parallel to the windrows and the conventional combine has a threshing cylinder that is perpendicular to the windrows. Often combines are modified to reduce mechanical injury by slowing conveyers and installing belt unloaders instead of using the more common unloading auger. A preferred combine for legumes is a rotary-type historically used for peanuts and adapted for bean seed harvest. It offers the advantage of seed harvest at higher moisture to minimize mechanical seed damage.

After harvest, it is important to preclean seed lots to remove excess debris, weeds, and soil before temporary storage prior to seed conditioning. Some seed companies moisturize excessively dry seed (< 10%) to 13–14% moisture content prior to conditioning to reduce mechanical damage. Immature and damaged seeds can be separated during post-harvest conditioning using air screen machines and gravity separation tables. Some seed lots are color sorted to remove low-quality off-color or wrong color seeds. Plant breeders have selected against the hard seed coat trait in legumes to improve germination of agricultural seeds. However, reducing seed hardness increases susceptibility to physical damage, such as seed coat cracking, during seed harvest, handling, and planting.

Peas

Soil temperatures of at least 5°–10°C (45°–50°F) are needed for good pea emergence which is best at 20°C (68°F). At temperatures > 25°C (77°F), germination decreases. Seeds are often treated with fungicides or biological treatments to minimize seedling diseases especially when planting in cold soils. Seeds should be planted in early spring to avoid high temperatures during bloom and seed development but late enough to avoid subfreezing temperatures during early

Fig. 13.9. Combining dried windrows of bean plants for seed harvest. (Courtesy of Tim Loeffler.)

development and flowering. Peas can be planted using grain drills but vacuum and belt seeders provide precision spacing. Determinate pea seeds are seeded relatively early in the spring at a rate of 150 kg/ha (134 lbs/acre) in rows 40 to 60 cm (16 to 24 in.) apart. Indeterminate cultivars are given more space to accommodate vine growth. The weight of seed needed for planting varies widely from 150 to 250 kg/ha (134 to 223 lbs/acre) depending on seed size, germinability, and growth habit (George, 2009). Higher populations reduce the number of fruits on each plant but increase overall pod yield. Higher plant populations also tend to concentrate seed maturity. Earlier cultivars generally produce flowers on lower nodes compared to taller and later genotypes.

Seed fields should be rogued during the growing season for genetically off-type plants and disease as described above. For large commercial seed production, high-quality genetic seed reduces the amount of rogueing necessary. Foundation or basic seed production should be planted with a enough space for easy inspection and rogueing activities. Integrated pest management (IPM) and proactive practices should be utilized to control diseases and insects before they become a problem.

Irrigation should be discontinued one to two weeks prior to harvest to allow complete seed maturation. Peas are harvested at storage moisture with a target seed moisture content of 12–13% either after the crop has been cut, windrowed, and cured or directly harvested after complete crop dry down. Direct combining is possible in areas where seeds dry sufficiently on the plants without dehiscing. Seed with > 16% moisture content must be dried to prevent mold growth and deterioration. Desiccants are sometimes used when mature plant and fruit drying is relatively slow. Drying agents are applied when pea plants begin to senesce, and the lowest pods have turned pale brown and paper-like. Desiccation application procedures vary with products. For example, 3 l (0.8 US gal.) of desiccant can be added to 100 l (26 gal.) of water and sprayed on foliage when random seed samples have a moisture content of 40% (George, 2009). Crops which have been treated with a desiccant can be cut and windrowed when the seed moisture content declines to approximately 28% or directly harvested when the crop reaches a moisture content of 14% or less (Fig. 13.10). Windrows are sometimes turned to speed drying, but some seeds will be lost to shattering each time.

The optimum moisture content for harvest corresponds to pods that appear dry, tan colored, and paper-like (Fig.13.10). At this maturity stage, further drying will not harm seeds. Peas are less sensitive to mechanical damage than beans. Precautions for harvesting and handling the crop to prevent mechanical damage should be followed for pea as well. Relatively low combine drum speeds help avoid pea seed damage (George, 2009). Threshing of windrowed plants should occur when seed moisture content is approximately 14% using rubber covered tines. Like bean, pea seed is usually precleaned and placed in temporary storage until the final conditioning is scheduled. Post-harvest conditioning of pea may include spiral separation, air screen cleaning followed by density separation, and color sorting if necessary.

Lima bean

Lima bean is less widely adapted compared to common bean. Lima bean requires slightly warmer temperatures for optimum growth. Favorable mean growth temperatures range between 15°C and 25°C (59°F and 77°F). Larger-seeded Limas generally prefer lower temperatures and higher humidity than small-seeded cultivars, especially for optimum pollination. For large-seeded Limas, temperatures above 30°C (86°F) and relative humidity < 60% may cause poor pollination and increase blossom drop. Moisture stress during flowering may also reduce fruit set. Environmental sensitivity restricts Lima bean seed production to specific climates such as coastal and central California, USA (Rubatzky and Yamaguchi, 1997).

Seeds germinate best between 15°C and 30°C (59°F and 86°F). Planting depths vary with seed size and range from 3 to 6 cm (1.2 to 2.4 in.), with large-seeded types planted deeper. Emergence is most rapid at about 25°C (77°F). Seeds are inoculated with *Rhizobium* before planting especially if other legumes have not been grown in rotation. Planting on raised beds improves drainage on heavier soils. Bush-type plantings are usually spaced with 60 to 90 cm (24 to 35 in.) between rows with in-row spacing of 10 to 15 cm (4 to 6 in.) (Rubatzky and Yamaguchi, 1997).

Fig. 13.10. Direct-seed harvest of pea seed crop in Idaho, USA. (Courtesy of Tim Loeffler.)

Lima bean is well suited for production on light-textured, warm, well-drained soils. A slightly acid soil with pH between 6 and 7 is preferred. Other than temperature and humidity restrictions, Lima bean seed production is similar to common bean (OSU, 2010a). Lima bean requires a longer growing season and consequently more moisture and nutrients compared to common bean. The crop is fertilized before planting based on soil test results. Supplemental fertilizer is often side-dressed after pod set. Although Limas are nitrogen fixers, they show a greater response to N fertilization compared to snap bean. Depending on soil type, 67–112 kg/ha (60–100 lbs /acre) of N is applied in split applications before planting, 40 days after planting, and sometimes at full bloom to meet crop needs (OSU, 2010a). Care should be taken not to over-fertilize. Excessive N early in the season will inhibit nodulation and *Rhizobium*. Too much N during establishment will delay flowering and increase disease. A Lima bean seed crop will remove about 20 lbs (22 kg) P, 100 lbs (112 kg) K, 10 lbs (11 kg) S, and 10 lbs (11 kg) Zn per acre (ha). Many of the same diseases and pests that affect common bean species also affect Limas.

Limas are harvested as the plants and seed pod transition from green to yellow. If harvest is delayed, the percentage of shattering pods increases. Lima seed crops are susceptible to mechanical damage. They are machine-harvested often with specialized threshers with two threshing cylinders operated at very low speeds to minimize mechanical damage. Cultivars prone to shattering can be cut and windrowed when seed moisture content is between 30 and 40%. Windrows are often turned to encourage uniform drying. Threshing or combining of windrows should occur when moisture content reaches approximately 15%. Large-seeded types are susceptible to mechanical damage especially at moisture contents < 15% and the risk increases at lower temperatures. Drop distances should be reduced when possible during harvesting, handling, and conditioning. Impact surfaces are often protected with rubber to minimize physical damage.

Seed Yield

Common bean seed yields range from 1500 to 2500 kg/ha (1338 to 2230 lbs/acre). The 1000 grain weight of *Phaseolus vulgaris* is about 250 g (8.8 oz) for large-seeded types to 600 g (21.1 oz) for smaller seeded cultivars.

The average yield of a pea seed crop is approximately 2000 kgs/ha (1784 lbs/acre). The 1000 grain weight varies from small- to large-seeded types, but generally is 330 g (12 oz) for the larger seed types and 250 g (8.8 oz) for smaller seeds.

Pest Management

Weeds

Weed control early in the season is essential for producing successful legume seed crops. Pea plants especially do not compete well with weeds. Selective herbicides are often used in conjunction with narrow row spacing for non-organic seed crops. Wider row spacing is needed for mechanical cultivation when herbicides cannot be used. Managing weeds in fence rows and fields adjacent to seed production, crop rotation, and stale bed preparations in conjunction with cultivation or herbicide applications are all parts of an effective weed control program.

Pathogens

Beans

White mold is caused by the fungus *Sclerotinia sclerotiorum* and develops as a cottony growth on the stem, stem branches, and pods. The fungus produces black, hard mats of sclerotia near the cotton-like growth, which overwinter. The sclerotia germinate the next season when the leaf canopy is complete, and the soil surface is cool, 15°C –18°C (59°F–65°F), and wet for ten to 14 days. White mold kills plants and infects pods, which severely reduces yield. High humidity and plant canopy temperatures from 16°C to 21°C (61 to 70°F) favor the spread of white mold (Meronuck *et al.*, 1993). Careful irrigation management is important because white mold is worse under moist conditions (Meronuck *et al.*, 1993).

Bean rust is a fungal disease caused by *Uromyces appendiculatus*. Symptoms first appear as small pale spots or lesions that turn yellow with small dark centers. Ten hours or more of constant moisture and temperatures between 17°C to 27°C (63°F to 81°F) cause infection and symptoms appear within 10–15 days. Early infection increases the potential for yield reduction. Crop rotations with unrelated crops for multiple years provide control. Crop residue is the primary source of inoculum and should be buried by plowing or burning. If the disease is identified early, fungicide may provide control (Meronuck *et al.*, 1993).

Common bean blight is due to *Xanthomonas campestris* pv. *phaseoli*, halo blight is caused by *Pseudomonas syringae* pv. *phaseolicola*, and brown spot is due to *Pseudomonas syringae* pv. *syringae*. These bacterial pathogens are restricted from most seed production areas. Common blight is favored by relatively high temperatures and appears as small, translucent, water-soaked spots on leaves. As spots enlarge, affected tissues die leaving brown spots with a narrow yellow margin. Lesions become large and have irregular shapes. Water-soaked sunken pod lesions eventually turn brownish-red. Bacteria may infect the vascular system and kill the main stem and branches (Meronuck *et al.*, 1993).

Halo blight infected plants exhibit symptoms similar to common blight. Halo blight lesions are surrounded by a large, pale-yellow circle 1.3 cm (0.5 in.) in diameter. Temperatures between 16°C and 20°C (60°F to 68°F) favor halo blight. The systemic plant infection produces stunted plants with small chlorotic trifoliolate leaves. Red or brown water-soaked lesions appear on pods (Meronuck *et al.*, 1993).

Brown spot lesions are similar to common blight. They have a narrow yellow border, but no halo, surrounding the lesions. Leaves also lack water-soaked symptoms. Brown spot infection is favored by relatively cool conditions. As lesions mature, the dead tissue in the center falls out resulting in a shot-hole. Infected pods may be twisted at the infection point (Meronuck *et al.*, 1993). Infected seeds spread and damp conditions favor the development of all three diseases. A three- to four-year rotation, incorporation of old bean debris, the reduction of volunteer beans, and the use of certified streptomycin treated seed helps to manage these similar diseases. Certain copper containing compounds may prevent the spread of causal bacteria (Meronuck *et al.*, 1993).

Fusarium solani f. sp. *phaseoli*, *Rhizoctonia solani*, and various *Pythium* spp. cause root rot and damping-off diseases in beans. Root-rotting fungi are present in the soil and live on decomposing vegetation. They attack plants whenever the population is large enough and the soil is cool, wet, and compacted (Meronuck *et al.*, 1993). Rotation is very important in controlling root rots. Rotations of three to four years between bean crops are effective. Poor drainage and waterlogged conditions can contribute to damping-off and root disease. Other crops such as sunflower, potato, sugarbeet, and soybean that are susceptible to or support the survival of these fungi should not be grown in rotation.

Several *Alternaria* spp. fungi cause leaf spot diseases of bean. Spots have an irregular shape and are gray-brown colored. When the weather is cool and humid, lesions coalesce, forming large contiguous areas. Infected leaves may tear and have a ragged appearance. *Alternaria* survives in infested crops and weed residue and may infect leaves through wounds. The disease is favored by wet and damp conditions and is found most prevalent on mature or senescent leaves. Satisfactory control measures have not been developed, although crop rotation and wider row spacings help to limit damage. No fungicide is effective (Meronuck *et al.*, 1993).

Angular leaf spot is a fungal disease caused by *Phaeoisariopsis griseola*. The disease affects the foliage and pods of beans and can be a problem when warm moist conditions accompany abundant inoculum from infected plant residues or contaminated seed. All above the ground tissues are susceptible. Leaf lesions are gray or brown with a chlorotic halo. After approximately ten days, lesions turn brown and necrotic and assume a distinct angular shape. Lesions may ultimately drop out, leaving holes in the leaf blade. Primary inoculum comes from seed or infested residue. Infection and disease occur between 16°C and 28°C (61°F to 82°F) with an optimum of 24°C (75°F). Planting disease-free certified seed and rotating two or three years between bean crops provide control (Meronuck *et al.*, 1993).

Anthracnose is caused by the fungus *Colletotrichum lindemuthianum* that affects all above ground tissues and is a regulated pathogen for several production areas. Lesions first appear as water-soaked angular areas that elongate, becoming brick red to purple before turning dark brown to black. Lesions appear on the petioles and on both the upper and lower surfaces of leaves. Pod lesions are tan to rust-colored and develop into sunken cankers outlined by black rings surrounded by a reddish-brown border. Tan to salmon-colored spores may form in the lesions. Temperatures of 13°C to 21°C (55°F to 70°F) favor the disease (Meronuck *et al.*, 1993).

Bean common mosaic virus (BCMV) stunts plants and causes malformed and mottled leaves. Leaves infected with BCMV are irregular shaped with light yellow and green patches. Infected leaflets are often narrower and longer than normal with a downward cupping. Early in the season, infected plants are yellowish, dwarfed, and spindly. Dark necrotic lesions and spots may occur on roots, petioles, pods, and leaves. The virus is vectored by aphids and seed. The only methods of control are resistant cultivars and non-infected seed. Other bean viruses include bean golden mosaic virus (BGMV), bean pod mottle virus (BPMV), bean southern mosaic virus (BSMV), bean yellow mosaic virus (BYMV), the bean strain of cucumber mosaic virus (CMV), the bean strain of sugarbeet curly top virus (SBCTV), and peanut stunt virus (PSV) (Rubatzky and Yamaguchi, 1997).

Peas

Although a few pea diseases are common to beans, many are unique. The following is a partial list of some of the most common diseases of pea: Aphanomyces root rot (*Aphanomyces euteiches*), anthracnose (*Colletotrichum pisi*), Ascochyta blight complex (*Ascochyta pisi, A. pinodella, Mycosphaerella pinodes*), bacterial blight (*Pseudomonas syringae* pv. *pisi*), black root rot (*Thielaviopsis basicola*), Botrytis (*Botrytis cinerea*), downy mildew (*Peronospora pisi, P. viciae*), Fusarium wilt and near wilt (*Fusarium oxysporum* f. sp. *pisi*, and other races), Fusarium root rot (*Fusarium solani* f. sp. *pisi*), leaf and pod spot (*Phoma medicaginis* var. *pinodella*), powdery mildew (*Erysiphe polygoni* and other *Erysiphe* spp.), Pythium root rot (*Pythium ultimum* and other *Pythium* spp.), rusts (*Uromyces pisi, U. fabae, U. trifolii*), Sclerotinia (*Sclerotinia sclerotiorum*), and Septoria blight (*Septaria pisi*).

Virus diseases are a major problem for pea producers. Most of the following list are vectored by aphids and other insects. Bean yellow mosaic virus (BYMV), pea strain, pea enation mosaic virus, pea streak virus, pea stunt virus, pea seed-borne virus, pea early browning virus, pea top yellow virus, common pea mosaic virus, and cucumber mosaic virus (CMV), bean leaf roll virus (BYMV) pea strain (CABI, 2023b).

Insects

Beans

Many insects are pests of common beans (CABI, 2023a). Lepidopterous caterpillars from the genera *Ascotis, Spilosoma, Amsacta,* and *Euproctis* all attack bean crops. Caterpillars feed on leaves, buds, flowers, and pods. Many predators and parasites attack the eggs of caterpillars including several species of *Trichogramma*. Most parasitized eggs turn black, but there may be a lag period before they do so. Generalist predators such as lacewings, minute pirate bugs, and damsel bugs feed on corn earworm eggs and small larvae as well. Sprays of *Bacillus thuringiensis* are an effective biological control. Chemical insecticide sprays may also be used for conventional seed production (UC IPM, 2023).

Several species of leafhoppers are found in dry beans with *Empoasca fabae* and *E. solana* being two of the more common (CABI, 2023a). Empoasca leafhoppers are small (3 mm, 0.13 in.) long, bright-green, wedge-shaped insects. The small, wingless nymphs or immatures are also wedge shaped and green. Adults and nymphs live on the underside of leaves. Leafhoppers cause leaf margins to turn yellow and become necrotic, and the entire leaf may turn yellow resembling virus symptoms. Insecticides may be used when leafhoppers first appear (UC IPM, 2023).

There are several types of leaf beetles from the family Chrysomelidae including the bean leaf beetle (*Ootheca bennigseni*) and redheaded flea beetles (*Systena frontalis*) that attack bean seed crops. Both beetles are oval-elongate shaped usually less than 13 mm (0.5 in.) long. Their color and shape are variable, making identification difficult. Adults feed on flowers and foliage while the larval forms generally attack roots (CABI, 2023a).

Darkling beetles (*Blapstinus* spp.) are from 3.5 to 6 mm (0.13 to 0.25 in.) long and vary from black or bluish black to rusty brown. Larvae are cylindrical, wireworm-like, soil-inhabiting, pale yellow to dark brown worms from 0.8 to 8 mm (0.03 to 0.33 in.) in length (UC IPM, 2023). Darkling beetles are numerous in spring and early summer. They run on the ground and are often found under clods or debris during the day. Darkling beetles girdle or cut off stems of young bean plants at or below the soil line (UC IPM, 2023).

Mexican bean beetle (MBB, *Epilachna varivestis*) adults are coppery brown with black spots. When adults migrate to a bean field, they lay yellow-orange egg masses on the underside of bean leaves. Eggs hatch into bright yellow, spiny, oval larvae that feed and molt repeatedly, pupating on the underside of leaves (UC IPM, 2023). Both adults and larvae feed primarily on leaves. Pod injury may occur if populations are high. The adult MBB overwinters and seeks out beans for feeding and reproduction. Overwintering adults colonize beans in June. There are two or three generations per season, usually increasing in numbers with each generation. The complete life cycle takes 30 to 40 days during the summer. *Pediobius foveolatus* is a commercially available biological control agent for MBB control (UC IPM, 2023). Chemical sprays are also available.

Some of the most serious pests of bean seed crops are weevils: cowpea weevil, *Callosobruchus maculatus;* broad bean weevil, *Bruchus rufimanus;* bean weevil, *Acanthoscelides obtectus*. Bean weevils and cowpea beetles are widespread, attacking filling and dried beans (CABI, 2023a). The adults are relatively small, 3.5 to 5 mm (0.13 to 0.2 in.) long, somewhat teardrop or triangular, and dull colored with white, reddish, or black markings (UC IPM, 2023). The eggs may be glued to the bean or the pod (in the case of cowpea weevil), glued to green pods (broad bean weevil), or laid loosely among beans or through cracks in the pods (bean weevil). The larval and pupal stages are spent inside the bean thus destroying them for commercial use. Infestations begin in the field as adults are introduced through discarded beans left in containers, harvesters, planters, etc. Broad bean weevil infestations also start in the field, but this pest does not create a problem in storage. As with the cowpea weevil, bean weevil will attack dried beans, and can be a serious pest consuming stored beans (UC IPM, 2023). Pupation occurs in the beans and adults emerge through a round hole in the seed coat. Damage is a combination of the feeding and contamination. Sanitation offers the most means of control. Because field infestations originate from beans, eliminate potential sources of weevils in production areas (UC IPM, 2023). Pyrethrins provide effective control of weevils.

Thrips (*Thrips tabaci, Heliothrips* spp.) feed on leaves and buds of young bean seedlings making plants look ragged. A sign of serious thrip infestation is distorted leaves that turn brownish around the edges and cup upward. Usually, plants outgrow the damage. Foliage-feeding thrips are effective natural predators of early-season spider mite (*Tetranychus urticae*) infestations of beans. Both adult and immature thrips may be found in spider mite colonies feeding on spider mite eggs. Biological control and unfavorable weather generally reduce thrips populations before treatment is necessary. Minute pirate bugs (*Orius tristicolor*) play a major role in controlling thrips populations. Thrip populations tend to build up on weeds. Cultivating nearby weedy areas before beans emerge will reduce the potential of thrips migrating when the weeds die. Destroying weedy areas after bean emergence will increase thrips problems (UC IPM, 2023).

The bean fly and stem fly (*Ophiomyia phaseoli, O. centrosematis*, respectively) are serious pests of many types of beans in Africa and Asia (CABI, 2023a). The most serious damage by adults occurs to plants in the first-true-leaf stage (UC IPM, 2023). Feeding soon after hatching produces larval mines on undersides of leaves just beneath the epidermis. Larvae in second and third instar mine into the cortex underneath the epidermis downward deeper into the plant and eventually the taproot. They pupate inside the stem close to the soil surface (UC IPM, 2023). Symptoms are feeding tunnels on the stems, yellow leaves, stunting, and plant mortality. There is no totally effective biological control. Carbaryl sprays may provide some control (UC IPM, 2023).

Bean aphids include: cowpea aphid, *Aphis craccivora*; bean aphid, *A. fabae*; pea aphid, *Acyrthosiphon pisum*; and green peach aphid, *Myzus persicae*. Aphids damage plants by:

1. sucking plant sap which causes curled leaves and stunted plants;
2. excreting honeydew causing sticky shiny leaves that turn black because of sooty-mold fungus development; and
3. spreading plant diseases (many viruses are vectored by aphids) (UC IPM, 2023).

Infestations frequently are localized with heavily infested leaves curling downward. Often natural parasites and predators prevent infestations throughout an entire field. Temperatures > 29°C (> 85°F) inhibit buildup of pea and green peach aphids. Some common predators of aphids in beans include lady beetles, syrphid or hover flies, and lacewings. Parasitic wasps attack each of the common aphid species. Insecticide may provide control, but prolonged use may lead to resistance (UC IPM, 2023).

The adult lygus bug (*Lygus hesperus, Chauliops fallax*) is about 6 mm (0.25 in.) long and about half as wide. It is generally brownish but varies from green to straw-colored, or light brown. The body is marked with a pattern of different shades of brown and occasionally yellow or red marks (UC IPM, 2023). A prominent V-shaped yellowish area is near the center of the body at the base of the wings. Lygus eggs are laid within plant tissue and are hard to see because only the cap is visible. Active green nymphs hatch from eggs. Lygus bugs may be present throughout the growing season and can be highly destructive. They have sucking mouthparts that pierce and consume plant tissue. The type of damage varies with plant age. During early bud and flowering stages, lygus bugs cause bud and flower loss, reducing yields. Lygus bug feeding on young, developing seed pods causes distortion, pitting, blemishes, and reduced germination in seed beans (UC IPM, 2023).

Lygus bug eggs are often attacked and killed by a small parasitic wasp, *Anaphesiole*. General predators, such as lacewings and damsel bugs, may prey on lygus bug nymphs as well. Avoiding the use of broad-spectrum insecticides will protect natural enemies. As a preferred host, alfalfa hay should be managed to suppress movement of lygus into dry bean fields by staggering alfalfa cuttings to preserve habitat. Leaving a small, uncut strip of alfalfa will limit the movement of lygus bugs into neighboring bean fields (UC IPM, 2023).

The most common nematode pests of beans are root-knot (*Meloidogyne* spp.) and lesion nematodes, *Pratylenchus* spp. Nematodes management in beans requires combined strategies such as cultivar selection, crop rotation, field sanitation, and fallow periods. Nematode-resistant bean cultivars are available such as: UC-301, White Ventura N. Maria, UC-90, UC-92, and Cariblanco N (UC IPM, 2023).

Peas

There are many insects, mites and nematodes that attack peas including: pea aphid (*Acyrthosiphon pisum*), lucerne bug (*Adelphocoris lineolatus*), black cutworm (*Agrotis ipsilon*), pea weevil (*Bruchus pisorum*), pea leafminer (*Chromatomyia horticola*), green looper caterpillar (*Chrysodeixis eriosoma*), white leafhopper (*Cofana spectra*), devil grasshopper (*Diabolocatantops axillaris*), flower thrips (*Frankliniella intonsa*), western flower thrips (*Frankliniella occidentalis*), red-legged earth mite (*Halotydeus destructor*), brown stink bug (*Halyomorpha halys*), spiral nematode (*Helicotylenchus dihystera*), cotton bollworm (*Helicoverpa armigera*), native budworm (*Helicoverpa punctigera*), pea cyst eelworm (*Heterodera goettingiana*), pea blue butterfly (*Lampides boeticus*), serpentine leafminer (*Liriomyza huidobrensis*), vegetable leafminer (*Liriomyza sativae*), American serpentine leafminer (*Liriomyza trifolii*), cabbage moth (*Mamestra brassicae*), bean flower thrips (*Megalurothrips usitatus*), peanut root-knot nematode (*Meloidogyne arenaria*), false root-knot nematode (*Nacobbus aberrans*), stemfly (*Ophiomyia centrosematis*), scarlet mealybug (*Pseudococcus calceolariae*), crown gall (*Rhizobium radiobacter*), bean bug (*Riptortus clavatus*), pea leaf weevil (*Sitona lineatus*), Costa Rican armyworm (*Spodoptera albula*), beet armyworm (*Spodoptera exigua*), cotton leafworm (*Spodoptera littoralis*), field thrips (*Thrips angusticeps*), honeysuckle thrips (*Thrips flavus*), and red flour beetle (*Tribolium castaneum*) (CABI, 2023b).

The pea weevil larvae can live and overwinter in pea seeds. This can be problematic for seed that is to be exported to other counties. Control practices are critical during the growing season and requirements to fumigate seeds are necessary in certain countries before export.

References

CABI (2023a) *Phaseolus vulgaris* (common bean). In: *Crop Protection Compendium*. CAB International, Wallingford, UK. https://www.cabidigitallibrary.org/doi/10.1079/cabicompendium.40626 (accessed 14 September 2023).

CABI (2023b) *Pisum sativum* (pea). In: *Crop Protection Compendium*. CAB International, Wallingford, UK. https://www.cabidigitallibrary.org/doi/10.1079/cabicompendium.41042 (accessed 14 September 2023).

Davis, J.G. and Brick, M.A. (2013) *Fertilizing Dry Beans*. Publication 0.539. Colorado State University Extension, Colorado. https://extension.colostate.edu/topic-areas/agriculture/fertilizing-dry-beans-0-539/ (accessed 14 September 2023).

Delahaut, K.A. and Newenhouse, A.C. (1997) *Growing Beans and Peas in Wisconsin: A Guide for Fresh-market Growers*. Publication A3685. University of Wisconsin Extension, Wisconsin. https://barron.extension.wisc.edu/files/2023/02/Growing-Beans-and-Peas-in-Wisconsin.pdf (accessed 14 September 2023).

Dickson, M.H. (1971) Breeding beans, *Phaseolus vulgaris* L., for improved germination under unfavorable low temperature conditions. *Crop Science* 11, 848–850.

Erker, B. and Brick, M.A. (2006) *Legume Seed Inoculants*. Publication 0.305. Colorado State University Extension, Colorado. https://extension.colostate.edu/topic-areas/agriculture/legume-seed-inoculants-0-305/ (accessed 14 September 2023).

Fofana, B., Vekemans, X., Du Jardin, P. and Baudoin, J.P. (1997) Genetic diversity in Lima bean (*Phaseolus lunatus* L.) as revealed by RAPD markers. *Euphytica* 95, 157–165.

Hochmuth, G.J. and Hanlon, E.A. (2010) *A summary of N, P, and K research with snap bean in Florida*. Publication #SL 313. University of Florida, IFAS Extension, Florida. edis.ifas.ufl.edu/publication/CV234 (accessed 14 September 2023).

Hofer, J., Turner, L., Moreau, C., Ambrose, M., Isaac, P., *et al.* (2009) Tendril-less regulates tendril formation in pea leaves. *Plant Cell* 21, 420–428.

Jones, K.M., Kobayashi, H., Davies, B.W., Taga, M.E. and Walker, G.C. (2007) How rhizobial symbionts invade plants: the *Sinorhizobium–Medicago* model. *Nature Reviews Microbiology* 5, 619–633. https://pubmed.ncbi.nlm.nih.gov/17632573/ (accessed 14 September 2023).

Kelly, J.D. (2010). *The Story of Bean Breeding*. Michigan State University, Michigan. www.canr.msu.edu/beanbreeding/_pdf/Story_of_Bean_Breeding_in_the_US.pdf (accessed 14 September 2023).

Li, X., Welbaum, G.E., Rideout, S.L., Singer, W. and Zhang, B. (2022) Vegetable soybean and its seedling emergence in the United States. In: Jimenez-Lopez, J.C. and Clemente, A. (eds) *Legumes Research: Volume 1*. IntechOpen. https://doi.org/10.5772/intechopen.102622 (accessed 14 September 2023).

Lorenz, O.A. and Maynard, D.N. (1988) *Knott's Handbook for Vegetable Growers*, 3rd edn. Wiley, Hoboken, New Jersey.

Makasheva, R.K. (1983) *The Pea*. Oxonian Press, New Delhi, India.

Meronuck, R.A., Hardman, L.L. and Lamey, H.A. (1993) *Edible Bean Disease and Disorder Identification*. University of Minnesota Agricultural Extension Service. https://hdl.handle.net/11299/205408 (accessed 14 September 2023).

Mohamed-Yasseen, Y., Barringer, S.A., Splittstoesser, W.E. and Costanza, S. (1994) The role of seed coats in seed viability. *The Botanical Review* 60, 426–439.

OSU (2010a) *Beans, Lima: Phaseolus lunatus*. College of Agricultural Sciences, Oregon State University. https://horticulture.oregonstate.edu/oregon-vegetables/beans-lima (accessed 14 September 2023).

OSU (2010b) Beans, Snap – Green, Romano, Yellow Wax. Oregon State University. https://horticulture.oregonstate.edu/oregon-vegetables/beans-snap-green-romano-yellow-wax (accessed 14 September 2023).

Stevens, P.F. (2023) Angiosperm phylogeny website. www.mobot.org/MOBOT/research/APweb/ (accessed 2 July 2013).

Rubatzky, V.E. and Yamaguchi, M. (1997) *World Vegetables: Principles, Production and Nutritive Values*, 2nd edn. Chapman & Hall, New York.

UC IPM (2023) Dry beans. Pest Management Guidelines. University of California, Integrated Pest Management. www.ipm.ucdavis.edu/PMG/selectnewpest.beans.html (accessed 14 September 2023).

Wang, F.B., Fu, J.F., Dong, L.F. and Zhu, Y.B. (2003) Tendril inheritance in semi-leafless pea and its utilization in breeding. *Hereditas* 25, 185–188.

Wilson, D.O. and McDonald, M.B. (1992) Mechanical damage in bean (*Phaseolus vulgaris* L.) seed in mechanized and non-mechanized threshing systems. *Seed Science and Technology* 20, 571–582.

14

Family Asparagaceae

Overview

Asparagus (*Asparagus officinalis*) has been a popular vegetable for centuries in many parts of the world (Rubatzky and Yamaguchi, 1997). Commercial production of asparagus as a food crop and for seed occurs in many locations including Europe, Asia, Australia, the Americas, and New Zealand. China is by far the largest producer of asparagus for food. Other major producers of edible asparagus in descending order are Peru, Mexico, Germany, Spain, Italy, USA, Japan, France, and Iran (Shahbandeh, 2023). Asparagus produces spears eaten when immature and tender after cooking. They can be preserved by pickling, canning, or freezing. Later in development after edible maturity, spears elongate, tips open, and produce stalks that are fibrous. At full maturity, spears develop into photosynthetic stem issues that produce flowers, fruits, and seeds.

Taxonomy

Asparagus was previously classified as a member of the families Liliaceae or Amaryllidaceae in older literature. Many plant scientists now place asparagus in the family Asparagaceae. Current taxonomy classifies Amaryllidaceae and Asparagaceae as separate families in the order Asparagales. *Asparagus officinalis* is the most popular of several edible species in the genus. Ornamental types are popular as house plants in some areas. *A. officinalis* is sometimes confused with *Ornithogalum pyrenaicum*, which produces edible shoots and is called "wild asparagus" in parts of Europe (BSBI Online Plant Atlas, 2020). There is no clear consensus on the native range of *A. officinalis*, but it generally includes most of Europe and western temperate Asia. Asparagus is believed to have originated in maritime saline habitats in this region.

Botany

Asparagus is a perennial, dioecious monocot that produces new above ground stems, commonly called "fern", each year from a fleshy underground "crown" (Figs 14.1, 14.2). The "crown" is a mass of fasciculate adventitious roots clustered with buds at the apex that give rise to above ground spears. The crown consists of unelongated basal internodes of old stems that increase in size with development of buds both apically and laterally, resulting in branching (Blasberg, 1932). When spears are not harvested as food and are allowed to grow, asparagus plants produce significant biomass 100–150 cm (40–60 in.) tall, with stout branching stems producing feathery foliage. The plant doesn't

DOI: 10.1079/9781789243260.0014

produce true leaves. The photosynthetic organs are needle-like cladodes (modified stems) in the axils of scale leaves. Cladodes are 6–32 mm (0.25–1.25 in) long and 1 mm (1/32 in.) broad, in clusters of four to 15, in a rose-like shape (Rubatzky and Yamaguchi, 1997).

Fig. 14.1. Asparagus "crown" with storage roots and old stems and clustered buds at the top.

The flowers are campanulate, greenish-white to yellowish, 4.5–6.5 mm (3/16–1/4 in.) long, with six tepals partially fused together at the base to form a bell-shape (Rubatzky and Yamaguchi, 1997). Flowers are produced singly or in clusters of two or three at junctions of the branchlets. Sex expression is usually dioecious, with separate male and female plants. Occasionally hermaphrodite flowers are found. Flowers on staminate plants have six stamens and functional anthers with a rudimentary stigma and ovary. Pistillate flowers have a functional three-lobed pistil, stigma, a three-carpel ovary, and rudimentary anthers. Flowers have nectaries and are insect pollinated. The fruit is a berry initially green becoming orange-red at maturity. Fruit size is 6–10 mm (1/4–13/32 in) in diameter (Fig. 14.3). The fruit have a tough skin that is difficult to rupture, contain sapogenins, and are inedible to humans. Fruits contain black seeds, two per carpel, that are round with a flat side (Rubatzky and Yamaguchi, 1997).

Types and Cultivars

Open-pollinated, F1, and F2 hybrid cultivars are available. Most modern cultivars are all male hybrids. Cultivar spear colors may be all green, green with purple tips, all purple, or white.

Fig. 14.2. Asparagus "fern" female plants have orange-red berries and male plants do not.

Fig. 14.3. Ripe asparagus berries attached to a female plant.

White asparagus is not a special cultivar, but rather spears grown in the dark under mounds of soil and black plastic film to exclude light to prevent chlorophyll development. UC 157 is a popular clonal hybrid cultivar producing both male and female plants developed at the University of California Riverside, USA in the late 1970s. Many seed companies sell second-generation F2 seeds of the original UC 157 F1 cultivar. Violetto D'Albenga and cultivars derived from it such as Purple Passion and Pacific Purple are open-pollinated and produce tetraploid, purple spears (Falloon and Andersen, 1999). Tetraploid cultivars tend to be less uniform when grown from seed because of their greater genetic complexity. Purple asparagus turns green after cooking because the water-soluble pigment anthocyanin leaches from the tissue. Some predominantly green cultivars have scale leaves and spear tips tinged with purple anthocyanin. This coloration is an important phenotype that helps identify certain cultivars.

Growth and Development

Buds on each crown can produce an upright stem, fern, and flowers. Growing buds/spears inhibit adjacent sprouting of new buds on the crown (Tiedjens, 1926). Each bud develops in the axil of the preceding one. Prior to elongation, buds are miniature spears. Most of the differentiation occurs prior to expansive growth, so spear emergence is largely due to cell elongation driven by water uptake.

Mature asparagus plants tolerate a wide range of environments, but newly emerging "spears" are damaged by frost. Seed crops are produced in climates where crowns remain quiescent until after the threat of subfreezing temperatures pass, so the first spears are not damaged or killed. Spears severely damaged by freezing cannot produce fern or flowers. Unproductive damaged spears waste storage reserves from the crown, potentially decreasing seed yield through reduced top growth. Uniform fern development improves the synchrony of seed maturation which is important for maximizing yield and seed quality of destructively harvested asparagus seed crops. Asparagus is considered a cool-season crop for food production since lower temperatures reduce growth rates and fiber development resulting in more succulent spears. Higher temperatures accelerate growth but do not inhibit asparagus development, so seed production is not affected by warmer climates that would be too high for acceptable edible spear quality.

New roots develop from the bases of actively growing buds and are very thick and fleshy, serving as the primary food storage organs of the crown. The large roots have a heavily suberized epidermis, but only a few root hairs are present. The bulk of the root mass is loosely fitting cortex, which through cell division considerably increases in diameter. This increase eventually causes crowding within crowns and between adjacent plants. This increase in crown size leads to decreased spear productivity over time necessitating new plantings. Planting of asparagus seed fields varies widely but in the absence of severe disease problems, a field should remain productive at least eight to ten years until crowding becomes an issue.

Soil

Drainage and soil fertility are major factors when selecting fields for asparagus seed production. Soils should be deep, well drained, and friable to allow development of an expansive root system. The field should not have been planted in asparagus for at least four years previously, especially if there is a history of Fusarium wilt or other soil diseases. Loams and sandy loams with organic matter and a pH between 6.5 to 7.0 are well suited to asparagus. Asparagus does not grow as well in acidic soils but tolerates salt better than other vegetables (Shannon and Grieve, 1999).

Establishment

Plantings are established from crowns, transplants, or direct seeding. Because of slow crown development from seed, commercial seed fields are often established from crowns or transplants and rarely direct-seeded. For production of crowns, seeds are sown usually in sterilized or fumigated seed beds with a linear spacing about 30 cm (1 ft) apart. The crowns are dug from the bed and transplanted to seed production fields when they are two years old although one-year-old crowns are sometimes used. "Crowns" are transplanted when quiescent in fall or winter, and the first shoots appear the following spring.

In shorter season areas, two-year-old crowns are established in furrows, so their tops are at least 15 to 25 cm (6 to 10 in.) below the surface. The bottom of the furrow should have sufficient loose soil to support deep root development. The furrow is gradually filled with soil as crowns develop. In longer season areas, seed crops are established using ten- to 15-week-old transplants. Transplants are grown in greenhouses using plug trays and transplanted in early spring. Transplants are planted into a furrow similar to crowns. The furrow is gradually filled in as the season progresses. Male and female asparagus plants are planted in separate rows for ease of seed harvest and plant inspection. Ratios of male to female rows vary between 1:1 and 1:3 depending on the cultivar and pollen production by the male line. If "supermales" (YY) are used as the pollen parent, all seeds harvested from the female rows will produce male hybrid plants regardless of the genetics of the female (see Genetics below). Female parents are selected based on their superior combining ability and other beneficial traits such as disease resistance and seed production.

Pollination

Since asparagus is dioecious, bees must transfer pollen between male and female plants. Honeybees are usually the primary pollinator and are attracted to nectaries at the base of the corolla in both male and female flowers. Growers should provide one to two colonies per acre to ensure adequate pollination of seed fields. Most pollination occurs in early mornings when pollen is first available and fully hydrated. Pollination declines in the afternoon as pollen begins to dry. There should be at least one male plant within 1.5 m (5 ft) of each female, at least one male for each six female plants (Huyskes, 1959). Asparagus hybrid seed fields should be isolated by 1.6 km (1 mile) from other asparagus including wild or escaped male plants. Asparagus seeds are sometimes dispersed by animals, especially birds, who consume berries and spread seeds that become weeds. Escaped or wild male plants may contribute unwanted pollen that contaminates asparagus seed fields.

Irrigation

Asparagus has a deep extensive root system. In areas with adequate rainfall, fields are not irrigated except during establishment or extreme drought. In areas with seasonal rainfall, the crop is irrigated periodically during the dry season to maintain growth and fruit development. Asparagus is more tolerant of saline irrigation water compared to other vegetables. Concentrations that reduce asparagus seed germination percentage by 50% (G_{50}) ranged from 6.3 to 14.3 dS/ml on filter papers saturated with NaCl + $CaCl_2$ (1:1 by weight) solution (Shannon and Grieve, 1999).

Rogueing

Since asparagus is a perennial, it is important to rogue fields aggressively before the first year of seed production. Critical times for rogueing include fern formation, flowering, and fruit set. When off-types and diseased plants are identified, entire plants including crowns should be removed to prevent regrowth. After initial seed harvest, rogueing should be conducted once each year at flowering to remove volunteer (escapes) and diseased plants.

Genetics

Sex expression is controlled by a single sex gene with two alleles (Y and X) and maleness (Y) is dominant (Sneep, 1953). Male plants sometimes bear bisexual flowers in addition to the male ones. Besides the stamens these bisexual flowers possess a functioning ovary with style and stigmas (andromonoecious plants). The rate of andromonoecious plants found in commercial fields fluctuates from 0 to 2% of male plants (Sneep, 1953). Perfect flowers allow selfing, which can be represented genetically by XY × XY. The same genotypes apply to cross-fertilization between andromonoecious and male plants, producing progenies of 1 YY, 2 XY, and 1 XX plants. Since the YY and XY plants are male, three quarters of the offspring is male and only one quarter is female. Males with YY alleles crossed with a normal female plant (XX × YY) produce XY seed that become all male F1 hybrid plants. Plants homozygous for maleness (YY) are sometimes called supermales. All male F1 hybrid asparagus cultivars have become popular and are the commercial vegetable industry standard because they produce more spears and often maintain productivity longer. Since male plants produce no seed, they store more carbohydrates in their crowns that are available for spear growth which is the reason for their greater productivity. Supermales are made by selfing male plants or making double haploids through anther culture. A doubled haploid plant can be regenerated from tissue culture. Elite parental plants can be cloned by tissue culture for breeding and seed production.

Production and Harvest

Harvest of a new seed production field usually begins the second year after establishment. Green berries develop bright orange-red color when ripe approximately 90 to 110 days post-anthesis. The small fruit adhere tightly to plants, so hand-harvest is not practical. The drying male fern rows are often removed first to eliminate the small number of seeds from andromonoecious or volunteer plants that were not removed during rogueing and to provide more space to harvest female rows.

Asparagus is a wet-seeded crop and fruits have a high moisture content at maturity. Fruit do not dehisce, so shattering is usually not a problem. The initial step of seed harvest is to cut the female plants and move them to a thresher that separates berries from plant biomass. Harvest usually occurs when fern begins to turn brown (Fig. 14.4). This ensures that seed development is complete and most seeds are fully developed. Female plants are hand-fed into the thresher that separates the fruit (Fig. 14.4). Isolated fruit are collected in bins as they exit the thresher (Fig. 14.5).

Water is often added to the bin contents. This allows fruit to be scooped out with a bucket or flow into the crushing machine depending on the feeding mechanism used. Added water also softens the hard berries, which dry out as the fern senesces. Maceration breaks the tough skins of asparagus fruit so a properly adjusted mechanical crusher will not damage seeds and is essential for efficient seed extraction (Figs 14.5, 14.6). The crushed fruit and water create a slurry of separated seeds and fruit pieces (Fig. 14.7).

Fig. 14.4. Asparagus fern cut and ready for berry harvest by threshing.

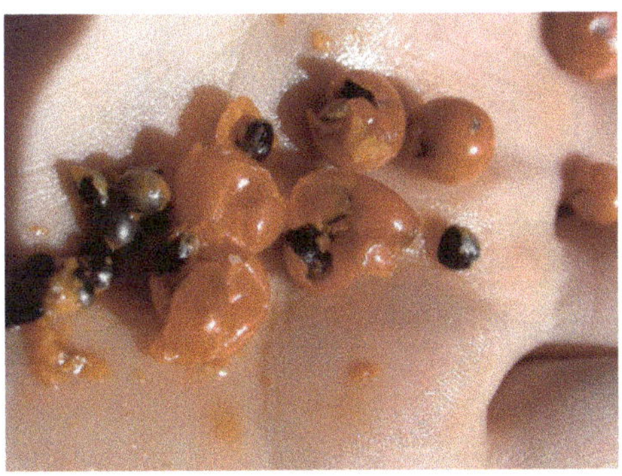

Fig. 14.5. Orange-red berries have tough skins and contain black seeds with a flat side.

Seeds can be separated from fruit debris by pouring the slurry in a container of water. The lighter fruit tissue floats to the top and can be easily separated from the seeds which sink to the bottom. Prolonged hydration may damage seeds so the separation should be done soon after the fruit are crushed. Alternatively, the slurry of fruit pieces and seeds can be washed through a plenum for fractionation so the lighter fruit debris flows to the end and the heavier seeds are collected earlier when they settle to the bottom. After separation, the predominate seed fraction is quickly washed to remove any adhering fruit tissue that may remain (Fig. 14.8). Plant debris after the separation and washing steps will be removed during seed conditioning after drying. Washed cleaned seeds are dried on forced-air screen trays or a revolving heated drier that allows dry air to circulate around seeds (Fig. 14.9). Drying is an important step for maintaining seed quality. Excessively rapid drying of seeds at high temperatures may damage seeds reducing both viability and vigor. Hydrated seeds can germinate precociously if they remain hydrated for too long especially at higher temperatures. It is important to not to over soak seeds and dry them as soon as they are clean.

Conditioning removes remaining plant debris from dry seeds. A spiral separator can effectively clean asparagus seed because their relatively round shape rolls faster to the outside chute at the bottom. In contrast the plant debris does not roll and moves slower, exiting the right or center chutes at the bottom where it is collected and discarded. Air screen machines and density separation tables also effectively clean dry asparagus seed. The dry, cleaned seeds should be stored under cold, low moisture conditions until final post-harvest treatments are applied and seeds are packaged for sale.

Fig. 14.6. A bin of harvested berries before mechanical maceration to release seeds.

Fig. 14.7. Slurry of berries, juice, water, and seeds after fruit maceration.

Declines in seed yield may be a sign of plant disease, nutrient deficiency, or crowding. There is no primary or secondary seed dormancy reported in cultivated asparagus. Washing seeds after harvest may improve germination (Rubatzky and Yamaguchi, 1997).

Seed Yield

Seed yields vary depending on both genetic and agronomic factors. A good seed yield would range from 1121 kg/ha (1000 lb/acre) to 1569 kg/ha (1400 lb/acre). One asparagus fruit may contain as many as six seeds. About 40 seeds weigh 1 g. There are approximately 1200 seeds in 1 oz.

Pest Management

Pathogens

Crop rotation should always be practiced. Fields that have grown asparagus in the past five years

Fig. 14.8. Seeds after fruit debris has been removed in a plenum and brief washing.

Fig. 14.9. Forced-air drying of seeds on screen trays to prevent damage from prolonged hydration.

should be avoided. If crop rotation is not possible soil fumigation or solarization may eradicate some diseases. Fusarium (*Fusarium oxysporum*) is a major disease of asparagus and the most prevalent species (Brizuela *et al.*, 2020). Some cultivars are more resistant than others. Asparagus decline syndrome (ADS) can be a major problem for asparagus seed production (Brizuela *et al.*, 2020). Eight *Fusarium* species belonging to seven species complexes have been identified in soils that cause ADS: *F. oxysporum*, *F. proliferatum*, *F. redolens*, *F. solani* sensu stricto, *F. equiseti*, *F. culmorum*, *F. compactum*, and *F. acuminatum*. *F. proliferatum*, *F. oxysporum*, and *F. redolens* showed pathogenicity to asparagus and were the main species associated with ADS (Brizuela *et al.*, 2020). *F. moniliforme* and *F. proliferatum* have been linked with internal and external root and crown rots of asparagus (Lori *et al.*, 1998). Crowns or transplants should be certified *Fusarium* free before planting.

Rust (*Puccinia asparagi*) is another major disease of asparagus. Rust causes dark brownish-red pustules (uredinia) on the foliage. These lesions release rust-colored spores throughout the summer. Rust can overwinter on plant debris. Infected foliage should be burned or buried to prevent reinfection in subsequent years. Phytophthora spear rot is characterized by soft, water-soaked lesions on shoots near the soil line. The lesions elongate rapidly and become light brown. Botrytis blight (*Botrytis cinerea*), also called gray mold, is a fungus that affects asparagus during cool rainy weather when temperatures are near 15°C (60°F). Several viruses affect asparagus and are widely distributed: asparagus virus I (AVI) (aphid-transmitted), asparagus virus II (AVII) (pollen- and seed-transmitted), and tobacco streak virus (an ilarvirus).

Insects

Asparagus beetle (*Crioceris asparagi*) is an important widespread pest of asparagus and the only plant it attacks. The beetle is 6.0 to 9.5 mm (0.25 to 0.4 in.) long and elongated. It is a distinctive metallic blue-black color with cream or yellow spots on its red-bordered elytra. Other insect pests include asparagus aphid (*Brachycorynella asparagi*) and the asparagus miner (*Ophiomyia simplex* Loew). The asparagus miner lives exclusively on asparagus plants. This insect is shiny black and is common in most major asparagus-producing regions of the world.

References

Blasberg, C.H. (1932) Phases of the anatomy of *Asparagus officinalis*. *Botanical Gazette* 94, 206–214.

Brizuela, A.M., De la Lastra, E., Marín-Guirao, J.I., Gálvez, L., de Cara-García, M., *et al.* (2020) *Fusarium* consortium populations associated with asparagus crop in Spain and their role on field decline syndrome. *Journal of Fungi* 6, 336. doi.org/10.3390/jof6040336 (accessed 14 September 2023).

BSBI Online Plant Atlas (2020) *Ornithogalum pyrenaicum* L. In: Stroh, P.A., Humphrey, T.A., Burkmar, R.J., Pescott, O.L., Roy, D.B. and Walker, K.J. (eds) *Plant Atlas 2020*, plantatlas2020.org/atlas/2cd4p9h.fv9 (accessed 22 September 2023).

Falloon, P.G. and Andersen, A.M. (1999) Breeding purple asparagus from tetraploid "Violetto D'Albenga". *Acta Horticulturae* 479, 109–114. doi.org/10.17660/ActaHortic.1999.479.13 (accessed 14 September 2023).

Huyskes, J.A. (1959) The value of comparative test progenies from open-pollinated female asparagus plants. *Euphytica* 8, 141–144.

Lori, G., Wolcan, S. and Mónaco, C. (1998) *Fusarium moniliforme* and *F. proliferatum* isolated from crown and root rot of asparagus and their association with asparagus decline in Argentina. *Plant Disease* 82, 1405. doi.org/10.1094/PDIS.1998.82.12.1405A (accessed 14 September 2023).

Rubatzky, V.E. and Yamaguchi, M. (1997) *World Vegetables, Principles, Production and Nutritive Values*. Chapman & Hall, New York, pp. 645–657.

Shahbandeh, M. (2023) Major asparagus producing countries worldwide 2021. Statistica. statista.com/statistics/279556/global-top-asparagus-producing-countries/ (accessed 14 September 2023).

Shannon, M.C. and Grieve, C.M. (1999) Tolerance of vegetable crops to salinity: a review. *Scientia Horticulturae* 78, 5–38.

Sneep, J. (1953) The significance of andromonoecy for the breeding of *Asparagus officinalis* L. *Euphytica* 2, 89–95.

Tiedjens, V.A. (1926) Some observations on root and crown bud formation in *Asparagus officinalis*. *Proceedings of the American Society for Horticultural Science* 23, 189–196.

15

Family Malvaceae

Overview

Okra is believed to have originated in Abyssinia or what is presently Ethiopia and the upper Nile River Region of Sudan (National Research Council, 2006). By the 12th century it was widely grown throughout Egypt. Over time, okra moved from North Africa to the eastern Mediterranean, India, and was introduced into North America via French colonists and/or from Brazil. The word okra first appeared on 1679 in the colony of Virginia. Today, India and Niger are the world's largest producers of okra, and the crop is widely grown across Africa, the Middle East, the Americas, Southern Asia, and China (FAO, 2023).

Taxonomy

Okra is in the family Malvaceae, also known as the mallow family, that consists of 244 genera with 4225 known species of flowering plants (Christenhusz and Byng, 2016). The most important vegetable crop in this family is okra, also known as gumbo, bhendi, or lady's finger. The Latin binomial for okra is *Abelmoschus esculentus* (L.) Moench and in older literature it is sometimes called *Hibiscus esculentus* (L.). Other economically important plants in the Malvaceae include cotton (*Gossypium* spp.), common mallow (*Malva sylvestris*), hollyhock (*Alcea rosea*), hibiscus (*Hibiscus syriacus*), cocoa tree (*Theobroma cacao*), and durian (*Durio zibethinus* L.).

Botany

Okra is a perennial in the tropics but is grown as an annual in temperate regions because it is killed by frost (George, 2009). Okra grows best in full sun with day temperatures from 27°C to 35°C (80°F to 95°F) and warm nights > 15.5°C (60°F). Okra is a fast-growing, erect dicot with many short branches. Leaves are large from 10–20 cm (3.9–7.9 in.) long and wide, have palmate venation, and five to seven lobes (Fig. 15.1).

The large perfect and self-fertile flowers are reminiscent of other members of Malvaceae such as cotton and mallow. Attractive flowers have five pale yellow, white, or pink petals (Fig. 15.1). The showy flowers can be 6 to 8 cm (2.4 to 3.1 in.) wide when fully open. Pistils and stamens are erect together in the center of each flower. The pistil is composed of multiple styles clustered beneath a prominent capitate stigma. A staminal tube with numerous short filaments and anthers form a sleeve surrounding the styles. (Purewal and Randhawa, 1947). The anthers dehisce 15 to 20 min after flowers open early in the morning. Some of the released pollen may self-pollinate the stigma, but flowers are mostly cross-pollinated by honeybees and bumble bees. Bees collect nectar from the base of the flowers

DOI: 10.1079/9781789243260.0015

Fig. 15.1. Okra plant architecture with buds and opening flower.

Fig. 15.2. Unopen flower bud (top right), fruit one day (middle) and three days (lowest) after anthesis. Aborted bud turns brownish yellow (top left) and unpollinated flowers rapidly senesce and drop within two days.

and in the process move pollen from the anthers to stigmas of different flowers. Okra pollen grains have many tiny pores and spines, are large, and relatively heavy (Purewal and Randhawa, 1947).

Each fruit is a capsule up to 25 cm (10 in.) long, pentagonal in cross-section, with from three to five fused carpels containing eight to 16 ovules distributed laterally in each. Okra is grown as an edible vegetable for its immature pod and seeds harvested within the first week after anthesis (Fig. 15.2). Pods rapidly become fibrous, woody, and inedible the first week after pollination. However, fruits and seeds fully develop to physiological maturity for seed production. The dried seeds may be used as a coffee additive or substitute. Dry okra seeds contain 18–20% oil and 20–23% crude protein.

Types and Cultivars

Standard cultivars are taller than 2 m (6.6 ft), while dwarf cultivars are less than 1 m (3.3 ft) tall. Cultivar external pod colors range from white through pale to dark green. There are red cultivars that may also be grown as ornamentals. Many commercial cultivars have ribbed fruit although some cultivars produce rounded fruit without ribs. Fruits are preserved by canning, dehydration, or freezing.

Many okra plants have stiff trichomes that irritate human skin on contact. Spineless cultivars have no or few trichomes on their pods and petioles. In West Africa, leaves, buds, and flowers of okra are also consumed as food. Dried stems are used as a source of paper pulp or fuel.

Growth and Development

Okra responds to constant warm weather and does not grow well if night temperatures are consistently below 15.5°C (60°F) (Rubatzky and Yamaguchi, 1997). Okra produces hydrocolloid mucilage that is an important characteristic of the vegetable (Gemede *et al.*, 2018). Compared to other vegetables, okra typically has a higher mucilage content that contains significant levels of protein, carbohydrate, neutral sugars, minerals, and other complex polysaccharides. The mucilage in the pod dries at crop maturity and doesn't hamper seed harvest or post-harvest processing. It is important to keep okra seeds dry during and after harvest to prevent mucilage rehydration that can interfere with processing.

Okra fruit develop acropetally so older fruit are at the bottom of the plant and newer fruit at the top. Sometimes older pods shatter before younger pods at the top of the plant fully mature. Topping plants after flowers at ten or 15 nodes have set, either mechanically or chemically, leads to more uniform seed development but may lower yields and increase labor. If pods are not harvested after fruit have set at several nodes, new flowers will fail to set due to the heavy fruit. So maturing fruit tend to inhibit new fruit development at a certain point in development.

Soil and Fertilization

Okra can be grown on a wide range of soil types ranging from loams, silt loams, or sandy loams with organic matter if they are well drained. Heavier soils are suitable if they are not waterlogged. Raised beds improve drainage on heavier soils. Okra is sensitive to soil acidity, so pHs of 6.0 to 7.0 are best. Fertilization rates vary dramatically with soil type and should be based on soil test results. Okra is considered a heavy feeder and will remove approximately 202 kg/ha N, 43 kg/ha P, and 188 kg/ha K (180 lbs/acre N, 38 lbs/acre P, and 168 lbs/acre K) per season depending on the soil type and cultivar. Okra generally responds well to fertilization when soil fertility is low. Split applications are common with row banding of a fraction of N and all P and K at planting, followed by side dressings of N two weeks before the start of flowering and again after approximately half of the fruits have set.

Establishment

Sowing rates vary from 5 to 8 kg/ha (4.5 to 7 lbs/acre) depending on whether a cultivar is dwarf or standard size. Okra is generally direct-seeded when soil temperatures exceed 16°C (60°F). Poor stand establishment can be a problem when soil temperatures are too cold for germination and early-season growth, or seed quality is poor. Commercial okra seed crops are often produced in warm deserts. For open-pollinated cultivar production, seeds are precision-planted in blocks with rows 0.8 to 1.2 m (2.6 to 4.0 ft) apart with in-row spacing of 0.3 to 0.5 m (1 to 1.6 ft) for dwarf cultivars and 0.6 to 0.8 m (2.0 to 2.6 ft) for larger cultivars. Hard seed coats may cause some cultivars to germinate poorly. Soaking seeds in hot water for one or two hours prior to planting may improve germination.

For F1 hybrid seed production, planting dates may be staggered to improve nicking between inbred lines. Transplanting can also be used to improve nicking. If seed quality is poor or the amount limited, transplanting is an option for producing breeder, foundation, and stock seed but is not a standard practice for commercial production. For F1 hybrid seed with male sterility incorporated into the female parent, rows are planted at a ratio of one male to two or three female rows, depending on the amount of pollen production by the male line.

Pollination

Plants are day-neutral, and flowering is developmentally regulated. The time it takes for okra to flower varies with both the cultivar and environment. First flowering for most cultivars ranges from 50 to 75 days after seeding under ideal conditions. Flowers generally appear at alternating leaf axils after the sixth to eighth node, but this varies with cultivar. Flowers are open only for one day shortly after sunrise until midday. The petals wilt in the afternoon and usually senesce and drop the following day. Seldom is more than one flower open on a given plant at one time.

Bee activity results in a high percentage of out-crossing because movement of the relatively heavy pollen is limited without insect assistance. Even pollen distribution across the stigma surface is required for full pollination. Poor pollination causes unfertilized ovules, poor carpel development, misshapen fruit, and fruit abortion.

Open-pollinated cultivars of okra are grown in much of the world. For open-pollinated production, plants of a specific cultivar are grown together and allowed to naturally intercross with no control of pollination. Okra flowers are attractive to bees, so the general recommendation of 5 hives/ha (2 hives/acre) applies for okra production. Recommended isolation distances among different cultivars vary with location, but a minimum isolation distance of 0.5 km (0.3 miles) is often used (George, 2009).

Irrigation

Okra has good drought tolerance, but yield is adversely affected when plants are water stressed during flowering and fruit development. Many okra seed crops are grown in warm desert areas, making irrigation a necessity. Flowering okra should receive approximately 4 cm (1.6 in.) of water per week to ensure fruit set and good seed development. At other times of development and when evapotranspiration is low, 2.5 cm (1 in.) per week may be sufficient on heavier soils depending on the environment, stage of growth, and cultivar. Sprinkler irrigation is popular to activate preplant fertilizer applications and promote uniform emergence. Drip irrigation is preferable to overhead sprinkler because the foliage is kept dry, reducing fungal diseases. In many areas more efficient drip irrigation has replaced furrow. Furrow delivery is inexpensive but uses larger amounts of water and can only be used on level ground.

Rogueing

Rogueing is important to remove off-type and diseased plants. Before flowering, plants should be checked to ensure they are true-to-type. This includes checking plant height, growth habit, pigmentation of leaves, and petioles' and stems' length and diameter (George, 2009). All off-type plants and those with disease symptoms should be removed. A second rogueing at flowering should check relative plant size, color, and intensity of flower pigmentation. Any off-type or diseased plants should be removed. The final rogueing is performed during fruiting to check trueness-to-type. Plants producing off-type fruit should be removed along with diseased plants, especially those showing virus symptoms (George, 2009).

Production and Harvest

After anthesis, it takes okra fruit an additional 30 to 50 days to reach full dry seed maturity for harvest (Santos *et al.*, 2019). The time from planting to once-over, destructive mechanical harvest of an okra seed field for an open-pollinated crop would vary widely from 130 to 160 days under optimum growing conditions (Fig. 15.3). The exact time will depend on the cultivar and environmental conditions.

Fig. 15.3. Open-pollinated okra seed field nearing harvest maturity.

Although open-pollinated cultivars are still widely grown and popular because farmers can save their own seeds, F1 hybrid cultivars are increasingly grown. Hybrids are popular in many areas because of their greater uniformity, vigor, and yields. In some areas, crossing flowers of inbred lines for F1 seed production is done by hand pollination after flowers are emasculated. For emasculation, buds likely to open the following day are ring cut at the base of the flower bud with a sharp blade. Cut petals and calyx sheath are removed so the staminal column and stigma are exposed. The undehisced anthers are removed with forceps. The emasculated buds are immediately hand cross-pollinated with male flowers collected and stored in bags under cool conditions from the previous day. The hand-pollinated flowers are marked with a thread, tag, or other marker for identification (Fig. 15.4). All other flowers on the female parent plants are removed to encourage development of the hand-crossed fruits and to prevent non-hybrid self-pollinated sibs from developing. Production of F1 okra is increasingly conducted in screen houses to exclude unwanted insects that transmit disease and prevent bees from pollinating maternal flowers that were not emasculated (Fig. 15.5). When hand-crosses are made in open fields with high insect populations, temporarily bagging hand-pollinated emasculated flowers will exclude unwanted foreign pollen until after the fruit has set. After fruit set bags are removed. Male rows are destroyed after pollination has concluded and well before seed harvest from female rows. Only marked fruit resulting from hand-crosses are harvested for seed.

Fig. 15.4. Peduncles of hand-crossed developing fruit marked with string on the female parent. All other fruit that did not result from crossing are removed.

Emasculation and hand pollination is time consuming and labor intensive. Genic-cytoplasmic male sterility has been developed in okra to control crossing for F1 hybrid seed production (Pitchaimuthu *et al.*, 2012). Male sterility is incorporated into the inbred line that will serve as the female parent, so self-pollination will not occur. The second inbred line is male fertile and serves as the pollen parent for the cross. Male and female parents are planted in alternating rows at a ratio of one male to three or four female rows, depending on the amount of pollen produced by the male parent. F1 hybrid seeds are harvested only from male-sterile, female parent rows. After all fruit have set, plants of the male parent are removed before seed harvest.

The traditional hand harvesting of mature dry pods is still conducted in tropical areas where there is sufficient labor. In these areas, generally multiple harvests are performed as the pods mature, senesce, and dry on the plant. For large-scale production, okra seed crops are combined with the gathering reel removed to reduce shattering. Combining begins when most fruits have dried before the oldest pods dehisce and lose a significant number of seeds.

Seed cleaning

Seeds are extracted when pods become dry and brittle. Seeds can be extracted from dried pods

Fig. 15.5. Hybrid seed is increasingly produced in screen houses. (Courtesy of Ananta Vegetable Seeds, Pvt, Ltd.)

with a stationary thresher. The most efficient method of extracting seeds by hand is to twist the pods open and shake out the dried seeds. Plant debris can be removed with a spiral separator because seeds are nearly round and roll faster than debris. Air screen machines can also effectively remove light debris. A density separation table can also remove underdeveloped light seeds.

Seed Yield

The number of seeds per fully pollinated okra fruit varies over a range from roughly 45 to 90 (Martin *et al.*, 1981). This equates to roughly 1000 to 1350 seeds per open-pollinated plant as a maximum yield. A relatively high seed yield for okra is 1900 kg/ha (1340 lbs/acre) achieved under optimal conditions in low humidity, irrigated, dry desert areas. In many tropical countries seed yields are lower, around 500 kg/ha (446 lbs/acre). Average seed weight is approximately 50 g (1.8 oz)/1000 seeds.

Pest Management

Pathogens

The most serious diseases of okra are fungal (Table 15.1). Foliar fungal diseases are worse when seed crops are grown at high humidity. The best environment for okra seed crops is warm, irrigated deserts. Several okra viral diseases have been characterized (Table 15.1). Since okra is related to cotton, many of the pests are common to both crops. Okra should not be rotated with cotton. If possible, cotton and okra seed fields should not be planted nearby for pest control.

Insects

Okra is affected by a diversity of insects (Table 15.2). Many okra insect pests also attack cotton so growing pest-free seed crops in cotton production areas can be challenging. Insect pests vary widely across regions of the world. Please check local recommendations for control options. Research on organic alternatives to pesticides has provided more treatment options in recent years (Table 15.2).

Table 15.1. Some important okra diseases. (From Kimura *et al.*, 1982; Venkataravanappa *et al.*, 2012; Texas Plant Disease Handbook, 2021.)

Common name	Scientific name	Comments
Fungal diseases		
Root rot	*Phymatorichum omnivorum*	Infected plants die suddenly from mid-summer to fall; leaves usually remain attached to the plant; control with crop rotation, soil fumigation
Blossom and pod rot	*Choanephora cucurbitarum*	Infected fruit and blossoms have "whiskery" appearance; infected tissues develop soft and rotten; favored by warm, humid weather; control with fungicide
Fusarium wilt	*Fusarium oxysporum* f. sp. *vasinfectum*	Wilting, followed by death; infected plants are stunted and yellow; fungus colonizes the vascular system; long crop rotations recommended
Verticillium wilt	*Verticillium dahliae*	Soil-borne disease causes gradual leaf yellowing, wilting, and dark brown discoloration of the vascular tissue; control with disease-resistant cultivars, crop rotation, fumigation
Powdery mildew	*Sphaerotheca fuliginea*	White, powdery spots on leaf surfaces, shoots, and fruit; control with cultural practices that promote dry foliage, resistant cultivars, fungicides
Leaf spot	*Alternaria* spp., *Ascochyta* spp., *Cercospora malayensis*, *Phyllosticta hibiscina*	Several organisms responsible; control with fungicide sprays, low humidity
Seedling disease	*Rhizoctonia* spp., *Pythium* spp.	Collapse of seedlings when okra is planted in cold wet soils; control with antifungal seed treatments
Charcoal rot/dry rot	*Macrophomina phaseolina*	Soil-borne disease is controlled by crop rotation or fumigation; infected plant debris should be destroyed at the end of the season
Southern blight	*Sclerotium rolfsii*	White mycelia and sclerotia, yellowing and wilting foliage as well as browning stems and branches; control with field sanitation, remove infected plants, fungicide treatments, reduce moisture
Boeremia leaf and fruit spot	*Boeremia exigua*	Brownish black, sunken cavities on the leaves and fruits
Gray mold	*Botrytis cinerea*	Water-soaked, brown or gray spots on fruit; control with fungicide sprays, water management, good air circulation, removed infected plant materials
Fusarium root rot	*Fusarium solani*	Yellowing wilting leaves, plant death, infected roots turn reddish brown; control with crop rotation, fumigation, soil fungicide drench
Virus		
Yellow vein mosaic disease (YVMD)	*Begomovirus* (family *Geminiviridae*)	Vectored by whitefly
Okra leaf curl disease (OLCD)	*Begomovirus* (family *Geminiviridae*)	Vectored by whitefly
Okra enation leaf curl disease (OELCD)	*Begomovirus* (family *Geminiviridae*)	Vectored by whitefly
Bacteria		
Bacterial leaf spot and blight	*Pseudomonas syringae* pv. *syringae*	Control by removing diseased plants, water management, copper spray

Table 15.2. Some important insect pests of okra. (From Mandal *et al.*, 2007; UC IPM, 2023.)

Common name	Scientific name	Symptoms and control
Insects		
Pink boll worm	*Pectinophora gossypiella*	Feed on pods; *Bacillus thuringiensis* (*Bt*) or insecticide sprays, irrigation management, crop rotations, manage alternative hosts
Corn earworm	*Helicoverpa zea*	*Bt* or spinosad sprays based on larve population scouting, appropriate chemical treatment may be used
Flea beetle	*Nisostra* spp., *Podagrica* spp.	Feed on immature fruit wall; treat with rock powders or powdered clay, pyrethrum extracts, or insecticide sprays
Aphids (green peach, potato, cotton, etc.)	*Myzus persicae* *Macrosiphum euphorbiae*	Insecticides only required for heavy infestations, plants generally tolerate low infestation, insecticidal soaps, neem, or canola oil may provide control
Stemborer	*Earias vitella*, *E. insulana*	Pheromone trapping, destruction of infected fruits, release of parasite *Trichogramma chilonis* or *Chrysoperla carnea*, *Bt* or insecticide sprays
Leafhopper	*Amrasca biguttula*	Feed on undersides of leaves cause leaf curling and desiccation; release of parasites, insecticidal soap, neem oil, or insecticides
Red spider mite	*Tetranychus telarius*	Feed on underside of leaves, under fine silken webs, yellow spots appear upper side of leaves causing curling and necrotic distortion; sulfur or insecticide spray
Whitefly	*Bemisia tabaci*	Parasitoids, insecticidal sprays, residue management
Nematode		
Root-knot	*Meloidogyne* sp.	Root becomes enlarged and distorted; fumigation, crop rotation

References

Christenhusz, M.J.M. and Byng, J.W. (2016) The number of known plants species in the world and its annual increase. *Phytotaxa* 261, 201–217. doi:10.11646/phytotaxa.261.3.1 (accessed 15 September 2023).

FAO (2023) FAOSTAT (Food and Agriculture Organization of the United Nations statistics). Crops and livestock products. Okra. https://www.fao.org/faostat/en/#search/okra (accessed 20 September 2023).

Gemede, H.F., Haki, G.D., Beyene, F., Rakshit, S.K. and Woldegiorgis, AZ. (2018) Indigenous Ethiopian okra (*Abelmoschus esculentus*) mucilage: a novel ingredient with functional and antioxidant properties. *Food Science & Nutrition* 6, 563–571. doi:10.1002/fsn3.596. PMID: 29876107; PMCID: PMC5980337 (accessed 15 September 2023).

George, R.A.T. (2009) *Vegetable Seed Production*. CAB International, Wallingford, UK.

Kimura, O., Ribeiro, R.D.L. and Robbs, C.F. (1982) Fruit rot and leaf blight of *H. esculentum* caused by *Pseudomonas syringae*. *Arquivos da Universidade Federal Rural do Rio de Janeiro* 5, 105–110.

Mandal, S.K., Sah, S.B. and Gupta, S.C. (2007) Management of insect pests on okra with biopesticides and chemicals. *Annals of Plant Protection Sciences* 15, 87–91.

Martin, F.W., Rhodes, A.M., Ortiz, M. and Díaz, F. (1981) Variation in okra. *Euphytica* 30, 697–705.

National Research Council (2006) Okra. In: *Lost Crops of Africa. Volume II: Vegetables.* National Academies Press, Washington, DC. doi.org/10.17226/11763 (accessed 15 September 2023).

Pitchaimuthu, M., Dutta, O.P. and Swamy, K.R.M. (2012) Studies on inheritance of geneic male sterility (GMS) and hybrid seed production in okra [*Abelmoschus esculentus* (L.) Moench.]. *Journal of Horticultural Sciences* 7, 199–202.

Purewal, S.S. and Rhandhawa, G.E. (1947) Studies in *Hibiscus esculentus* (ladyfinger) (okra) I. Chromosome and pollen studies. *Indian Journal of Agricultural Science* 17, 129–136.

Rubatzky, V.E. and Yamaguchi, M. (1997) *World Vegetables: Principles, Production, and Nutritive Values*. Chapman & Hall, Hoboken, New Jersey.

Santos, R.F.D., Gomes-Junior, F.G. and Marcos-Filho, J. (2019) Morphological and physiological changes during maturation of okra seeds evaluated through image analysis. *Scientia Agricola*, 77. doi.org/10.1590/1678-992X-2018-0297 (accessed 15 September 2023).

Texas Plant Disease Handbook (2021) Okra. Texas A&M AgriLife Extension. plantdiseasehandbook.tamu.edu/?s=okra (accessed 15 September 2023).

UC IPM (2023) Cotton. Pest Management Guidelines. University of California, Integrated Pest Management. ipm.ucanr.edu/agriculture/cotton/ (accessed 15 September 2023).

Venkataravanappa, V., Lakshminarayana Reddy, C.N., Swarnalatha, P., Devaraju, *et al.* (2012) Molecular evidence for association of *Cotton leaf curl Alabad virus* with yellow vein mosaic disease of okra in North India. *Archives of Phytopathology and Plant Protection* 45, 2095–2113. doi: 10.1080/03235408.2012.721682 (accessed 15 September 2023).

16

Family Poaceae

Overview

Maize, or corn as it is called in some locations, is native to the western hemisphere. Many believe maize was domesticated in Tehuacan Valley of southern Mexico over 7000 years ago (Ranere *et al.*, 2009). As a basic food plant, maize spread to the north from its center of origin to the United States and south into South America hundreds of years before the arrival of Europeans. Wild maize pollen, 60,000–70,000 years old, has been found on lake bottoms near Mexico City (Beadle, 1981; Sears, 1982).

A close progenitor of maize is not found in the wild. Modern cultivated maize has evolved from a wild, annual grass, perhaps with a terminal flowering structure with male flowers above and female flowers below. Another theory is that the original plant may have had a terminal male spikelet with several small female spikelets 19 to 25 mm (3/4 to 1 in.) long at the nodes immediately below (Goodman, 1988). Teosinte (*Z. mays* subsp. *mexicana*) may be similar to the wild plant related to modern maize. Starting with a 2.5 cm (1 in.) wild pod, Native Americans transformed maize over 2000 years by systematically collecting and cultivating plants. This domestication resulted in larger cobs and more rows of kernels, improving maize as a food crop (Galinat, 1992; FAO, 2002). More recently, plant breeding and genetic engineering have improved yield, nutritional quality, insect resistance, and made some cultivars resistant to herbicides. Modern cultivars of maize can no longer exist in the wild because of their resistance to lodging and seed release. Sweet, pop, and ornamental corns are the primary focus of this chapter. Sweet corn did not become a major food crop until about 140 years ago and is consumed primarily in North America. It is becoming more popular in parts of South America, Europe, and Asia.

Botany

There are many forms of maize, and they are sometimes classified in subspecies or botanical varieties based on different phenotypes rather than genetic differences. This chapter will focus on sugar accumulating types (sweet corn, *Zea mays* var. *saccharata*), ornamental corn (multicolored often variegated maize ears sometimes classified among different subspecies), and popcorn (*Zea mays* var. *everta*) (Fig. 16.1). All are warm-season, frost-intolerant, annual monocots from the family Poaceae (George, 2009).

The rigid stems (axis or culm) of maize range in height from 1.3 to 2.5 m (4.3 to 8.2 ft) depending on the cultivar (Fig. 16.1). The culm is enveloped by alternating leaf sheathes originating from each node. The long alternating opposed leaves extend along the entire stem.

DOI: 10.1079/9781789243260.0016

Fig. 16.1. "First early" sweet corn (right row) and ornamental corn (center and left rows) show genetic variation in color and height between types. Corn is monoecious with male flowers (tassels) at the top of the plant and female flowers (ears) lower on the main stem.

At the ligule, each leaf sheath angles away from the stem to form a long, broad, curved leaf with prominent midrib. The large leaf blades are alternate, wavy, and grass-like in appearance. The long leaves are generally uniform in width and have many parallel veins running along their length (Rubatzky and Yamaguchi, 1997).

The nodal root system is composed of whorls of secondary adventitious roots that develop at the basal nodes of the stem and grow laterally. These lateral roots branch profusely, spread widely, and penetrate deeply into the soil, though the depth of penetration depends on soil characteristics. These relatively shallow, branched secondary roots absorb nutrients from the soil. Besides nodal roots "prop or brace" roots help to anchor the plants in the soil and provide additional nutrient absorption. Brace roots are limited to the first two or three nodes of the stalk above the soil line (Gardner *et al.*, 1985).

Maize is monoecious with separate male and female flowers on the same plant. Male flowers are borne at the top of each plant (tassels) while female flowers (ears) are borne separately lower on the main stalk. Ears develop above a few of the lower leaves in the midsection, between the stem and leaf sheath. Female flowers are terminal inflorescences protected from desiccation and predation by thick ear leaves called husks. The carpels are anchored to a cob in a multiple fruit structure. Elongated stigmas, called silks, extend beyond the whorl of husk leaves at the top of each ear (Figs 16.1, 16.2). Stigmas range in color from whitish green to pale yellow to burgundy, are 18 cm (7 in.) in length, and are reminiscent of hair (Figs 16.1, 16.2; Abendroth *et al.*, 2011). For successful pollination, pollen grains must move by air from the anthers on the tassel and land on silks protruding through the husks. Sweet corn hybrids are estimated to

Fig. 16.2. Sweet corn silks extending past the husks at the top of an ear.

produce 8,200,000 to 14,960,000 pollen grains per tassel (Fonseca *et al.*, 2003). The amount of pollen shed is affected by temperature, wind, and cultivar. Pollen is shed from three to ten days. At the end of each stigma is an ovule, which may develop into a "kernel" if fertilized by pollen via a pollen tube (Abendroth *et al.*, 2011). For fertilization, the pollen must germinate, and each tube grow the length of the silk to fertilize an ovule. The pericarp of the fruit is fused with the seed coat and referred to as a caryopsis, typical of grasses. The fruit pericarp and seed coat are not easily distinguished and exist as a single structure. The entire kernel is usually referred to as a "seed" by agriculturalists despite being a caryopsis (Gardener *et al.*, 1985). For seed production, each caryopsis must fully develop before harvest. Plants of vegetable sweet corns and popcorn cultivars are smaller than starchy kernel grain types. The tassels of some sweet corn cultivars are distinct with larger size and lighter color than grain corn (Fonseca *et al.*, 2003). Ornamental plants may have purple culms and husks but some cultivars are indistinguishable from agronomic grain corn (Fig. 16.1).

Growth and Development

There is no seed dormancy reported in commercial grain, sweet, ornamental, or popping types of corn. Corn kernels must imbibe approximately 30% of their weight in water to germinate. Repeated wetting and drying cycles after planting can cause poor stands. Germination is faster in warm soils with adequate soil moisture than in cooler or drier soils, which delay emergence.

Like other monocots, the protective coleorhiza is the first tissue to emerge from a kernel. Radicle emergence occurs through the ruptured coleorhiza adjacent to its tip. The coleoptile or "spike" emerges next from the embryo side of the kernel within one-to-many days after the radicle, depending on soil temperature and moisture (Abendroth *et al.*, 2011). The coleoptile is a rigid plant tissue that completely encloses four or five embryonic leaves (plumule) that form during caryopsis development. The coleoptile initially moves toward the end of the kernel away from the radicle by elongation of the mesocotyl. The plumule leaves slowly enlarge eventually causing the coleoptile to split open near the soil surface (Gardener *et al.*, 1985).

The primary seed root anchors the seedling. Seminal (seed) roots originate from the scutellar node located within the embryo. Seed roots are composed of a radicle and lateral seminal roots (Gardener *et al.*, 1985). The nascent seminal root system helps sustain seedling development through water uptake. A corn seedling survives primarily on the energy reserves from the kernel's endosperm until the nodal root system develops later. Sweet corn endosperms are reduced in size and contain higher quantities of soluble sugars than starch. Soluble sugars can leach upon hydration, attracting microbes to the rhizosphere around the seedling. Loss of nutrient reserves, particularly in supersweet cultivars (*sh_2* genotypes), may reduce seedling vigor, which is detrimental to establishment under adverse conditions. Modern cultivars are selected for higher seedling vigor than some of the early supersweet (*sh_2*) cultivars. After seedling emergence, the seminal roots growth rate slows as the nodal root system begins development above the mesocotyl (Gardener *et al.*, 1985).

A new leaf should appear around every three days during early seedling growth, while later leaves may appear in one or two days under warmer conditions (Abendroth *et al.*, 2011). Main-season sweet corn hybrid cultivars may produce up to 20 leaves. Earlier maturing hybrids will produce fewer leaves (Fig. 16.1). Development may be faster than described under higher-than-normal temperatures and slower under cooler conditions. Maize development is very closely related to heat exposure. Because of the strong correlation, growing degree days (GDD) can predict maize development and maturation of seed crops. While growing degree days are often associated with scheduling sequential plantings of corn for fresh market and processing, there is utility in tracking GDD as an indicator of seed corn maturity. The equation to calculate GDD is: GDD = [(Maximum temperature + Minimum temperature)/2] – 10°C. So, for example if the daytime temperature is 20°C and the nighttime temperature is 10°C, you would add 20° + 10° = 30°, then divide 30°/2 = 15°, and then subtract 15° – 10° = 5° GDD. Negative numbers are not counted (NDAWN, 2023).

Corn development can be divided into discrete vegetative and reproductive stages. Reference points for vegetative development are: VE = emergence, V1= first collar leaf, V2 = second collar left, V(*n*) = *n*th collar leaf, VT = tasseling (Pioneer Agronomy Sciences, 2023). Reproductive stages are: tassel (VT) = bottom-most branch of tassel completely visible without silk emergence, silking (R1) = silks extend beyond the husks, blister (R2) = kernels white on outside, clear liquid endosperm inside, milk (R3) = kernel colored outside, milky white endosperm fluid inside, dough (Fig. 16.3) (R4) = kernel endosperm thick/pasty, cob pink, whitish-green, or red, maturing (R5) = kernels beginning to wrinkle, physiological maturity (R6) = shriveled kernels, black layer formed and maximum dry weight (Pioneer Agronomy Sciences, 2023). At physiological maturity, a distinct black layer typically forms at the base of the kernel isolating it from the cob; however, some parent lines of sweet corn may form a more indistinguishable brown layer.

Fig. 16.3. R3 stage white kernel sweet corn cultivar ear development with husks pulled back. Pale green silks attach to each ovule (kernel). A single yellow kernel in the middle of the ear was fertilized by pollen from a yellow plant due to xenia. A corn earworm insect larva is feeding on the tip of the ear.

Through generations of selection and breeding, corn cultivars have been developed for long or short days at different latitudes. The response of corn to day length is not as pronounced as for other crops like soybean, where cultivars are adapted to specific latitudes. With corn, the period from emergence to flowering is reduced by short days and increased by long days (Jenkins, 1941). An early-maturing northern sweet corn cultivar developed for southern Canada adapted for long days will flower earlier if grown in Mexico under shorter days. Average yield of the same cultivar at higher latitudes exceeds that in the tropics by four times in the field and by 2.5 times in experimental plots largely because of day length and other environmental factors (Chang, 1981). Longer days maximize photosynthesis and cooler night temperatures reduce respiration. High night temperature reduces the corn yield of the same cultivar in the tropics by some 30% by increasing respiration compared to higher latitudes

(Chang, 1981). In much of North America, China, Argentina, southern Brazil, Uruguay, and Europe, long-day, exceeding 14 h 30 min, corn hybrids are grown. Seed of long-day types is produced between 30 and 45° N and S latitudes. Conversely, a short-day cultivar developed for sub-Saharan Africa would grow vegetative longer when produced under long days at higher latitudes in Europe. This means that seed corn cultivars developed for temperate climates are best produced in temperate regions, while tropical cultivars should be produced in the tropics.

Sweet corn is often divided into different maturity classes based on time to harvest from direct seeding (Tracy, 2000). For example, "first early" cultivars reach edible maturity in 65–74 days and approximately 85–94 days for seed corn maturity after planting. First early cultivars usually produce small plants, small ears, and have poorer quality compared to later maturity classes (Fig. 16.1). First early cultivars produce short plants with approximately ten to 13 leaves, only 12–14 rows of kernels per ear, and shallow kernels. Edible maturity of "early" cultivars is reached at 75–80 days and seed corn maturity at approximately 95–100 days. Early types produce taller plants with more leaves (14–16), more rows of kernels per cob, and deeper kernels compared to first early cultivars. "Main-season" cultivars mature mid-summer and are grown for processing and fresh market. They reach edible maturity in 81–90 days with seed corn maturity in 101–110 days. Plants in the main season class are large with approximately 17–20 leaves. Ears in this class have about 18 rows, deep kernels, and excellent eating quality. "Late"-maturing cultivars may have over 20 leaves, reach edible maturity in > 90 days, > 110 days for seed corn maturity, and produce very high-quality, large ears with deep kernels (Tracy, 2000). Sweet corn seed crops are sometimes divided into different maturity classes based GDD maturity. It takes 250 GDD°C (450 GDD°F) for first early cultivars to mature a seed crop and 350 GDD°C (630 GDD°F) for a main-season cultivar to reach full seed maturity (NDAWN, 2023).

Most modern grain cultivars are predominately single-stalked plants without sucker shoots; whereas sweet corn often produces sucker shoots especially at low plant populations (wider spacing). Some heirloom sweet, pop, and ornamental corn cultivars develop sucker shoots, adventitious shoots, or root sprouts at their bases. These small plants are detrimental for seed production and should be removed. Sucker plants do not develop productive ears but compete with the mother plant for nutrients and photosynthates.

Soil

Seed corn can grow on a wide range of soil types ranging from sandy to clay loams. Irrigation is recommended for lighter soils because they have lower moisture-holding capacity. Good drainage is important regardless of the soil type. Highest yields occur at a pH range of 6.0 to 6.8 although the crop tolerates a wider pH range from 5.5 to 7.5. Acid soils with pH below 5.5 should be avoided or lime applied before planting to adjust the pH into the acceptable range (George, 2009). A crop rotation cycle with a minimum of two years is best practice to reduce risk from diseases, insect pressure, volunteer corn, and other factors.

Fertilization

Sweet corn has a relatively high nutrient requirement. Sufficient nutrients should be available early in the season to ensure stand establishment and rapid growth of both male and female lines to encourage simultaneous flowering. Seed corn responds well to nitrogen. Nitrogen fertilization varies widely depending on soil type and cropping history. Generally, more nitrogen is needed on light sandy soils than heavier silt or clay loam soils. Side dressing with nitrogen is common with corn particularly after plants reach knee height. Many growers row-band nitrogen at planting and then side dress with anhydrous ammonia later in development although other forms of N can be used. Phosphorus and potassium fertilization is usually based on soil test results. Often, all the potassium and phosphate are row-banded at planting. Micronutrients are only applied when soil tests show a need.

Inbred lines for F1 hybrid production often lack vigor. Inbred lines do not respond to fertilizer, and particularly nitrogen, in the same way as an F1 hybrid. It is easy to over-fertilize inbreds trying to compensate for their lack of vigor. Proper fertilization, especially of inbred lines, is essential to maximize seed production without wasting fertilizer.

Pollination

Pollination of corn occurs after pollen grains are released from tassels, fall or are blown through the air, and land on the silks of the same or a different plant. Fertilization is the union of the male (pollen) and female (ovule) gametes. Technically, pollination is usually successful (i.e. the pollen usually reaches the silks). However, unsuccessful fertilization (i.e. pollen tube failure, silk failure, pollen death) is more likely cause of poor kernel set and yield losses.

Silks begin to elongate from the ovules ten to14 days prior to growth stage R1 or approximately at the V12 leaf stage described above. Silk elongation begins from the basal ovules at the bottom of each cob, then proceeds sequentially upward to ovules at the tip. Because of this acropetal sequence of silk elongation, silks from the basal ovules typically emerge from the husk first. Silks from ovules near the tip of each ear generally emerge last. Silks grow by expansion of existing cells. Silks lengthen up to 3.8 cm (1.5 in.) per day as they initially emerge from the husk (Fig. 16.2). Complete silk emergence from an ear generally occurs within four to eight days (Nielsen, 2020). The elongation rate slows as silk cells reach maximum size. Silks are less receptive to pollen over this time. Silks remain receptive to pollen grain germination for up to ten days after emergence then senesce if there is no pollination. Silk elongation stops shortly after pollen is captured, germinates, and penetrates the silk (Nielsen, 2020).

Pollen grain germination occurs within minutes after a pollen grain lands on a receptive silk. A pollen tube develops and grows inside the silk and fertilizes the ovule within 24 h (see Chapter 2, this volume). Pollen grains can land and germinate anywhere along the length of an exposed receptive silk. Many pollen grains may germinate on a receptive silk, but typically only one will successfully fertilize the ovule. Most successful ovule fertilization events occur during the first four to five days after silking (Nielsen, 2020).

Several factors may interfere with successful fertilization. Collapse of silk tissue during senescence restricts growth of the pollen tube. Silk emergence usually occurs close to the time of pollen shed in most cultivars. However, when two inbred lines are crossed, pollen shedding and silk growth may not be synchronized. It is critical that pollen is shed by the male parent when the silks are exposed and receptive on the female parental line. To improve the synchronization of the parent lines, nicking trials are conducted to determine the best time to plant each parent for successful hybrid seed production. When the female or the male parents are planted at different times, it is called split planting. Often the male parent is planted at two different times to assure plenty of pollen is available when the silks of the female are receptive. Poor nicking increases the potential for pollen contamination from other corn fields and an increase of self-pollination from tassels that were not removed from the female parent.

To further complicate seed production, the synchrony of pollen shed and silk formation can be altered by environmental stressors in production fields. Silks have the highest water content of any corn tissue, so they are sensitive to moisture deficits. Silk emergence may be delayed or incomplete because of severe drought stress. If the delay is sufficiently long, pollen shed may finish before receptive silks are available, greatly reducing seed set (Nielsen, 2020). Drought also accelerates pollen shed, so if silks are not yet developed, the timing of pollination is disturbed. Some genotypes developed for drought tolerance have rapid silk growth rates, meaning that silk elongation is more drought tolerant. Excessively long silks are often symptomatic of poor pollination (Nielsen, 2018). Under normal growing conditions, genotypes expressing drought stress genes may silk four or five days prior to pollen shed decreasing pollination (Nielsen, 2020). Severe leaf roll early in the day suggests occurrence of adverse moisture conditions that may reduce pollination. Pollen loses viability once temperatures are > 34°C (> 93°F) especially at low humidity. Pollen shed usually occurs during early to mid-morning at lower temperatures, so daytime high temperatures alone may not be responsible for pollination problems (Nielsen, 2020). Any stress such as inadequate water, low soil fertility, or too high plant densities may delay silking two or more weeks until pollen is no longer available (Nielsen, 2020).

Poor pollination may be caused by insects feeding on silks, called "clipping". Insects that cause severe prolonged clipping are the corn rootworm (*Diabrotica* spp.) and the Japanese beetle (*Popillia japonica*). Occasionally, silks fail to emerge because elongation is twisted or coiled into a ball inside the husk (Nielsen, 2000). One possible explanation for silk "balling" is a physical restriction of silk elongation because of "tight" or long husk leaves in certain genotypes. Unusually cool nights during silk elongation may contribute to balling (Nielsen, 2000).

Corn pollen is wind-borne and isolation distances of 213–600 m (700–1969 ft), depending on endosperm type, kernel color, natural barriers, wind direction, and other conditions, are recommended. It is extremely important to ensure maximum isolation between sweet corn seed fields and other types of *Zea mays* grown as fodder or for grain production (Darrah *et al.*, 2003). Example minimum isolation distances for hybrid sweet corn fields include:

- From field corn or other sweet corn cultivars should be at least 201 m (660 ft).
- From corn of different colors or popcorn should be least 402 m (1320 ft).
- Bicolor hybrids should be 201 m (660 ft) away if the adjacent corn is the same color as the bicolor male parent. This distance should be twice the above amount if the bicolor male parent is a different color.
- A minimum distance of 1609 m (5280 ft) from transgenic sweet corn (*Bt*) seed production.
- Isolation distances for parent seed increases are typically twice the amount for each of the applicable situations above.

Volunteer *Zea mays* plants must be removed from the isolation area around seed production fields. If a seed crop must be produced in proximity to another corn type, additional "trap" rows of male pollinator for F1 hybrid production may be planted as a buffer around rows harvested for seed. The buffer rows produce desirable pollen and are not harvested as part of the seed crop.

Irrigation

Sweet corn has a shallow fibrous root system with most roots in the top 0.6 m (2 ft) of soil. Keeping adequate moisture in this topsoil layer is a priority especially during stand establishment and early-season growth. Pollination is a critical time to prevent drought stress which interferes with silk growth, pollen shed, pollen germination, and pollen tube growth. Drought stress will cause poor ear tip fill, delay maturity, cause uneven harvest, and reduce yield.

Supplemental irrigation or rainfall must provide at least 2.5 cm (1 in.) of water per week for uninterrupted growth. More irrigation is needed on sandy soils. Corn seed crops are often irrigated with overhead sprinkler systems such as in the Columbia Basin of Washington state, USA. Furrow irrigation is limited to areas where the ground is relatively level and water is plentiful and is still common practice in the Treasure Valley in Idaho, USA. In areas where water is limited, furrow irrigation has been replaced by more efficient methods such as surface and subsurface drip. Although the range is broad and dependent on several factors, to fully mature a corn seed crop requires in the range of 550 to 750 mm (22 to 30 in.) of water per season.

Rogueing

Both parents must be inspected when F1 hybrid seed is produced. Rogueing criteria are applied according to the detailed description of each parent line provided by plant breeders. Rogueing is normally performed when contaminating plants are at their most visible stage of development (George, 2009). Field rogueing must be completed before male or female inflorescences are fully developed. An additional check for trueness-to-type, especially characters such as seed color, are made after harvest when the kernels are clearly visible prior to drying and conditioning.

Genetics

Parent lines of seed corn should be grown in climates similar to where the hybrid seed production will be ultimately grown to prevent genetic drift. Sweet corn genetics are complex with different alleles affecting sugar content and kernel color. Sweet corn has evolved through a series of natural genetic mutations that have been maintained as new cultivars. F1 hybrid corn was first produced commercially in the 1930s by crossing inbred lines. Today almost all commercial sweet corn, pop, and grain corn cultivars are F1 hybrids. However, a few open-pollinated ornamental and specialty heirloom cultivars remain in production commercially (Darrah *et al.*, 2003). Transgenic sweet corn expressing the *Bacillus thuringiensis* (*Bt*) and herbicide resistance genes for insect and weed control, respectively, have been developed and may be grown only if regulations allow in areas such as the USA and Canada. Transgenic sweet corn production requires additional isolation distances to prevent contamination of non-transgenic seed production crops.

Sweet corn genes can be divided into three groups. The oldest type is due to the recessive su_1 (O22637, mutant allele of isoamylase) that causes sugar rather than starch to accumulate in the endosperm of each kernel (Tracy *et al.*, 2006). This is "standard" sweet corn (Table 16.1). After the harvest of standard sweet corn, sugar is very rapidly converted to starch, if not refrigerated to slow the conversion.

Another major sweetness allele is the sugary enhancer (se_1). This recessive mutation causes a higher sugar content than su_1 cultivars (Table 16.1). In se_1 cultivars, the conversion of sugar to starch after harvest occurs more slowly. Sugary enhancer cultivars lose moisture at a slower rate than normal sweet corn as well, extending their harvest period as a food crop.

Supersweets carry the shrunken-2 (sh_2) mutation (P55241, for glucose-1-phosphate adenylyltransferase) (Table 16.1). Conversion of sugar to starch is very slow, causing the supersweets to have highest sugar content (Darrah *et al.*, 2003). Studies show that supersweet cultivars can maintain their sweetness on the stalk for up to 7 days and off the stalk for up to 30 days in cold storage (Welbaum, 2015). Supersweet (sh_2) cultivars have a tender pericarp in addition to a higher sugar content compared to *su* and *se* types. The distinctive supersweet phenotype is extremely shriveled kernels caused by its small endosperm (Darrah *et al.*, 2003). Some supersweet cultivars germinate poorly and have low seedling vigor making establishment more difficult, particularly in cold soils. Vigor of sh_2 seed corn has improved since its introduction because of improved inbred lines and post-harvest conditioning.

Recently sweet corn cultivars called "synergistic hybrids" (sy) combine both sugary enhancer (se_1) and supersweet (sh_2) genes in kernels on the same ear to produce a unique cultivar with tender kernels, high sugar content, improved eating quality, and long shelf life (Cummins and Motsenbocker, 2022). Similar to supersweet cultivars, these hybrids also can have lower germination and seedling vigor. Other cultivars that include both types of starch synthesis mutations are "augmented supersweet hybrids" (shA) where (se_1 or su_1) mutant alleles are combined with a (sh_2) mutant allele. The kernels of these hybrids are even more shriveled, fragile, and have lower germination and seed vigor than sh_2 hybrids. ShA hybrids need higher optimal soil temperatures for germination.

"Xenia" refers to the immediate and recognizable effect of a pollen parent genetics on the phenotype of the developing kernel (Rubatzky and Yamaguchi, 1997). Corn is one of the few vegetable crops where pollen genes determine the phenotype of the fruit. The two most common xenia effects are kernel color and starch content. Understanding xenia is important for maintaining seed quality during seed production. Characteristics of the pollen parent become visible on the developing ear almost immediately after pollination. Controlling pollination of sweet corn is especially important when sweet corn is grown near a field of agronomic (grain) corn because many of the field corn pollen traits are dominant and expressed over the desirable recessive traits of sweet corn including starch content and kernel color (Table 16.1, Fig. 16.4).

It is also important to isolate supersweet (sh_2) corn from su_1 and se_1 types of sweet corn. If isolation is not maintained and cross-pollination occurs, then both the supersweet and the su_1 and se_1 of sweet corn may produce starchy kernels (Table 16.1). Both su_1 and se_1 genotypes can cross-pollinate and maintain at least su_1 levels of sugar (Table 16.1; Darrah *et al.*, 2003).

In determining sweet corn kernel color, yellow is dominant to white (Figs 16.3, 16.4). When pollen from a yellow pericarp plant fertilizes an ovule with white pericarp, the resulting kernels have yellow endosperm because of the Mendelian dominance of yellow over white. When yellow corn is pollinated by white corn, the kernels are lighter yellow and often capped with white. There is no genetic linkage between kernel color and sweetness because each trait is independently inherited (Darrah *et al.*, 2003).

Table 16.1. Effects of pollination on kernel sugar content of grain corn (Su_1) and su_1, se_1, and sh_2 sweet corn cultivars.

	Pollen genotype			
	Sweet corn			Starchy grain
Genotype of kernel	su_1	se_1	sh_2	WT
su_1 sweet corn	sweet	segregates 1:2:1	starchy	starchy
se_1 sweet corn	su_1 level of sweetness	sweet	starchy	starchy
sh_2 sweet corn	starchy	Starchy	sweet	starchy
Su_1 grain corn	starchy	Starchy	starchy	starchy

Fig. 16.4. Su_1 sweet corn ear with some kernels cross-pollinated by agronomic starchy "dent" grain corn (expanded, bright yellow). The dominant starchy genes from grain corn pollen cause the developing kernel to be pale yellow and starchy because of xenia rather than shriveled and opaque like the sweet corn.

Production and Harvest

Corn should be planted when soil temperatures exceed 20°C (68°F) for the best emergence after the threat of frost has passed. Emergence is slow when soil temperatures are below 15°C (59°F). Planting densities are determined by mature plant and seed sizes. Larger seed cultivars are generally sowed at a density up to 15 to 30 kg/ha (13.4 to 26.8 lbs/acre) for small and large seeds, respectively (George, 2009). Between-row spacing ranges from 70 to 100 cm (2.3 to 3.3 ft) apart depending on the machinery used (George, 2009). In-row spacings range from 12 to 25 cm (5 to 10 in.) apart depending on germination and vigor test data. Plant populations can vary from 74,000 to 123,000 plants/ha (30,000 to 50,000 plants/acre) depending on the parent lines and their yield potential. Typical row ratios are 1 male:4 female, or 2:4 for poor pollen producers. The amount of pollen from a typical healthy male parent is sufficient to pollinate ten female plants. In some seed production areas, two male rows are planted at different times on the same bed to improve nicking by providing an extended supply of pollen for the female parent to maximize seed yields.

Since spacing is important to optimize pollination for seed production, precision belt or air seeders are preferred for establishment. Sweet corn seeds are more fragile than other types of corn, so planters with "finger" pickup seed selection mechanisms may break fragile sweet corn seed unless encrusted seeds are used. Sweet corn and especially sh_2 cultivars are sometimes encrusted to prevent kernel breakage and to improve singulation during planting.

Inbred corn lines crossed to produce F1 hybrid seed are sometimes weak and poor seed producers, especially in the early days of hybrid development. To increase seed yield, a double-cross hybrid system was developed. As the name implies, producing a double-cross hybrid requires two stages of crossing involving two pairs of inbreds (Fig. 16.5). In the first cross, two pairs of hypothetical inbreds, *C* and *D* and *W* and *Z*, are each crossed to produce single-cross hybrids, *CD* and *WZ*. In the second cross, the two single-cross hybrids produced in Cross 1 are crossed to produce double-cross hybrid seed (Fig. 16.5). Typically, both pairs of breeding lines in Cross 1 are closely related but neither pair is as closely related to the other. In other words, *CD* and *WZ* are *not* closely related. Heterosis is achieved in Cross 2 because of this genetic diversity (Fig. 16.5).

Fig. 16.5. Diagram of double-cross hybrid seed corn production.

Unlike a single-cross hybrid, however, plants of a double-cross hybrid are less uniform. Three-way cross hybrids are produced by crossing a hybrid with a pure line. In most cases, single-cross hybrids produce the most vigorous seed compared to the best double-cross or three-way cross hybrids. Over time as corn breeders improved the performance of inbreds through selection, commercial production has shifted back to the more desirable single-cross hybrids. Modern single-cross hybrids produce the greatest heterosis and satisfactory yields of seed corn without resorting to double or triple crossing. Sometimes two closely related inbred lines are crossed (A1 × A2) and is used as the female parent crossed to a male parent line (B) to produce a modified single cross (A1A2 × B). The advantages of these hybrids include better seed quality and higher seed yields.

Parental lines for F1 hybrid seed production along with instructions on optimum ratio of male to female parental lines to ensure adequate pollination and maximum yield are normally provided by the seed company. A taller male parent will distribute pollen more efficiently than a small one and may require a lower male-to-female ratio. Cytoplasmic or genic male sterility is sometimes incorporated into the F1 hybrid female parent to ensure crossing occurs only between the male and female inbred lines without physically detasseling the female parent. Although male sterility can reduce the labor involved in corn seed production, susceptibility to certain diseases is increased. Male-sterile inbred lines also require a restorer line to maintain. The southern corn (*Zea mays* L.) leaf blight (SCLB) epidemic of 1970–71 was blamed on over reliance on cytoplasmic Texas male-sterile (cms-T) lines in hybrid seed production (Bruns, 2017). The leaf blight epidemic occurred because of a favorable environment for the disease, combined with > 85% of the hybrids using susceptible cms-T male-sterile genetics. Hybrid corn seed production is much less reliant on cms-T male sterility after 1971 as hybrid seed production returned to detasseling of the female parent to control crosses (Fig. 16.6). The SCLB epidemic reminds commercial seed producers of the danger of creating cultivars with a narrow genetic background (Bruns, 2017).

Without satisfactory genetic or chemical control of pollen production, the female parent line must be detasseled to prevent selfing. It is important that all the tassels of the female parents and their tillers, if produced, are completely removed before any pollen is shed and silks are exposed on the female parent. Detasseling methods include pulling the tassel by hand, primarily by walking or by workers perched on a mobile platform. Machines with rotating automated cutters are sometimes used but still require late developing tassels to be removed by hand. Excessive cutting and leaf removal may reduce yields. Typically, a field is detasseled several times during the flowering period and additional pulls may be required if a crop lacks uniformity. Seed companies usually employ inspectors to monitor activities during the flowering period to assure timely and proper removal of tassels and proper quality control. Male parents are often removed from seed fields after pollenation, so seeds from the male parental line do not contaminate F1 hybrid seeds from the female parent (Fig. 16.7).

Sweet corn seed production is best produced in irrigated areas where there is low relative humidity and low rainfall to reduce disease pressure especially during seed maturation. In North America, most seed production for temperate hybrids of sweet corn takes place in Treasure Valley of Idaho and the Columbia Basin of Washington state, USA. Temperate cultivar production also occurs in Chile and New Zealand for local use as well as counter-season production for the northern hemisphere. Hungary and France also produce seed for regional use.

Fig. 16.6. Female rows of a sweet corn seed field in Idaho, USA have been detasseled to prevent selfing during F1 seed production. Planting ratio is one male row to four female rows. (Courtesy of Tim Loeffler.)

Fig. 16.7. Hybrid sweet corn seed field in Hungary with the male rows removed so non-hybrid seed will not be accidently mixed with hybrid seed rows. (Courtesy of Tim Loeffler.)

Tropical sweet corn seed production occurs in Brazil, Thailand, and countries with similar climates. Major popcorn seed production takes place in Idaho, Nebraska, Indiana, Illinois of the USA.

Seeds for ornamental and popcorn are frequently harvested with a combine configured for corn picking and shelling (Fig. 16.8). Popcorn should be combine-harvested at a field moisture of between 14 and 18%, with an optimum of 16 to 17%. Combine settings are different for popcorn than other types of corn. Slower cylinder speeds and wider concave clearances reduce kernel damage. A combine working at or near its rated capacity will produce less kernel damage than when operated at low throughput (Ziegler *et al.*, 1989).

Plots of sweet corn seed are only hand-harvested for small-scale or specialty production. Larger scale commercial production is harvested with mechanical ear pickers because kernels are fragile when harvested as whole ears, when kernel moisture content can be 45% or higher depending on the genotype. Sweet corn should be harvested at about 35 to 40% moisture for su_1, 45–40% for se_1, and 50% for sh_2 genotypes. Often the sugary enhanced and sh_2 cultivars are harvested at higher seed moisture levels due the presence pathogenic diseases and pregermination. Visual appearance of harvest stage kernels is slightly wrinkled, and translucent with either a more indistinguishable brown or more distinguished black layer (George, 2009). After harvest, husks are removed and naked whole ears are placed in slotted containers, bins, or cribs with good air circulation or forced-air dried. Ears are dried at 35°C–38°C (95°F–100.4°F) to a moisture content of 9 to 11%. Quality of sh_2 seed is significantly improved by a slow drying at temperatures < 38°C (100.4°F).

Removing kernels from cobs is called shelling (Fig. 16.9). The dried whole ears flow on to a revolving cylinder with round peg projections that separates seeds from the cob within the shelling machine. The sheller must be adjusted correctly to reduce seed breakage. Modifications often made to shellers to accommodate brittle sweet corn include rubber padding and finger pegs for gentle separation. After shelling, kernels are conditioned using an air screen machine to remove pieces of husk or cob debris followed by a

Fig. 16.8. Corn ear picker harvesting whole ears. Harvested whole ears are final dried and shelled as a separate process. (Courtesy of Tim Loeffler.)

Fig. 16.9. Dried sweet corn seed ears ready for shelling followed by conditioning. (Courtesy of Tim Loeffler.)

density separator. The clean seed lot is often graded to separate different sized kernels.

Seed Yield

The seed yield depends on the cultivar. For hybrid sweet corn seed, the ratio of female plants cultivated per unit area determines yield. Yields of 1500 kg/ha (1338 lbs/acre) are generally obtained for sweet corn. Specialist producers may obtain yields of up to 2500 kg/ha (2230 lbs/acre). Typical popcorn yields are 3922 to 5043 kg/ha (3500 to 4500 lbs/acre). The 1000 grain weight of the sh_2 (supersweet) cultivars is about 100 g (3.5 oz) and roughly 210 g (7.4 oz) for *su* and *se* cultivars.

Pest Management

Pathogens

Bacterial (Stewart's) wilt (*Pantoea stewartii*) is vectored by flea beetles. The bacteria colonize the vascular system disrupting plant hydration and producing wilt symptoms. Sweet corn and popcorn are particularly susceptible, although some resistant cultivars are available. Southern corn leaf blight (SCLB, *Bipolaris maydis*) has three races: Race O, Race C, and Race T. SCLB symptoms vary depending on the infectious pathogen's race but generally cause elongated leaf necrotic lesions. Race T is infectious to corn plants with the Texas male-sterile cytoplasm. Common corn rust fungus (*Puccinia sorghi*) is a frequently occurring disease in North America. Several aphid-transmitted viruses infect corn including maize yellow dwarf virus, barley yellow dwarf virus, sugarcane mosaic virus, and cucumber mosaic virus, maize dwarf mosaic virus.

Insects

Insects, and particularly worms, are a major problem for seed corn production. Prolonged and severe silk feeding by corn rootworm, the larval form of the cucumber beetle (*Diabrotica* spp.), and Japanese beetles (*Popillia japonica*) interferes with the success of pollination. *Carpophilus lugubris* is the most common sap beetle pest of sweet corn. They also feed on the silks and attack ear tips causing usually superficial damage. European corn borer (*Ostrinia nubilalis*) is a worm that can bore directly through the husk to eat developing kernels. Corn earworm (*Helicoverpa zea*) attacks through the silk end of the ear and eats developing kernels under the husks (Fig. 16.3). Cutworm (*Agrotis ipsilon*) larvae are large, 2.5 to 5.0 cm (1 to 2 in.) long when fully developed, smooth, dull-colored caterpillars, which curl tightly when handled. They hide in the soil during the day and feed at night on the base of small emergent corn plants. Dense populations of corn leaf aphid (*Rhopalosiphum maidis*) cause direct damage through the removal of photosynthates. Large amounts of honeydew deposited by aphid feeding on maize tassels can decrease seed yields by reduce pollen shed (Cerena and Glogoza, 2004). Several damaging viruses are transmitted by *R. maidis*.

References

Abendroth, L.J., Elmore, R.W., Boyer, M.J. and Marlay, S.K. (2011) *Corn Growth and Development.* Publication PMR 1009. Iowa State University Extension and Outreach Publication https://store.extension.iastate.edu/Product/Corn-Growth-and-Development (accessed 12 September 2023).

Beadle, G.W. (1981) Origin of corn: pollen evidence. *Science* 213, 890–892.

Bruns, H.A. (2017) Southern corn leaf blight: a story worth retelling. *Agronomy Journal* 109, 1218–1224.

Cerena, M.J. and Glogoza, P. (2004) Resistance of maize to the corn leaf aphid: a review. *Maydica* 49, 241–254.

Chang, E.H. (1981) Corn yield in relation to photoperiod, night temperature, and solar radiation. *Agricultural Meteorology* 24, 253–262. https://doi.org/10.1016/0002-1571(81)90049-2 (accessed 12 September 2023).

Cummins, D. and Motsenbocker, C. (2022) *Sustainable Gardening for School and Home Gardens: Sweet Corn.* LSU AgCenter, Baton Rouge, Louisiana. https://repository.lsu.edu/susgard/18 (accessed 12 September 2023).

Darrah, L.L., McMullen, M.D. and Zuber, M.S. (2003) Breeding, genetics, and seed corn production. In: Ramstad, P.E. and White, P. (eds) *Corn: Chemistry and Technology.* American Association of Cereal Chemists, Minneapolis, Minnesota, pp. 35–68.

FAO (2002) *The Role of Women in the Conservation of the Genetic Resources of Maize – Guatemala.* Food and Agriculture Organization of the United Nations and the International Plant Genetic Resources Institute, Rome. www.fao.org/publications/card/en/c/98c35da2-a63f-5856-b337-bca1843a6e1c/ (accessed 12 September 2023).

Fonseca, A.E., Westgate, M.E., Grass, L. and Dornbos, D. L. (2003) Tassel morphology as an indicator of potential pollen production in maize. *Crop Management* 2, 1–15.

Galinat, W.C. (1992) Evolution of corn. *Advances in Agronomy* 47, 203–231.

Gardner, F.P., Pearce, R.B. and Mitchell, R.L. (1985) *Physiology of Crop Plants.* Iowa State University Press, Iowa.

George, R.A.T. (2009) *Vegetable Seed Production.* CAB International, Wallingford, UK.

Goodman, M.M. (1988) The history and evolution of maize. *CRC Critical Reviews in Plant Science* 7, 197–220.

Jenkins, M.T. (1941) Influence of climate and weather on growth of corn. In: Hambidge, G. (ed.) *1941 Yearbook of Agriculture: Climate and Man.* US Government Printing Office, Washington, DC, pp. 308–341.

NDAWN (2023) *Corn Growing Degree Days (GDD).* North Dakota Agricultural Weather Network, North Dakota State University, North Dakota. https://ndawn.ndsu.nodak.edu/help-corn-growing-degree-days.html (accessed 12 September 2023).

Nielsen, R.L. (2000) *Scrambled Silks, Anyone?* Corny News Network, Purdue University, Indiana. www.kingcorn.org/news/articles.00/SilkBalling-0718.html (accessed 12 September 2023).

Nielsen, R.L. (2018) *Unusually Long Silks in Corn.* Corny News Network, Purdue University, Indiana. http://www.kingcorn.org/news/timeless/LongSilks.html (accessed 12 September 2023).

Nielsen, R.L. (2020) *Tassel Emergence & Pollen Shed.* Corny News Network, Purdue University, Indiana. www.kingcorn.org/news/timeless/Tassels.html (accessed 12 September 2023).

Pioneer Agronomy Sciences (2023) *Staging Corn Growth.* Corteva Agriscience. https://www.pioneer.com/us/agronomy/staging_corn_growth.html (accessed 12 September 2023).

Ranere, A.J., Piperno, D.R., Holst, I., Dickau, R. and Iriarte, J. (2009) The cultural and chronological context of early Holocene maize and squash domestication in the Central Balsas River Valley, Mexico. *Proceedings of the National Academy of Sciences USA* 106, 5014–5018.

Rubatzky, V.E. and Yamaguchi, M. (1997) *World Vegetables.* Chapman & Hall, New York.

Sears, F.B. (1982) Fossil maize pollen in Mexico. *Science* 216, 932–934.

Tracy, W.F. (2000) Sweet corn. In: Hallauer, A.R. (ed.) *Specialty Corns*, 2nd edn. CRC Press, Boca Raton, Florida, pp. 155–199.

Tracy, W.F., Whitt, S.R. and Buckler, E.S. (2006) Sugary1 and the origin of sweet maize. *Crop Science* 461, 49–54.

Welbaum, G.E. (2015) *Vegetable Production and Practices*. CAB International, Wallingford, UK, pp. 151–157 and pp. 248–266.

Ziegler, K.E., Ashman, R.B., White, G.M. and Wysong, D.S. (1989) *Popcorn Production and Marketing*. Publication CR-2104. Division of Agricultural Sciences and Natural Resources and Oklahoma State University. http://corn.agronomy.wisc.edu/Management/pdfs/NCH05_CR-2104web.pdf (accessed 12 September 2023).

17

Family Solanaceae

TOMATO

Overview

Tomato is indigenous to western South America. Wild species are found in the Andean region and adjacent areas which are considered centers of origin (Welbaum, 2015). Germplasm from this region is used for genetic improvement. The wild cherry tomato, *Solanum lycopersicum* var. *cerasiforme* (Dunal), is present throughout tropical and subtropical America and is a progenitor of cultivated tomatoes.

Tomato was unknown to Europeans until early explorers introduced collected specimens. Upon its arrival in Europe, tomato was considered poisonous. Tomato leaves and immature fruit do contain hazardous quantities of the alkaloid tomatine. The tomato was grown in the early American colonies as a curiosity called "love apple" and not widely used as food until after 1800. Other cultures accepted tomato as a food crop earlier. Tomatoes are now a very significant global food crop.

Botany

S. lycopersicum plants are short-lived perennials in their center or origin, but in temperate regions are grown as annuals. Plant stature ranges from highly indeterminate to strongly determinate. Growth habits vary from erect, to semi prostrate, to vining. Plant height varies from 0.15 to 10.0 m (0.5 to 33 ft) with solid, thick stems. In deep soils, strong taproots sometimes grow to depths of 1.8 m (6 ft). Leaves are compound pinnate, coarsely toothed, and curled or flat. Small glandular hairs on stems, leaves, and peduncles produce a distinctive odor.

Flowers are produced in clusters opposite and between leaves. Some genotypes have 30 or more flowers per cluster, but commonly the range is four to 12 on a broad, flat raceme. Flowers are perfect, about 2 cm (0.8 in.) in diameter. Yellow anthers fuse to form a pendent-shaped tube, often called the pollen tube when viewed longitudinally, with a yellow, star-shaped corolla (Fig. 17.1).

Self-pollination is most common. Flowers do not produce nectar. Cross-pollination, usually by bees, occurs to varying degrees. Pedicels, or fruit stems, of some genotypes have an abscission zone visible as a swollen "joint" a short distance above the fruit. Some cultivars, like those developed for mechanical harvesting, are "jointless" without an abscission layer on the pedicel. Therefore, jointless cultivars must separate cleanly from pedicel at the fruit junction with no stem attachment.

The tomato fruit is a fleshy berry with a slightly hairy surface when very young but smooth at maturity. Fruit of most cultivars are globose, but other diverse shapes such as elongated, plum, and pear-like exist as well. Noticeable

DOI: 10.1079/9781789243260.0017

Fig. 17.1. Open tomato flowers. Petals and fused anthers form a pollen tube that surrounds the ovary, stigma, and style. The pollen released by air movement drops onto the stigma inside the enclosed pollen tube resulting in self-pollination. (Courtesy of Tim Loeffler.) Also see Fig. 17.12.

lobes are present on fruits of some cultivars, an indication that they have multiple carpels. Ripe fruit can have a wide range of colors ranging from red, pink, tangerine, orange, yellow, purple, white, or green. Most fruit have solid colors but striped cultivars also exist. Red color is due the carotenoid pigment lycopene. Other carotenoid pigments cause orange color. Xanthophyll (oxygenated carotenoid) accumulates in smaller quantities and contributes to the yellow color in ripe fruit. Intermediate colors occur because of differing ratios of these pigments as well as anthocyanin (purple), and differences in skin color. Red tomatoes have a yellow skin and red flesh (pericarp), while pink cultivars also have red flesh, but because of a recessive gene, their skin is colorless. Yellow flesh, the result of another recessive gene, when overlaid by yellow skin produces bright yellow fruit. If yellow flesh is combined with colorless skin, the fruit is pale yellow (Rubatzky and Yamaguchi, 1997).

Types and Cultivars

Tomato genetics are well understood largely because of the late Dr Charles Rick at the University of California, Davis, who spent 60 years collecting and characterizing tomato germplasm. More recently, the tomato genome was sequenced and extensively studied by many.

For years, tomatoes used for both processing and fresh market were hand-harvested, large-fruited indeterminate cultivars. Beginning in the 1960s, processing and fresh market tomatoes diverged into distinct types through plant breeding. Small-fruited bush cultivars with concentrated fruit set were developed for once-over, destructive mechanical processing in anticipation of farm labor shortages at that time (Schmitz and Seckler, 1970). Bush plants produce many small, thick-skinned fruit with few and small locules resistant to impact and compression damage. In addition, red-ripe fruit remain attached to the vine until harvested. Vine storage of ripe fruit was a critical trait needed for high yields. Fruit did not abscise naturally but were easily detached by harvest equipment (Grandillo *et al.*, 1999).

Fresh market tomato fruit range from small to large and are predominately hand-harvested. Plants of fresh market cultivars are semi-determinate or indeterminate and often trained to a support system. Mechanical harvesting of mature, green, small or medium-sized fruit for fresh use occurs in some markets. Hand harvesting continues to predominate in most markets because fruit, particularly after the onset of ripening, are easily damaged.

Fresh market tomato cultivars are often divided into large, also known as Beefsteak or Beef type, medium-fruited which includes the TOV (tomato-on-the-vine) as well as saladette or salad types, and the small-fruited cherry and grape types. There is also an intermediate-fruited tomato with a size between medium- and cherry-sized tomatoes, known as cocktail types, that are primarily produced in greenhouses. Small-fruited, cherry, pear, or grape cultivars are popular because they are less prone to damage during harvesting, transport, and have become very popular as healthy snack foods. They can also be shipped at a more advanced stage of maturity with less damage. Some medium- and small-fruited cultivars are developed to produce clusters of fruit that ripen together. Fruit clusters, with the stems attached that can be harvested like bunches of grapes, have become popular with consumers and increase harvest efficiency. Fresh market tomato cultivars are classified and developed for open-field and protected growing environments. Field cultivars are predominantly determinate if unsupported or semi-determinate for string weave systems. Field cultivars have a shorter harvest period based on the season and cultivar grown. Protected cultivars are predominately indeterminate, are grown on high wire and string systems, and require extensive pruning and crop management. Protected cultivars are often grafted to provide root disease resistance and promote vine strength as the production season can be as much as nine months with vines growing as long as 10 m (33 ft). Protected growing environments range from unheated plastic or netted houses to heated glasshouses and the length of the harvest season varies. Outdoor cultivars require resistance to common field diseases. Breeding cultivars for protected culture focuses on resistance to diseases common to controlled environments.

To increase the firmness of vine ripe fruit, the Flavr Savr genetically engineered tomato was one of the first transgenic vegetables developed in the early 1990s. Flavr Savr contained a second copy of the tomato polygalacturonase gene inserted backwards (antisense direction) into the genome. The polygalacturonase enzyme produced by this gene degrades pectin, a component of the tomato cell wall, which causes fruit softening. When the antisense gene is expressed, it interferes with the production of the polygalacturonase enzyme, delaying fruit softening.

Other slow-ripening tomatoes have been developed by preventing or reducing the production of ethylene, the gaseous hormone that naturally triggers fruit ripening in tomato and other climacteric fruit. The transgenic cultivar Endless Summer expressed a truncated version of the ACC synthase gene. This interfered with the endogenous ACC synthase, impairing ethylene biosynthesis. Endless Summer and Flavr Savr were briefly marketed commercially; however, lack of consumer interest and acceptance of transgenic vegetables in North America and Europe prevented these cultivars from being successful. Neither are sold commercially.

Tomato plants are relatively easy to genetically engineer using *Agrobacterium tumefaciens*, so new transgenic cultivars can be expected in the future with improved environmental stress tolerance, pest resistance, improved nutrition, improved taste, etc. (Ruf *et al.*, 2001; Goyal *et al.*, 2007).

Naturally occurring genetic mutations that slow fruit ripening have been used by breeders to extend storage life (Matas *et al.*, 2009). The ripening inhibitor mutant gene *rin* limits ethylene production, and the non-ripening *nor* gene prevents ethylene biosynthesis. Another mutant gene, *ale*, prolongs the storage of harvested fruit, and has less adverse effects on color development. These mutants have been bred into cultivars such as the Extended Shelf Life (ESL) group that ripen slowly for long-distance shipping of fresh market tomatoes (Plunkett, 1996; Matas *et al.*, 2009). The ESL cultivars are another approach to developing tomatoes that can withstand the rigors of shipping at more advanced stage of ripeness. Since the *rin*, *nor*, and *ale* are natural mutations and not genetically engineered, consumers are more accepting of their use.

Tomatoes are increasing grafted to improve resistance to soil-borne pathogens and improve top growth. Cultivars often used as rootstocks include F1 Maxifort and Beaufort. Seed production of root stocks is the same as for other cultivars except that their end use is for grafting. Small seedling rootstocks are top or side grafted to production scions when the stems are about 2 mm in diameter and the seedling has two or three true leaves.

Growth and Development

Tomato cultivation is adaptable to varied environments. Production ranges from high elevations near the equator to temperate regions at higher latitudes. Tomatoes do not grow well in the humid tropics because of disease problems and may require grafting. Tomato is grown annually because it cannot survive temperatures at or below freezing. Tomato is chilling sensitive and damaged by prolonged exposure > 10°C (50°F). Chilling injury is a physiological disorder caused by exposure to low, non-freezing temperatures that significantly reduces fruit and seed production, particularly when exposure occurs before fruit maturity. Tomatoes grow optimally up to about 27°C (81°F). Pollen fertility may decline in some genotypes above this threshold particularly if there is little fluctuation between day and night temperatures. Some cultivars, such as Solar Set, are better adapted to setting fruit at high temperatures than others. Low night temperatures may help offset effects of high daytime temperatures (Willits and Peet, 1998).

Tomato is day-neutral, so flowering is not dependent on photoperiod. Tomato prefers high light intensities. High light intensity tends to accelerate flowering, whereas low light intensity reduces vegetative growth and delays flowering. There is a tendency for flower truss initials to abort at relatively low light levels at higher latitudes and in winter greenhouses (Kinet and Peet, 1997). Also, low light intensities and temperatures less than 10°C (50°F) or greater than 30°C (86°F) can cause green fruit drop, reducing seed yield (Rubatzky and Yamaguchi, 1997). An appropriate level of vegetative growth should be achieved before flowering begins, so plants are able to support maximum development with little fruit drop. Once reproductive growth begins, the fruit becomes the major photosynthetic "sink" with proportionally less fixed carbon directed to vegetative growth. Underdeveloped flowering plants typically have poor fruit and seed yields.

Soil Fertility and Nutrient Management

Tomatoes are grown successfully on a wide range of soil types, from sandy to fine-textured clays, as well as soils with high organic matter content. Organic or muck soils may produce excessive nitrogen through organic matter decomposition leading to overly vegetative "leggy" plants with poor fruit set. Tomato plants generally have extensive root systems, but most develop within the upper 60 cm (24 in.) of soil. Taproots may extend 6 ft (1.8 m) deep when not restricted by hard pans or high water tables. Deep root development affords some drought tolerance (Rubatzky and Yamaguchi, 1997).

Tomato grows best with uniform moisture and good drainage. Tomatoes are intolerant of waterlogging, especially during emergence and near fruit maturation. Excessive moisture promotes damping-off and root rot diseases. Raised-bed culture is recommended on moist heavy soils to improve drainage. Crop rotation is strongly recommended to help minimize soil-borne disease (Le Strange *et al.*, 2000). A crop rotation cycle of several years between other Solanaceous crops is necessary if pathogen pressure is high and fumigation or solarization is not used.

Tomatoes tolerate slightly acid to moderately acid soil conditions over a pH range 5.5–6.8. Nutrients are often applied in a split application, at planting and after fruit set. To avoid root burn starter fertilizer should be banded a short distance from transplants or seeded row. Nutrient applications are based on soil test results prior to establishment. Some N-P-K is applied at planting or transplanting to ensure optimum seedling growth. Liquid starter solutions are sometimes poured around the root ball at transplanting to provide nutrients for rapid uptake and water to alleviate transplant shock. Optimal mid-season applications by side dressing or fertigation can be accurately made based on petiole or foliar analyses.

Nitrogen is very important for vegetative growth. Fertilization with a higher ratio of NO_3^- compared to NH_4^+ encourages rapid early-season growth particularly when soils are cool and little nitrification occurs. Nitrogen is important for developing sufficient canopy size to maximize photosynthesis and support flowering, fruiting, and seed development. However, excess nitrogen fertilization may cause excessive vegetative growth, delaying fruit set and seed harvest (Rubatzky and Yamaguchi, 1997). The overall nitrogen fertility management goal for tomato is initially high availability to stimulate vegetative growth, lower levels at flowering, and increasing nitrogen again after fruit set. Fertigation is the most efficient application method for nitrogen particularly for hand-harvest plots where drip lines will not interfere. Side dressing is frequently used as well, particularly for mechanical harvesting (Le Strange *et al.*, 2000). Predicted soil nutrients removed when producing a tomato seed crop in kg/ha are: N: 70 to 125 (62 to 112 lbs/acre), P_2O_5: 150 to 225 (134 to 201 lbs/acre), K_2O: 150 to 225 (134 to 201 lbs/acre).

Sufficient phosphorus is needed for early plant development and flowering. A relatively high level of available phosphorus and potassium is important for a satisfactory seed yield and quality. Calcium is important for cell wall development and helps avoid ripening disorders and blossom end rot discussed below under Physiological Disorders (Le Strange *et al.*, 2000).

Pollination

Tomato flower structure favors self-pollination because anthers dehisce internally from fused anthers that form a pollen tube. Natural flower movement releases pollen which lands on the enclosed stigma without need for insects. When open-pollinated seed production occurs in protected structures or areas with little wind, daily flower agitation may improve pollen transfer from anthers to stigmas (Fig. 17.2). Stigmas remain receptive to pollination for four to seven days. Style elongation usually coincides with pollen release. Depending on temperature, fertilization occurs within 48 h after pollination.

In some cases, cross-pollination of tomato flowers may occur. Most modern tomato cultivars were developed to have short styles that remain inside the pollen tube. However, some cultivars produce flowers with elongated styles, which favors cross-pollination. Stylar elongation results in the stigma sticking outside of the pollen tube, so it doesn't receive pollen from the same flower. In some cultivars excessive style elongation is a result of high temperatures (Rubatzky and Yamaguchi, 1997). During F1 hybrid seed production, insects in open-field, unprotected environments may introduce foreign pollen after flowers have been emasculated for hand-crossing. Insect attraction to tomato flowers is more likely when seed crops are grown in Central or South America near their original habitat (George, 2009). Most hybrid seed production is now grown under net or plastic houses, greenhouses, or glasshouses to reduce production variables negatively impacting the crop and seed quality. High temperatures interfere with viable pollen production, pollen dispersal, and ovule viability (Ho and Hewitt, 1986). Hot, drying winds negatively affect pollination. Cool/cold temperatures and high humidity and/or low light intensity limit pollen shedding as well. The most temperature-sensitive periods for fruit set are about five to ten days before anthesis, and two or three

Fig. 17.2. Open-pollinated tomato seed production in a plastic covered enclosure that modifies the environment and keeps disease and insects out. (Courtesy of Tim Loeffler.)

days following pollination. Temperatures at or below 15°C (54°F) inhibit pollen formation and function. Fertilization is favored by day temperatures between 21°C and 30°C (70°F and 86°F), and night temperatures between 15°C and 21°C (59°F and 70°F). For many cultivars, day temperatures above 32°C (90°F) reduce fruit set, which is negligible at 40°C (104°F). The period from fruit set to seed maturity coinciding with fruit full physiological maturity ranges from 32 to 60 days depending on the cultivar and environment.

The minimum isolation distance between different tomato open-pollinated cultivars for seed production is relatively short because of the high level of self-pollination. Most experts specify separation distances of 30 to 200 m (98 to 656 ft). The isolation distance should be sufficient to avoid accidentally mixing fruit from different plant genotypes at harvest. Distances between two parental lines grown for F1 hybrid seed production may be as close as 2 m (6.6 ft). Even closer spacing is possible in greenhouses (George, 2009).

Irrigation

A continuous supply of water either from rain or irrigation is recommended to produce tomato seed crops. Water requirements for tomato seed crops are about 25–30 mm (1–1.2 in.) weekly. This recommendation applies to tomato seed production, although more may be required for sandy soils with little organic matter. On a hot dry day, evapotranspiration may exceed 10 mm (0.4 in.) in desert production areas. Mid-season crop coefficients that relate water needs to evapotranspiration generally range from 1.05 to 1.25 for tomato (Hanson and May, 2006). Although irrigation frequency and amounts vary, tomato seed crops grown may require 600–900 mm (24–35 in.) of water over an entire season.

Tomato is susceptible to several serious foliar diseases, so keeping foliage dry is a priority. Surface or subsurface drip irrigation is used for tomato seed production with the benefits of using water efficiently while keeping foliage dry. Drip systems often require less labor, utilize less water, reduce diseases, and restrict between-row weed growth. Other advantages include the ability to fertigate. Sprinkler irrigation is not allowed for most hybrid seed production and is rarely used due to increased foliar diseases. Furrow irrigation is inexpensive, uses large quantities of water, requires flat land, should be used with raised beds, and may be difficult to control and is less frequently used.

Rogueing

Rogueing is usually conducted in three stages. The first stage is prior to flowering. Plants should be carefully checked for desirable characteristics such as growth habit, leaf morphology, stem characteristics, and color such as anthocyanin typical of the cultivar. Diseased plants should be recorded, sampled, removed, and destroyed.

The second stage is at early flowering and first set before significant fruit growth. Inspect plants for general plant morphology and characters as defined for stage one. Check immature fruit for the correct phenotype according to the cultivar description.

Stage three of rogueing is prior to harvest and will include fruit quality, morphology, color, relative size, and diseased plant removal. Crops testing positive for bacterial canker and other quarantined pathogens may need to be destroyed based on local laws. More benign diseases may require additional testing, sanitation, and disinfection treatments before the seed can be sold.

Genetics

Tomato cultivars can be F1 hybrid, open-pollinated, or transgenic. Use of F1 hybrid cultivars has increased over the past 30 years. Hybrids are widely used for both fresh market and processing because of their extra earliness, improved uniformity, vigor, and increased yield, and disease resistance compared to open-pollinated cultivars. Crossing two inbred parental lines by hand produces F1 hybrid seeds and requires emasculation of the female parent and pollen collection from the male parent (Fig. 17.3). Male sterility is available and used by some companies for hybrid tomato seed production. The extra labor and development make hybrid tomato seed more expensive compared to seed from open-pollinated cultivars (Fig. 17.3). Transgenic cultivars have been developed but currently are not widely grown because of a lack of consumer acceptance and

Fig. 17.3. Cross-pollination by hand for F1 hybrid tomato seed production. (Courtesy of Tim Loeffler.)

the high cost of development/regulation. When producing hybrid seed, lots should be tested by polymerase chain reaction (PCR) to ensure the proper hybrid cross was made, particularly when genes code for traits without an obvious phenotype.

Production

Hybrid cultivars dominate commercial markets. Much of the F1 hybrid tomato seed is grown in Peru, Chile, India, Thailand, and China where labor costs are lower and exported to a central processing facility and then re-exported for sale. Most hybrid seed production takes place in net houses, greenhouses, or glasshouses depending on the production area to control the environment, exclude foreign pollen, and control diseases and pests (Fig. 17.4). Tomato seed production has evolved significantly over recent years with an increasing emphasis on seed-borne disease management, control, and prevention. In general, the more a growing environment is protected, the higher the level of control required in seed production. Typically, open-pollinated and heirloom cultivars for outdoor field production are grown in the open with little or no control over sanitation practices. Controlled environment tomato vegetable production has increased in acreage, is highly developed in the Netherlands, France, and Canada, and is rapidly expanding in many other areas of the world. Growth in greenhouse vegetable tomato production has increased demand for high-quality seed of tomato cultivars developed for controlled environments. Many countries require Good Seed Production Practices (GSPP) certified seeds be used for greenhouse tomato production (GSPP, 2023).

GSPP-certified seeds used in heated glasshouses require completely closed seed production systems because the threat of diseases such as bacterial canker is great. GSPP certification requires extensive control and sanitation procedures, testing, and documentation at all steps of production (GSPP, 2023). Seed production of glasshouse cultivars occurs in temperature-controlled greenhouse structures with high mounted support systems using rockwool, coconut coir, or similar substrates.

For disease control, seed production should not occur in the same space where tomatoes or related Solanaceous crops were grown in the previous two or three years with some seed companies requiring up to five-year rotations.

Fig. 17.4. Greenhouse tomato seed production with supplemental lighting. (Courtesy of Tim Loeffler.)

Growing tomato seed crops in rotation with legumes, rice, wheat, and Brassicas may reduce diseases and nematodes. With soil fumigation using steaming, solarization, or a suitable sterilant, successive or related seed crops can be grown with little or no rotation time in some cases.

Establishment

Tomato plants for seed production are usually established from plug transplants (Fig. 17.5). In most cases, plug transplants, produced in disease-free soilless media, have replaced bare-root transplants produced outdoors in soil (Styer and Koranski, 1997). Almost all seed companies require their stock seed to be tested and free from seed-borne diseases prior to planting. Planting contaminated seed can infect other nearby tomato transplants or seed crops. Transplants improve uniformity, control insects, and reduce diseases prior to field or greenhouse establishment.

Before planting, fungicides or biologicals are added as seed treatments to reduce damping-off disease. Seeds are planted in plug trays and placed in a "germ room" for a few days at ideal temperatures and high relative humidity to encourage uniform germination. After emergence, trays are transferred to a greenhouse to complete development prior to transplanting at the five- to seven-true-leaf stage.

Planting densities depend on whether a cultivar is indeterminate or semi-determinate grown on a support system or a bush type. Traditional spacing for supported plants is 90 to 120 cm (3 to 4 ft) between rows with in-row spacing of 30 to 60 cm (1 to 2 ft). For protected seed production, populations range from 15,000 to 30,000 plants/ha (6000 to 12,000 plants/acre) depending on the cultivar, the growing environment, pruning methods, number of stems and support systems to be used.

Plants are established by hand or using transplanters on bare soil or raised beds. Often plastic film mulch is installed over the rows to help modify soil temperatures, conserve moisture, and for weed control. Tomatoes seed crops can be grown in open-field locations with at least four months of warm, frost-free weather, with average temperatures greater than 16°C (61°F). Vegetative and reproductive growth at lower temperatures is slow, and an extended period of exposure to 10°C (50°F) or less causes poor stands and chilling injury. Later in the season, chilling injury increases membrane leakage and metabolic abnormalities in affected fruit (Zhao

Fig. 17.5. Tomato plug tray transplants growing in a greenhouse before transplanting. (Courtesy of Tim Loeffler.)

et al., 2009). Consequences of chilling injury are failure to ripen and develop full fruit color, irregular (blotchy) color development, premature softening, surface pitting, browning of seeds, and increased decay (Snowdon, 2010). A wide diurnal difference between day and night temperatures tends to improve flowering, foliage growth, and fruit quality. Fruit set is best between 21°C and 30°C (70°F and 86°F) with night temperatures between 15°C and 21°C (59°F and 70°F). For many cultivars, day temperatures above 32°C (90°F) reduce fruit set, which is negligible at 40°C (104°F). Night temperatures are more critical than day temperatures for good fruit set (Rubatzky and Yamaguchi, 1997).

Training plants to stakes or overhead strings keeps plants upright and easier to work with. The plants are pruned to establish the number of main stems, to remove side shoots, and to remove basal leaves to balance plant growth. Typically, determinate plants have three or four stems and indeterminate plants can have one to four stems per plant.

F1 hybrids

Hybrid seed production requires crossing and maintaining of two compatible parental inbred lines that serve as male and female parents. The male serves as the pollen parent and is normally sown two to four weeks earlier than the female plants to ensure adequate pollen for crossing. The ratio of male to female plants depends on the flowering habits of the individual parental lines. A male-to-female ratio of about 1:4, is a general guideline. Before crosses are made, plants from both lines must be checked for trueness-to-type and inspected for seed-borne disease symptoms. Off-types, undesirable, or suspect plants should be removed as explained for the first stage of rogueing. Suspect plants with certain seed-borne diseases require proper documentation, sampling, and testing to confirm presence to decide the outcome of the crop and quarantine requirements. Flowers on the female line must be emasculated during the late bud stage before the flower opens in preparation for cross-pollination. Emasculation removes anthers before pollen is shed so selfing cannot occur and is a separate operation prior to the hand pollination (Fig. 17.6).

During emasculation, the anther cone is cleanly removed from the intact flower by pulling the base upward with forceps. Emasculation takes practice and is a skill that workers must acquire thorough training and practice. The stigma must not be damaged during emasculation, so successful pollination and fertilization can occur (Fig. 17.6). The timing of emasculation

Fig. 17.6. Emasculated tomato flower with pollen tube (fused anthers) removed by hand. The pale green ovary, style, and stigma, yellow petals, and green sepals are visible with unopen buds in the background. This flower will be the female parent for F1 hybrid seed production. (Courtesy of Tim Loeffler.)

is related to the rate of flower development. Environmental conditions determine how long it takes for flowers to open and when emasculation should be conducted. Emasculation is often performed in early morning but can be done anytime during the day in some areas and can be the same day as hand pollination. The timing of emasculation can occur every other day for some female parents and daily for other female parents. For greenhouse seed production, flowers are emasculated up to two days in advance of pollination. Male-sterile lines are sometimes used as the female parent, but are less common than emasculation and hand pollination. When male-sterile lines are used, bees can transfer pollen from male to female flowers. Since pollination may be unsuccessful since tomato flowers lack nectaries to attract bees, hand pollination may be the most reliable method to obtain high seed yields when using male sterility. Male sterility only prevents unwanted selfing. Seeds of the male-sterile line are produced by crossing with a male fertility restorer line.

Pollen is collected from the male line for crossing either by detaching individual flowers or shaking open flowers and collecting pollen. Detached flowers are dried and shaken in a sealed container to release and collect pollen. Drying is usually for one to two hours at about 25°C (25°F), 30 to 40% relative humidity. Flowers are discarded after pollen collection. The pollen should be light yellow. Usually, the pollen is stored overnight and is used the next day, or the female line can be pollinated immediately. Alternatively, pollen can be stored for future use. Tomato pollen lasts longest if frozen and stored at low humidity. Frozen tomato pollen can be stored for 12 months, although the practical limit appears to be closer to six months. Another technique for collecting pollen is by oscillating each open flower with a vibrator, like a modified electric toothbrush, to release pollen into a tube or small sealable vial placed at the open end. The collected pollen can be applied with a fine brush or finger to each stigma of the female parent flower (Fig. 17.7). Hand-pollinated flowers may be tagged, or a lobe of the calyx removed with small clippers to clearly confirm that a hybrid cross was made immediately after pollination (Figs 17.6, 17.7). In a seed production field, emasculation, pollen collection, and hand pollination usually take place three times a week.

For greenhouse seed production, flower buds are checked daily or on alternate days, depending on temperature and day length, to determine if emasculation is needed.

Open-pollinated production

For open-pollinated seed production, plants are allowed to self and are carefully rogued for off-types.

Fig. 17.7. Greenhouse tomato seed production. Indeterminate plants are trained vertically. Strings are tied around tomato stems to mark hand-pollinated fruit. (Courtesy of Tim Loeffler.)

When maintaining stock or foundation seed and parental lines for future crossing, care must be taken to clearly label and separate different genotypes. This will ensure fruits of different identities remain separate and are not accidently mixed at harvest. It is important to physically separate and properly label plants, harvested fruit, and seeds of different genotypes, so accidental mixing cannot occur.

Fruit ripening

Tomato fruit are climacteric, and ripening is naturally triggered by the gaseous hormone ethylene (Brady, 1987). Depending on the cultivar, most tomato fruit mature 32 to 60 days after anthesis. Temperature greatly influences the rate of fruit ripening with faster ripening occurring at higher temperatures. Optimum temperatures for fruit maturation and color development are between 20°C (68°F) and 24°C (74°F). At favorable temperatures, the red pigment lycopene develops during ripening even in the dark. However, light accelerates color development and intensity. Lycopene synthesis is inhibited to a greater extent than other carotenoids above 32°C (90°F) or below 10°C (50°F). At these extremes, mature fruit tend to develop a yellowish to orange-red color instead of deep red. At temperatures greater than 40°C (104°F), fruit tend to remain green because chlorophyll isn't broken down (Brady, 1987).

Seed Harvest

For best seed quality, fruits should be harvested when fully mature and ripe. Harvest color can vary with cultivar. Highest seed quality occurs when fruits are at the red-ripe stage just beyond the point of edible maturity. Seeds harvested from immature fruit at the mature green or breaker stages are not physiologically mature and have lower vigor. Immature seeds require additional dry after ripening to obtain full vigor. Over maturation and delayed harvest can cause accelerated aging of seeds while in fruit. Color development and fruit softening occur in response to ethylene (Brady, 1987). Ethylene application 100–150 ppm can accelerate ripening when supplied at or slightly beyond the mature green stage, but this doesn't enhance seed development and is not used for seed production.

Precocious seed germination in fruits is prevented because of embryo abscisic acid levels and the osmotic potential of the fruit, which prevents full hydration (Berry and Bewley, 1992). Plant breeders should select against precocious seed germination during cultivar development. Nevertheless, it still occurs on occasion during seed production particularly if harvest is delayed (Fig. 17.8). Seed after-ripening can occur in fruit

Fig. 17.8. Precocious seed germination in tomato fruit.

when exposed to high temperatures for prolonged periods. Just as in dry stored seeds, after-ripening of hydrated seeds in fruit increases seed vigor as abscisic acid levels decline.

Fruits of some open-pollinated cultivars are collected by a destructive mechanized harvester in a single operation. Once-over harvesting combines seeds of different stages of fruit development, contributing to variable seed quality. Destructive mechanical harvesting is less common due to the increased use of F1 hybrid seed. Hand-pollinated hybrid fruit are hand-picked from the female lines whether in the field or under protective cover in containers usually one or two times each week. This ensures that only healthy fruits that have been cross-pollinated are used. Marking fruit, by labeling or calyx modification, confirms that a cross was made and is extremely important to exclude selfed fruit from hybrid seed lots. Fruits can be extracted the same day or kept overnight under cool temperatures to allow after-ripening before extraction.

Post-harvest processing includes seed extraction followed by cleaning, washing, seed sanitation, and drying. Seed extraction can be performed manually or mechanically. For manual extraction, the fruits are cut equatorially and the seeds and surrounding gelatinous material are squeezed or spooned into containers or the fruits can be placed into a bag and crushed (Fig. 17.9). Mechanical extraction feeds hand-picked fruit directly into a mobile fruit crusher positioned in a field or at a seed-processing facility. The crusher macerates the fruit into a purée of mucilage, seed, juice, and fruit pieces. This mixture of fruit tissue and seeds is next passed through a revolving cylinder screen that allows the seeds and juice to pass through the mesh where they are collected below. Fruit debris is conveyed through the sloping cylinder screen where it falls back the field or is captured in a bin for recycling. The isolated seeds with gelatinous material are next processed using one of the seed separation techniques described below.

Seed separation

After seed extraction from fruit, additional wet processing is required. Seed separation isolates clean seeds from fruit juice and mucilage before final washing and drying. Tomato seed mucilage is the outer layer of the seed. Several seed-cleaning technologies can remove gelatinous tissue that tenaciously adheres to the seed coat without damaging viability. The method of separation will depend on the target market (protected v. open-field), quantity of seed, need to control seed-borne pathogens, phytosanitary requirements for seed movement, and environmental temperatures. The following are commonly used separation technologies.

Fig. 17.9. Ripening tomato fruit in cross-section. Seeds are embedded in a thick gelatinous material that must be removed during processing. Seed harvest should occur after fruits are red-ripe slightly beyond edible maturity but before severe fruit decomposition. (Courtesy of Tim Loeffler.)

Fermentation

Fermentation relies on natural microbial activity to break down the gelatinous mucilage around the seed. The pulp containing the extracted tomato seed is left to ferment for up to three days over a temperature range from 20°C to 35°C (68°F–95°F) (Fig. 17.10). The rate of fermentation is dependent on the ambient temperature and can take as long as five days at lower temperatures. In warmer areas, natural fermentation may be completed in as little as 24 h.

For natural fermentation, the mixture must be stirred several times daily to maintain uniformity and to avoid seed discoloration. Fermentation containers are often covered with cloth to reduce insect activity. Frequent inspections will determine when the seeds' gelatinous coating has sufficiently broken down. A downside to fermentation is that it must be carefully managed to avoid seed damage from the heat generated and over hydration. Hydration after the mucilage is digested can cause premature germination or accelerated aging if seeds remain hydrated at high temperatures for too long. It has been claimed that fermentation reduces seed-borne bacterial canker of tomato (George, 2009); however, other more effective seed disinfection methods are now used.

Pectinase enzyme separation

A newer method for separating tomato seed mucilage uses pectinase to enzymatically degrade it. Pectinase, available commercially, can be added to the seed mucilage and juice solution or after fermentation as a final wet cleaning step. Pectinase treatment is popular because it effectively and rapidly cleans seed without damage. Pectinase is most effective if the pH of the tomato crude tomato extract or final aqueous cleaning solution is adjusted to 3.0 to 6.0 with hydrochloric acid. The treatment temperature is usually under ambient conditions. An aqueous pectinase solution is prepared by mixed 3 g/l of pectinase to water. The most effective concentration of enzyme solution may vary with the composition of the crude tomato extract, purity of the enzyme, solution pH, and incubation temperatures. The enzyme solution is added to the fermentation or final cleaning solution and stirred periodically. After as little as two hours and not more than 24 hours, the seeds are rinsed clean from mucilage and other debris. This procedure shortens the time seeds are hydrated after extraction. Enzyme treatment also prevents exposure to high temperatures and prolonged hydration, preventing damage by premature germination or excessive heat.

Fig. 17.10. Buckets of fermenting tomato seed and pulp after extraction from a tomato seed field in the Sacramento Valley of California, USA.

Separation with sodium carbonate

This method is relatively safe and is effective for small quantities of seed in cooler areas as an alternative to natural fermentation. The juice, mucilage, and seed slurry is mixed with an equal volume of 10% sodium carbonate solution. The mixture is incubated for up to two days at room temperature. At the end of the treatment period, the seed is washed in a sieve and dried. Sodium carbonate separation tends to darken the seed coat and is not normally used for commercial seed lots. Plant breeders and seedsmen use sodium bicarbonate for maintaining breeding material and inbred parental lines.

Seed washing after separation

After separating the mucilage, the isolated seed is thoroughly washed to remove residue and any remaining debris before disinfection. The seeds can be washed in a water trough or flume for larger seed volumes. Small quantities of seeds are often washed in buckets.

Seed disinfection

Disinfection of tomato seeds is a critically important step to reduce or eliminate seed-borne pathogens from the surface of the seed. After washing, the seeds are usually disinfected with hydrochloric acid and often other chemicals such as trisodium phosphate (TSP) before drying. Alternatively, the seeds are dried and may go through a seed disinfection step later after the crop arrives at a processing facility. Most hybrid seeds are exported from the country of production to another country for seed cleaning and packaging and eventually sold and re-exported to many other countries for sales. There can be extensive phytosanitary requirements for seeds to be shipped from one country to another and seed disinfection can be required for export for certain target pathogens, especially quarantined diseases.

Disinfection with hydrochloric acid

Commercial producers long favored this method because it kills bacteria and produces light-colored clean seed. Hydrochloric acid treatment also meets the export and import requirements for shipping seeds to other countries for additional processing. Various dilutions and durations of treatment are used commercially. Hydrochloric acid treatment is usually applied after separation and seed washing but is sometimes used in conjunction with fermentation. One recipe for HCl treatment of small tomato seed lots uses 575 ml of 37% hydrochloric acid mixed with 10 l of seed and pulp for 30 min. The acid should be added to the water and seed solution. Adding acid to water generates less heat and reduces the likelihood that acid will splatter onto workers. Workers handling hydrochloric acid solutions must wear appropriate face shields and protective clothing. Sodium bicarbonate and the hydrochloric acid separation methods reduce seed coat transmission of virus disease (George 2009). Hydrochloric acid treatments may remove pubescence on some tomato seeds allowing them to flow better in bulk. Hydrochloric acid treatment also helps control seed-borne transmission of bacterial canker disease (Dhanvantari, 1989). Over-exposure or excessive concentrations of hydrochloric acid may damage seeds. After the treatment process is complete, the seeds are thoroughly rinsed and centrifuged before they are subjected to other disinfection agents or are dried.

Disinfection with trisodium phosphate and calcium hypochlorite

Another disinfection treatment to kill seed-borne diseases and particularly viruses is TSP. Treatment is carried out immediately after separation and washing but before drying. TSP is also used following hydrochloric acid and fermentation treatments. Wet seeds are soaked in 10% trisodium orthophosphate solution for 30 to 60 min. and then immediately rinsed in clear water, centrifuged, and dried. Calcium hypochlorite is used as an optional disinfection agent when certain bacterial pathogens may be present.

Washing and drying

After the final seed separation treatments described above, it is essential that tomato seeds be thoroughly rinsed to remove residues. Even tiny amounts of residue can cause clumping and interfere with seed flow during planting. There are several ways to wash large quantities of seeds efficiently. A series of increasingly smaller pore sieves can filter debris while allowing clean seeds to pass through. Large quantities of extracted seeds are washed in a long water trough, also called a sluice or flume, with a fall of about 2.5 cm per 125 cm trough length (1 in. per 50 in.). The trough has ripples at various intervals, so the denser "good" seed sinks while undeveloped seeds and other debris floats out a drain near the end. Operating a seed washing trough effectively comes with experience. It is wise to have a suitably sized catch screen over the wastewater drain to protect against accidental seed loss (George, 2009).

Seeds should be dried immediately after final cleaning, because prolonged hydration can initiate germination or damage seeds due to accelerated aging. Care should be taken not to dry seeds too rapidly or at temperatures exceeding 35°C (95°F) because either action can damage seeds. Seeds can be dried using different technologies.

To rapidly remove excess surface moisture, tomato seeds are placed in strong mesh bags and spun in a centrifuge to remove loosely bound surface water (Fig. 17.11). Seed can be dried in several ways. Precision rotary dryers with temperature and relative humidity control are the preferred drying system for F1 hybrid, high-value seeds. An alternative method is drying tables with forced air and is common in most seed production areas. Sun drying in warm climates with low humidity can be accomplished by spreading seeds on cloth or screen and brushing occasionally to expose both sides if precision drying equipment is not available.

Post-harvest Seed Conditioning

After the seeds are dried, they are precleaned using air screen cleaners or air column separators to remove chaff, light seeds, and deformed seeds, sampled for initial testing, and packaged for export or shipment to a seed-processing plant. Samples are submitted for germination, seed health, and genetic testing. Most tomato seeds are exported from the country of seed production to a regional or central global location for seed processing, packaging, and distribution. For seeds to be exported, additional seed health testing may be required to meet phytosanitary requirements before the seed can be shipped.

Upon arrival at the seed-processing location, additional sampling occurs to confirm preliminary results and to perform more rigorous testing. Based on the results for seed health tests, additional seed disinfection may be required. Clean dry seeds are further upgraded using seed polishers, aspirated screen cleaners, seed density separation tables, and liquid density separation

Fig. 17.11. Tomato seeds are placed in strong mesh bags and spun in a centrifuge to remove excess water. (Courtesy of Tim Loeffler.)

as needed for final seed cleaning. Seed samples are collected of the final cleaned seed and again tested for seed health, seed viability, and physical purity. Dry seed is held in bulk containers, bags, or packed ready to ship in moisture-proof packaging and held in controlled cold seed storage until shipment. Prior to packaging, seeds may be primed and/or pelleted depending on market requirements and demand.

Seed Yield

Tomatoes fruit contain 150 to 300 or more light cream to brown-color, flat seeds. Seeds are typically 2 to 3 mm (0.08 to 0.12 in.) long and range in pubescence from sparce to heavy. There are approximately 300 to 350 seeds/g (8500 to 10,000 per oz) (Rubatzky and Yamaguchi, 1997). It is difficult to provide an average since seed yield varies significantly among geneotypes and production areas. Yields vary according to fruit type, type of plant, e.g. determinate or indeterminate, number of fruit trusses per plant, and plant density. There are three main methods of estimating tomato seed yield: according to weight of fruit; according to unit area of plants; and counting seeds per fruit.

For greenhouse seed production, 1 kg (2.2 lbs) of fruit will produce approximately 4 g (0.14 oz) of seed (approximately 1200 seeds). Large-fruited greenhouse tomato cultivars tend to have larger seeds with a 1000 seed weight of 3.3 g (0.12 oz) more than field or determinate cultivars (George, 2009). Cherry or grape types have much smaller seeds. A more accurate determination is to count the number of seeds per fruit and multiply by historical seed weight. By comparison, field and determinate cultivars have a 1000 seed weight of about 2.5 g (0.09 oz) or 240 to 400 seeds per g (0.04 oz). Seed weight is 1% of fruit weight, e.g. 1 kg (2.2 lbs) of fruit will produce 10 g (0.35 oz) of seed. The expected tomato seed yield in the USA is between 250 and 400 kg seed/ha (223 and 357 lbs/acre). In Africa, yields range from 10 to 50 kg/ha (8.9 to 44.6 lbs/acre) (George, 2009). A multiplication factor of 200 can be used for indeterminate cultivars, while for field-grown determinate cultivars a multiplication factor of 50 is more accurate (George, 2009).

Physiological Disorders

Tomato is subject to several ripening disorders that should not be confused with genetic off-types or pathogenic diseases during rogueing. True-to-type fruit should be firm and not easily deformed when overripe (Le Strange *et al.*, 2000). Non-pathogenic fruit disorders are sometimes characteristic of certain cultivars and usually induced by moisture stress, nutrient stress, and temperature extremes. A fully pollinated quality tomato fruit should appear smooth, with small blossom- and stem-end scars. In addition, normal fruit should be free from growth cracks, cat-facing, zippering, sun-scald, insect injury, and mechanical injuries. Physiological disorders include blossom end rot, which is the formation of a necrotic black sunken area on the fruit. Blossom end rot is a water-stress-induced localized calcium deficiency that tends to be much worse in some cultivars than others (Snowdon, 2010). Ensuring that tomato seed fields are properly limed and adequately watered during the season can prevent blossom end rot.

Blotchy ripening is a failure of sections of fruit wall to color uniformly. The tissue tends to remain firm and may be white, pink, or yellow but sometimes turns brown. The disorder seems to be associated with mineral stress and cool conditions. Internal browning is caused by discoloration of vascular tissues in the fruit wall. Gray wall is observed as areas of discoloration within the fruit wall. One form of non-uniform ripening is attributed to a toxin produced by whitefly feeding (Snowdon, 2010).

Additional fruit physiological disorders are zipper, a long, thin brown scar tissue resembling a zipper, which usually runs from the stem end to the blossom end caused when an anther or other flower part remains attached to the newly forming fruit. Fruit puffiness is caused by inadequate pollination resulting in partially filled locules.

Cat-face is also caused by poor pollination and results from low or high temperatures at pollination (Fig. 17.12). Symptoms of cat-facing include non-uniform locule formation and a large, irregular, corklike blossom scar that resembles a cat's face on the bottom of the fruit. Some cultivars like Beefsteak or Ponderosa tend to produce a high percentage of cat-faced fruit because their stigmas naturally extend beyond the pollen tube making incomplete pollination likely. Recent studies have shown that the *SlLst* gene plays a crucial role in stigma exertion. Low light intensity may also contribute to cat-facing in some cultivars. Radial cracking is a genetically controlled trait that is a problem in some heirlooms. Concentric fruit cracking is caused by water collecting around the fruit stem. Most modern cultivars were developed with genetic resistance to cracking disorders. Green shoulder is a condition where the top of the fruit fails to turn red and remains green and hard even in ripe fruits. Many modern cultivars contain the uniform ripening gene that prevents green shoulder formation. Bleached tissue with pale white or yellowish coloration is a symptom of sun-scald. Direct sunlight causes sun-scald when high temperatures kill fruit tissue due to poor foliage cover.

Chilling injury caused by exposure to temperatures below 10°C (50°F) causes membrane leakage and prevents proper ripening, irregular color development, premature softening, surface pitting, browning of seeds, and increased secondary infection, especially black mold caused by *Alternaria* spp. Freezing injury is initiated at about −1°C (30°F) for a few hours. Symptoms include a water-soaked appearance of tissues, excessive fruit softening, and desiccation of locular gel.

Pest Management

Weeds

Weed competition is most damaging early in the growing season. Popular weed control strategies include stale bed, mulching, fumigation with approved chemicals, solarization, cultivation, and selective herbicides (Stall and Gilreath, 2002). Soil solarization controls many soil-borne diseases, nematodes, and weed pests. Removing weeds in ditches, fence rows, and surrounding fallow areas helps reduce populations in seed fields. Weeds also harbor disease pathogens and insects that transmit them. A stale bed prepared ten days to two weeks before seeding or transplanting often allows the first flush of weeds to be killed with shallow cultivation or non-selective herbicides. When tomato seedlings are about 10 cm (4 in.) tall, shallow cultivation can effectively create a dry layer of soil along the row to prevent weed seeds from germinating and to smother small weed seedlings (Stall and Gilreath, 2002). Two parasitic plants, dodder *(Cuscuta* spp.) and broomrape *(Orobanche ramosa)*, are major pests of tomato in some seed production areas that are difficult to control. Weed management is easier with transplanted crops, because clean cultivation can be practiced until establishment. Often seed crops are transplanted through plastic mulch to control weeds. Crop rotation can effectively

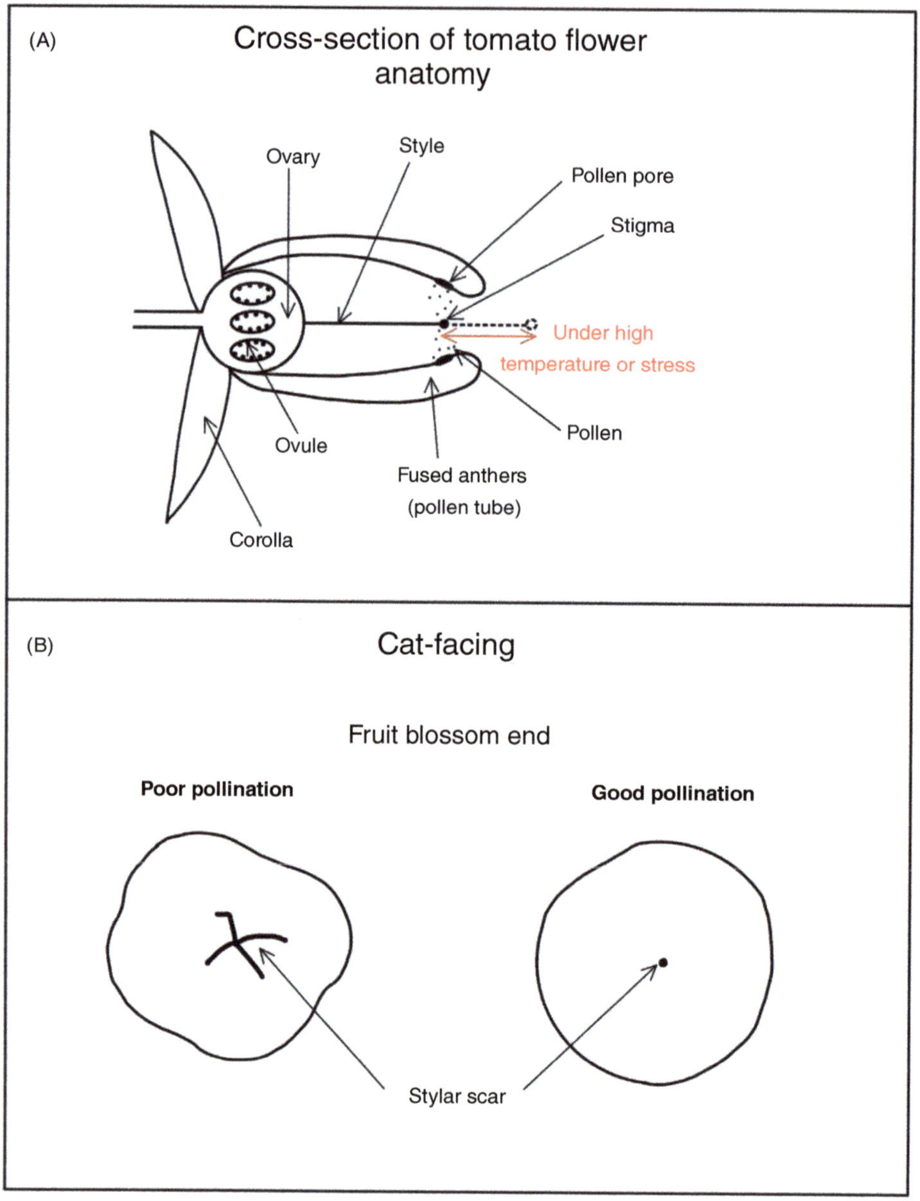

Fig. 17.12. (A) Cross-section of tomato flower anatomy. In some genotypes, the style naturally elongates beyond the pollen tube, resulting in poor pollination and encouraging cross-pollination by insects. Other genotypes experience style elongation during stressful conditions like high temperature. (B) Poor pollination reduces seed yield and causes the style to adhere to the ovary because of slow flower senescence, resulting in a blossom end scar called "cat-facing." (Courtesy of Stephanie Welbaum.)

reduce difficult weed problems by altering the environmental conditions that favor a particular species or by allowing alternative control methods. For example, alfalfa (*Medicago sativa*) is a good rotation crop with tomato because its frequent cutting cycle keeps certain weeds from reproducing.

Maize (*Zea mays*) is also effective in rotation with tomato because some maize herbicides control weeds related to tomato such as *Solanum nigrum* and *Solanum carolinense* as well as field bindweed (*Convolvulus arvensis*). Maize is not a host for dodder (*Cuscuta* spp.) (Stall and Gilreath, 2002).

Pathogens

Many diseases affect tomato. Seed production areas where critical seed-borne diseases occur may not be used until disease management and control issues are resolved. Many countries have quarantined pathogens that require extensive reporting and resolution if present in seed production. For example, positive confirmation of one prohibited diseased plant in a seed or vegetable production greenhouse in some countries requires complete removal and destruction of the crop. GSPP-certified seed with a focus on producing clean seeds at all steps of production, processing, and handling is a requirement by many greenhouse seed producers (GSPP, 2023).

Bacteria diseases of tomato include bacterial canker (*Clavibacter michiganensis* subsp. *michiganensis*), a seed-borne disease that spreads rapidly and is one of the most serious greenhouse diseases. Quarantine requirements in many countries require that bacterial canker contaminated seed lots must be destroyed. Planting clean seeds, growing disease-free transplants, and managing all steps of production are the most important measures for control. Bacterial wilt (*Ralstonia solanacearum*) is a serious disease in hot and humid areas. There are no effective chemical options, but long-term crop rotation, wilt-resistant cultivars, and sanitation practices provide control. Bacterial spot (*Xanthomonas vesicatoria*, *X. campestris*) are serious seed-transmitted diseases during cold, wet weather. *Xanthomonas* spp. damage blossoms, foliage, and stems but are difficult to detect on mature fruit. Control measures include copper fungicides, resistant cultivars, disease-free seed, and healthy transplants. Cultivar resistance is unavailable for some strains of bacterial spot. Other diseases include bacterial speck (*Pseudomonas syringae* pv. *tomato*), bacterial soft rot (*Erwinia carotovora* subsp. *carotovora*), pith necrosis (*Pseudomonas corrugata*), and syringae leaf spot (*Pseudomonas syringae* pv. *syringae*) (APS, 2016).

Important fungal diseases of tomatoes are early blight (*Alternaria solani*), black leaf mold (*Pseudocercospora fuligena*), leaf mold (*Mycovellosiella fulva*), powdery mildew (*Leveillula taurica*), southern blight (*Corticium rolfsii*), Botrytis blight (*Botrytis cinerea*), and target spot (*Corynespora casiicola*). Fusarium (*Fusarium oxysporum*) and Verticillium (*Verticillium dahlia*) wilts are soil-borne and attack vascular tissue. Fungal diseases are controlled by crop rotation, fumigation, and disease-resistant cultivars where available. Phytophthora root rot (*Phytophthora parasitica* and *P. capsici*) can infect plants throughout the entire season. Preventing soil saturation for extended periods provides the best control. Corky root (*Pyrenochaeta lycopersici*) is problematic in some areas, and a control strategy is to avoid early planting in cold soils. Late blight (*Phytophthora infestans*) is a concern particularly in high humidity, occurs during rainy periods, and during late summer and fall (APS, 2016).

There are many virus diseases that affect tomato. Virus damage can cause negligible to substantial yield loss, and some can be seed-borne. Depending on the virus, transmission is through direct contact or through insect vectors such as aphids, whiteflies, and thrips. Some of the most common viruses are curly top, cucumber mosaic, tobacco mosaic, tobacco etch, and potato Y. Tomato-spotted wilt and tobacco streak are common in some areas. Other viruses affecting tomato may include whitefly-transmitted geminivirus, tobacco leaf curl virus, and tobacco ringspot virus. Viruses tested for presence in seed include the tobamoviruses and pepino mosaic virus (PepMV). A recent virus that has serious impact on greenhouse production is tomato brown rugose virus (ToBRFV). It is a tobamovirus that is easily transmitted mechanically and can be seed-transmitted. In many countries, quarantine requirements dictate that ToBRFV contaminated seed lots must be destroyed. Disease-resistant cultivars are being introduced. Growing virus-resistant cultivars, if available, is the best control method. Disease-free stock seed prevents spread as well. Early control of insect vectors and general field sanitation is beneficial (APS, 2016). Seed disinfection agents like TPS can remove viruses from tomato seed coats.

Insects and nematodes

Insect pests of tomato are numerous and vary with location (Kennedy, 2003). The following is a partial list of some major pests of tomato. The tomato fruit worm (*Helicoverpa armigera* and *Helicoverpa zea*) and various armyworms (*Spodoptera* spp.) are destructive pests, causing significant losses due to fruit boring. Naturally occurring beneficial insects like *Trichogramma* spp. egg parasites, the larval parasite *Hyposoter exiguae*, and predators such as big-eyed bug and minute pirate bug are important for the biological control of tomato fruit worm. Sprays of *Bacillus thuringiensis* and the Entrust formulation of spinosad are effective against many worms that feed on tomato and are generally acceptable for organic seed production. Pyrethroids and/or insect growth regulators may provide effective control for armyworms and stink bugs (*Euschistus conspersus* and *Nezara viridula*). However, pyrethroids are a broader-spectrum chemistry and may inadvertently kill beneficial insects. Integrated pest management (IPM) monitoring programs have been developed in many areas for determining treatment thresholds for initiating fruit worm and armyworm controls. Tomato seed crops should not be planted near other alternative insect host plants, such as maize and cotton, that serve as refugia.

Cotton aphid (*Aphis gossypii*) is a major pest during the dry season and vectors cucumber mosaic virus. The silverleaf whitefly and sweetpotato whitefly (*Bemisia argentifolii* and *B. tabaci*) are both serious pests that also vector tomato yellow leaf curl virus. Thrips are also a problem pest for glasshouse seed crops, especially *Frankliniella occidentalis*, which vectors tomato-spotted wilt virus. Other important tomato fruit pests include tarnished plant bug (*Lygus hesperus*), potato tuber moth (*Phthorimaea operculella*), and tomato pinworm (*Keiferia lycopersicella*). Some important foliage pests are beet leafhopper (*Neoaliturus tenellus*), green peach aphid (*Myzus persicae*), tobacco hornworm (*Manduca sexta*), tomato hornworm (*M. quinquemaculata*), leafminers (*Liriomyza sativae* and other species), looper caterpillars (*Autographa californica* and *Trichoplusia ni*), potato aphids (*Macrosiphum euphorbiae*), tomato russet mite (*Aculops lycopersici*), flea beetles (*Epitrix hirtipennis* and others) and wireworms (*Limonius* spp.) (Kennedy, 2003).

The primary insect pests of tomato seedlings are flea beetles (*Epitrix* spp.), darkling ground beetles (*Blapstinus* spp.), and cutworms (*Peridroma* and *Agrotis* spp.), but they do not typically cause major problems during transplant production or in larger field plants. Occasionally, the garden centipede (*Scutigerella immaculata*) can damage young transplants. Root-knot nematodes (*Meloidogyne incognita* and other species) will cause gall formation. Crop rotation, resistant cultivars, or fumigation are the best controls for root-knot nematodes.

PEPPERS

Overview

Capsicum sp. peppers should not be confused with black pepper (*Piper nigrum*), a flowering tropical vine native to Asia in the family Piperaceae cultivated for its dried fruit which is used as a spice (McGee, 2004). The genus *Capsicum* is strictly of New World origin. Pepper samples discovered in Peruvian burial sites and seed remnants from 5000 BC near Tehuacan, Mexico, suggest early cultivation as a food crop. *Capsicum* species were likely domesticated in different areas: *C. annuum*, Central to Southern Mexico; *C. frutescens*, Central America, probably Guatemala; *C. chinense*, Ecuador and southern Columbia; *C. baccatum*, eastern Bolivia, and *C. pubescens*, mountainous Bolivia and southern Peru (Pickersgill, 1997). Today, wild plants are found from the southern United States to southern South America. European explorers took specimens to Europe where they were eventually adopted as food crops and quickly moved around the world.

The most notable feature of peppers is flavor, which ranges from mild to strongly pungent. The non-pungent peppers are more popular in Europe. The term chili pepper (also spelled chile or chilli pepper) refers to the fruit of plants from the genus *Capsicum*. In some cultures, chili is used alone without the word pepper. Also, the name chili implies elongated fruits that are often but not always pungent, rather than blocky "bell" types.

Pungent peppers have antimicrobial properties. This is probably why their consumption is traditionally higher in tropical regions where they are used to help fight disease (Cichewicz, and Thorpe, 1996).

Botany

Peppers belong to the family Solanaceae, like tomato and eggplant. They are warm-season, frost-sensitive, tropical perennials grown as annuals in short season temperate climates. Peppers are herbaceous but eventually become woody and shrub-like at maturity. Most peppers are erect and highly branched reaching 0.5 to 1.5 m (1.6 to 4.9 ft) tall when mature. Root systems are extensive and anchored by a taproot. Stems are smooth and leaves relatively simple and broadly lanceolate to ovate with few trichomes (Rubatzky and Yamaguchi, 1997).

All domesticated *C. annuum* cultivars are predominately self-pollinated, although some out-crossing occurs (Rubatzky and Yamaguchi, 1997). Single or multiple fruit form at each node for *C. annuum* (Fig. 17.13). The bell-shaped calyx usually enlarges and at least initially covers the top or much of the immature fruit (Fig. 17.13).

Botanically, the fruit is indehiscent and does not abscise (Fig. 17.14). Like the tomato, the fruit of the pepper is a berry, but its structure is considerably different. When the fruit is small it is very much like tomato but as it grows, the outer wall, or pericarp, grows more rapidly than the placenta. This results in fruit filled with large air-filled cavities or locules. Inside the fruit, the placenta is soft white tissue beneath the calyx to which developing seeds attach. Narrow white bands of placenta form the barriers of each carpel but usually don't extend all the way to the blossom end. With maturity, the texture of the outside pericarp wall becomes smooth and glossy, while the inside has a blistered rough texture composed of large watery cells (Fig. 17.14; Rubatzky and Yamaguchi, 1997). Fruit of

Fig. 17.13. Senescing and immature flower buds of *C. annuum*. Several flowers are visible near the apical meristem associated with compressed nodes. The brown senescing petals are two- or three-days post anthesis. The immature buds near the top of the plant will open in one to three days depending on the temperature.

Fig. 17.14. Mature orange bell pepper nearing seed harvest. The pedicel lacks an abscission zone and must be cut or snapped off the plant without breaking surrounding branches.

domesticated species are variable in pungency, which is related to their capsaicin content.

C. annuum is the most widely cultivated and economically important species and includes both sweet and pungent fruited cultivars of various colors, shapes, and sizes. Among the different Capsicum species, corolla colors vary from white to greenish white, and lavender to purple. Anther colors are blue, purple, or yellow. *C. frutescens* has blue anthers, greenish-white corollas, and commonly has two or more fruit developed per node and is believed native to the lowlands of tropical America. Southeast Asia is a secondary area of diversity. Wide variations in flavor and pungency exist among the many accessions of this perennial. *Capsicum chinense* produces some extremely pungent fruit. Except for a distinct ring-like constriction at the base of the calyx, *C. chinense* resembles *C. frutescens* and *C. annuum*. Fruit is smooth and shapes are as diverse as those of *C. annuum*, but often with puckered fruit walls. *C. chinense* has a characteristic citrus-like aroma. Both *C. baccatum* and *C. pubescens* are grown primarily in South America (Pickersgill, 1997).

Types and Cultivars

Open-pollinated and F1 hybrid cultivars of *C. annuum* are available. Immature *C. annuum* fruit colors are highly variable including green, yellow, or even purple. Mature colors include red, orange, yellow, or a mixture of these colors (Fig. 17.14). Green is due to chlorophyll, red and purple due to carotenoids, and purple is due to anthocyanin. As with color, there are many fruit shapes ranging from blocky, conical, round, to pencil shaped, or combinations. Fruit walls differ in thickness. Fruit lengths vary from 1 cm (0.4 in.)

to more than 30 cm (11.8 in.) and from 1cm (0.4 in.) to about 15 cm (5.9 in.) in diameter.

The systematic classification of pepper is challenging because of the wide range of fruit and plant types that fit multiple groupings. Fruit shape is frequently used for horticultural classification. Peppers are often divided into pungent and non-pungent types. A general convention among plant breeders is to develop bell types that are non-pungent and elongated types are pungent.

Non-pungent peppers

Bell peppers: historically, large, blocky, four-lobed "bell peppers" are popular in North America and Europe (Fig. 17.14). The highest quality bells, like the open-pollinated cultivar California Wonder, have a thick fruit wall, four locules (fruit chambers), and both ends of the fruit are generally flattened. California Wonder reaches green maturity in about 95 to 100 days and red physiological maturity, when the seeds are fully developed, after approximately 125 days. Similar cultivars are often referred to as California Wonder or Bell types and most are F1 hybrids. Cultivars that mature in 65 to 75 days are available but have poorer quality with three locules rather than four, thinner fruit walls, less blocky shape, and smaller size.
Paprika: a spice made by grinding dried, usually red, pepper fruits from a wide range of cultivars.
Pimento (or pimiento): cultivars that produce conical or tomato-shaped fruit with thick walls. The fruit are sliced and used fresh or canned as a condiment to add color.

Pungent peppers

Pungency is caused by capsaicin (pronounced cap-say-i-sin), an odorless, colorless, and flavorless vanillyl amide of isodecylanic acid chemically related to vanilla (Guzman *et al.*, 2011). Capsaicin induces a warm feeling by stimulating chemoreceptor nerve endings in the skin and can be a powerful irritant if it contacts mucous membranes of the intestinal tract and eyes. Capsaicin is concentrated in glands located in the pepper's placenta, the inner soft white tissue attached to developing seeds.

Pungent cultivars tend to be smaller and elongated but this is a generalization. There is no genetic relationship between fruit shape and pungency. Although pungency may increase with pepper maturity, the increase is gradual with development and does not dramatically increase with color change or physiological maturity. In other words, peppers are pungent throughout development due to their genetic makeup. There is no transition from sweet to pungent at a particular stage. A few important types of peppers are described below.

Tabasco (*C. frutescens*): grown in warm-temperate and tropical regions. Tabasco fruit have an irregular shape, are 5 cm (2 in.) long, being small, elongated fruit that are moderately pungent. Fruit is pickled or processed into a popular spicy red seasoning sauce.
Cayenne (*C. annuum*): also known as the Guinea spice and is named for the city of Cayenne in French Guiana. Cayenne is called a cow-horn type because it has an elongated curved shape. Cayenne is pungent and red at maturity. Mature fruits are dried and ground into powder for flavoring and medicinal use.
Banana (*C. annuum*): has a shape and immature color resembling a banana. It is an elongated "wax" type that turns red at maturity. "Wax" is a general term for yellow fruit. Wax peppers may be either pungent or sweet. Cultivars of banana peppers can be either mild or slightly pungent. They are used fresh or processed.
Jalapeño (*C. annuum*): bullet shaped, 5–9 cm (2–3½ in.) medium long, deep green when immature, red at maturity, and somewhat pungent. Non-pungent cultivars with the same shape have been developed for consumers who enjoy the distinctive flavor but without pungency. The name comes from Xalapa, Veracruz, Mexico, where jalapeño is traditionally grown.
Cherry peppers (*C. annuum*): named for their resemblance to the tree fruit and are round green when immature and turn intense red at maturity. Cherry peppers are small and round, may be pungent or sweet depending on the cultivar. Cherry peppers are consumed fresh or pickled.
Scotch Bonnet (*C. chinense*): a pungent cultivar popular in the Caribbean islands, Guyana, the Maldives, western Africa, and increasingly North America. The name is derived for its resemblance to a Tam O'Shanter hat (Andrews, 1998). Scotch Bonnet fruit change from green to colors ranging from pumpkin orange to scarlet red as they advance in maturity.

Habanero (*C. chinense*): (habañero in Spanish) describes fruit with similar appearance, pungency, and flavor to Scotch Bonnet. Although the names "habanero" and "Scotch Bonnet" are sometimes used interchangeably, they are distinct cultivars. Habanero fruit are slightly elongated and pointed compared to the compressed appearance of Scotch Bonnet. A typical mature habanero fruit is 2–6 cm (0.8–2.4 in.) long. Immature habaneros are green and transition to orange-red at maturity.

Tepin: also called 'chiltepin', 'chiltepe', and 'chile tepin', is often listed as the botanical variety *glabriusculum* of *Capsicum annuum* that is native to southern North America and northern South America (Singh, 2006). The name "tepin" is derived from a Nahuatl word meaning "flea". The plant is a shrub that usually grows to a height of about 1 m (3.3 ft), but may reach as high as 3 m (9.8 ft) (Richardson, 1995). The tiny tepin fruits are extremely pungent, red to orange-red, round or slightly ellipsoidal, and about 0.8 cm (0.31 in.) in diameter.

Thai chili: also called bird's eye chili, is a very pungent pepper from Southeast Asia. The name bird's eye chili is generic and describes several small-fruited types including the tepin that are spread by birds. They are perennial, producing small, tapered fruits, 2 to 3 cm (0.8 to 1.2 in.) in length, similar in appearance to tabasco, with two to three fruits per node. Red is the most common mature fruit color. Taxonomically, bird's eye chili types were traditionally classified as *C. frutescens* L. but they are also classified as *C. chinense* by some (DeWitt and Bosland, 1997).

Ornamental cultivars have green or variegated foliage and many small, multicolored fruit that develop above the canopy. The fruit are edible and adapted to container gardens in urban areas as both ornamentals and vegetables/spices (Fig. 17.15). Ornamental fruit may be pungent or non-pungent (Bosland, 1996).

There is considerable variation in plant characteristics among pepper cultivars. Like eggplant and tomato, pepper plants can be classified as determinate, semi-determinate, or indeterminate depending on how many nodes are produced before vertical growth is terminated with a flower cluster. Determinate plants are small and bushy, while indeterminate plants are tall with continuous vegetative growth. Semi-determinate plants are intermediate between the two growth habits. Other distinguishing features that differentiate cultivars include leaf cover, which is an important characteristic to protect fruits from sunburn, fruit size, and position of fruit bearing, e.g. some cultivars develop fruit in the center of the plant, complicating harvest since pepper branches are brittle and break easily.

Fig. 17.15. Early-maturing, determinate, ornamental, non-pungent, edible chili pepper cultivar suited for container culture. Yellow is the immature fruit color, orange is intermediate, and red is mature.

Parthenocarpy refers to the development of an ovary into seedless fruit independent of pollination or fertilization. Parthenocarpy can occur naturally or is induced by environmental conditions. Natural parthenocarpy is genetically controlled (Nsabiyera *et al.*, 2013). Male-sterile pepper cultivars such as Angello develop parthenocarpic fruits from ovary growth without fertilization. In parthenocarpic cultivars, fruit development and initiation are uncoupled from fertilization. If flowers of a parthenocarpic cultivar receive viable pollen from a fertile male plant, seeds develop normally and are used to perpetuate the cultivar.

Growth and Development

Peppers do not grow well under cold conditions especially when night temperatures drop below 10°C (50°F) for extended periods. Temperatures ranging from 0°C to 10°C (32°F to 50°F) cause off-flavor, poor growth, and poor color development because plants and fruit are susceptible to chilling injury. Since early-season growth is slowed by cold weather, planting should be delayed until soil and air temperatures warm in spring. Mean day temperatures of 20°C to 25°C (68°F to 77°F) are ideal for growth of bell peppers.

Peppers are more tolerant of high temperatures than tomatoes because of superior pollination. Pungent types tend to be more high temperature tolerant and set fruit better compared to bell peppers. In general, small-fruited pepper cultivars are more tolerant of both high and low temperature extremes. Pepper flowers generally set fruit poorly below 15°C (59°F) or above 32°C (90°F) because of poor pollen production. Cool temperatures can sometimes result in environmental parthenocarpic fruit development. Pepper pollination and fertilization are optimum from 20°C to 25°C (68°F to 77°F). Plants are day-neutral with no photoperiod sensitivity. Flowering usually begins between one and two months after planting with fruit achieving full size about a month after anthesis depending on the cultivar (Rubatzky and Yamaguchi, 1997).

Pepper plants have extensive root systems that are moderately deep-rooted with taproots reaching a depth of more than 1 m (3.2 ft). Although peppers are generally drought resistant, intermittent periods of moisture and/or nutritional stress can dramatically reduce plant growth, fruit size, and yield. Flowers and young fruit may drop if water stress occurs at flowering and fruit set. Unlike tomatoes, peppers are not climacteric. Green bell and many chili peppers do not ripen after harvest when treated with the gaseous plant hormone ethylene. Bell pepper ethylene production rates are very low in the range of 0.1–0.2 μl/kg·h at 10°C to 20°C (50°F to 68°F) (Cantwell, 2013).

Fertilization

Fertilization should occur in response to soil testing before establishment and foliar testing throughout the season. Fertilization in response to actual soil and plant needs ensures healthy growth and reduces waste. Soil pH should be between 6 and 6.5. Lime application prior to soil preparation can raise pHs if below this range.

Peppers do not remove as many nutrients from soil as tomatoes. A pepper seed crop will remove roughly 134 kg N, 13.4 kg P, and 134 kg K per ha (120 lbs of N, 12 lbs of P, and 120 lbs of K per acre) (Maynard and Hockmuth, 1996). Peppers respond to nitrogen fertilization, which is often applied prior to establishment by row banding. A liquid starter solution high in P is often applied to the root zone when crops are established by transplanting. A standard starter solution recipe is 1.4 kg (3 lbs) of 8N-24P-8K or 10N-52P-17K dissolved in 200 l (50 gal) of water with 118 ml (½ cup) applied to the root ball of each transplant to reduce shock and stimulate early growth (Maynard and Hockmuth, 1996). Up to two thirds of the nitrogen is often banded as top dressing at planting. After flowering and fruit set, the remaining third is applied (George, 2009). High N favors vegetative over reproductive growth thus delaying seed and fruit maturity. Excessive nitrogen wastes costly fertilizer and may damage the environment. Calcium can help prevent blossom end rot which affects some cultivars during water stress.

Pollination

C. annuum and *C. frutescens* are generally self-pollinated. Like tomato, peppers require physical movement of flowers either by wind or artificial agitation in greenhouses to release pollen from anthers surrounding the stigma. Despite having perfect flowers, some crossing occurs among plants within both species. Bees visit pepper flowers to collect nectar, and in the process move pollen to stigmas of flowers on other plants. High temperatures may decrease pollen viability and reduce selfing. Cross-pollination is increased by large populations of pollinators. Delayed dehiscence after flowers open encourages crossing by pollinators. An extreme case of 68% cross-pollination was reported in India (George, 2009).

For open-pollinated seed, fields are isolated by at least 0.5 km (0.3 miles) to reduce potential for crossing. Isolation of 1.6 km (1 mile) or more is typically recommended between pungent and non-pungent types or when producing "foundation" quality seed for genetic preservation.

For F1 hybrid seed production, most production occurs in greenhouses, net houses, or tunnels. Pollen is collected from the male parent and stored until inbred lines are crossed (Fig. 17.16). Prior to emasculation for F1 hybrid production, female plants are "cleaned" by removing open flowers, fruit, and flower buds below a certain

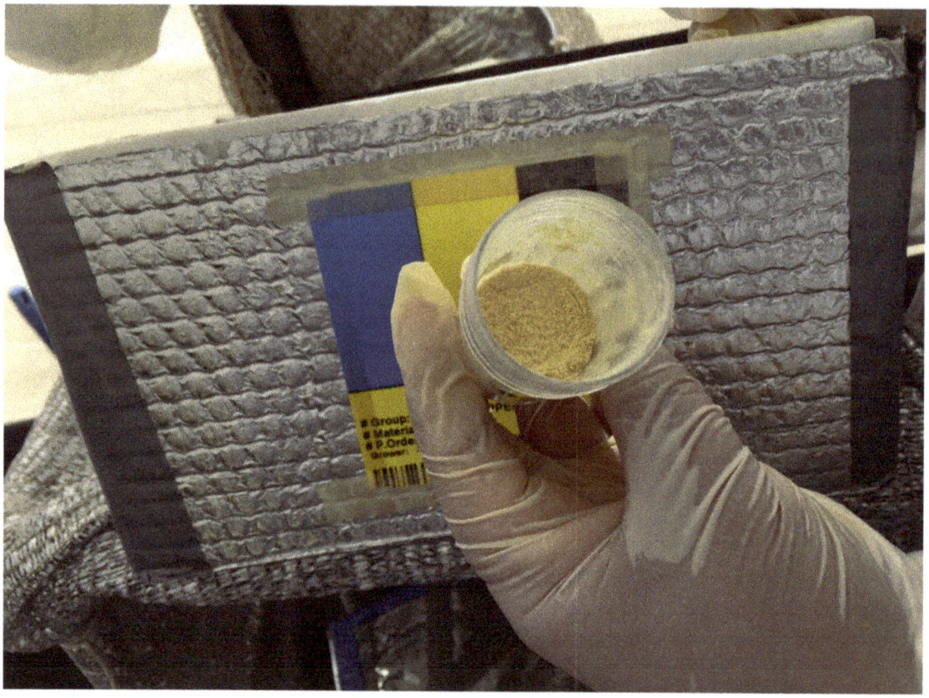

Fig. 17.16. Vial of pollen collected from the male parent for hand-crossing with emasculated flowers of the female parent. (Courtesy of Tim Loeffler.)

node. Only flowers above this node will be emasculated and pollinated for hybrid seed production. Flowers for crossing are emasculated before the pollen is ripe and the buds are still closed. Stigmas are receptive to pollen before flowers open. Emasculation occurs over several days. Floral buds targeted for emasculation are marked with twist tie or string usually with a unique color for a particular day (Fig. 17.17). Depending on the hybrid, there is a certain number of flowers crossed per plant since only a limited number of developing fruit can be supported all the way to seed maturity by each plant.

An alternative to emasculation is male sterility (Nsabiyera *et al.*, 2013) and is used in some hot pepper cultivars primarily for the Asian market. In most male-sterile genotypes, anthers fail to fully develop and do not produce viable pollen. Therefore, no emasculation of the female parent is required, and this labor- and time-consuming step can be skipped in advance of crossing with pollen from the male parent. When male-sterile female lines are used, flowers are often bud-pollinated.

Pollen must be harvested from the male parent in advance of making F1 hybrid crosses. There are two techniques for harvesting pollen.

- Floral buds harvest method: Buds are taken from the male line and stamens are removed and dried. After drying (12 to 15 h), the pollen is shaken out and screened like tomato pollen isolation described above. Pollen is then stored until the next morning, or in some cases, it is used immediately for pollination after isolation.
- Vacuum pollen harvest method: A specially made device or the vibrating end of an electric toothbrush is adapted to induce pollen shed. The vibrating device is held below the anther to release pollen. Pollen is collected in a tube or small container. The pollen collected is used immediately or stored.

The length of pollen storage is very temperature and relative humidity dependent. Pollen should

Fig. 17.17. Hand-crossed pepper fruit for F1 hybrid seed production color coded with string to identify crossing dates. (Courtesy of Tim Loeffler.)

be stored at 10°C (50°F) or less if it is used within a few hours and less than 5°C (41°F) and 50% relative humidity if kept until the next day. Cryopreservation of pollen enables overcoming asynchronous flowering and genetic improvement by conservation of alleles until needed. Pollen can be frozen for 30 days at –18°C to –20°C (0°F to –4°F). For successful storage of frozen pollen, moisture content should be reduced to < 10% using silica gel or other desiccants. Dried pollen can be stored cryogenically below –196°C (–320°F) in liquid nitrogen for months or years and is an important tool for long-term germplasm preservation.

For successful pollination, the stigma should be sticky and receptive. Pollination is more successful in that morning when relative humidity is higher, and temperatures range from 20°C to 25°C (68°F to 77°F). Pollination will not be successful if stigmas are dried or shriveled. To ensure successful pollination, pollen should be at optimum maturity and the pistil receptive. To prevent foreign pollen from contaminating the stigma after hand cross-pollination, the female flower is covered with a gel cap or small piece of aluminum foil for outdoor production. This also prevents desiccation. When seed production is in a greenhouse, protection from foreign pollen is less critical because insect access is limited. Fertilization occurs about 24 to 36 h after pollination. If pollination, fertilization, and fruit set are successful, rapid ovary enlargement occurs. Fruit development is usually apparent three to five days after pollination. A target with hand-pollinating bell peppers is to achieve five to eight fruits per plant with 200 seeds per fruit and to avoid pollinating many fruits with less seeds per fruit.

Irrigation

Pepper seed production occurs in desert areas with low humidity to minimize disease pressure, so irrigation is required. If the crop is to be grown in an area that is not consistently rain fed, sufficient irrigation water must be applied to prevent flower drop and poor fruit set. Water requirements for pepper seed production range from 40 to 100 cm (16 to 39 in.). The broad range reflects differences in cultivar and environmental conditions. A general irrigation recommendation is for a minimum of 2.5 cm (1 in.) per week to ensure continuous growth. On sandy soils, more water may be required. Drip irrigation provides uniform and efficient water application directly to the root zone. Furrow irrigation may be an option if sufficient water and infrastructure are available (Fig. 17.18). Some greenhouse seed production occurs using rockwool or peat moss bags and utilizes specialized drip irrigation. For production in greenhouses or

Fig. 17.18. Pepper seed field on raised beds using furrow irrigation. Red fruit are hand-harvested for wet seed extraction.

other protected structures, the water requirement for a mature bell pepper plant is around 1 l per plant/day. The EC of irrigation water should be less than 1.25 milliequivalent per litre (1 mEq/l).

Rogueing

Seed producers should be provided with a list of defining traits of the seed crop cultivar. This list is invaluable for identifying off-type plants. Prior to transplanting, young plants should be inspected for uniformity, leaf morphology, pigment, and diseases. Before flowering, check each plant for characteristics typical of the cultivar including growth habit, vigor, foliage, pigment, and leaf morphology. Any plants with irregular characteristics should be removed from the field and destroyed. At the time of first flowering and fruit set, check for the same characteristics listed for the pre-flowering stage. In addition, also check for seed-borne diseases and any suspected plants should be removed and tested to determine if any further actions are needed. All genetic off-type plants must be removed and appropriately discarded. When the first fruit reach market maturity, check for the same characteristics listed for stages above as well as fruit colors and morphologies.

Genetics

Pungency is a genetically heritable trait. Capsaicin is the compound most closely linked to pungency although other compounds may also be involved. Genetic resistance has been incorporated into various cultivars of *C. annuum* by traditional plant breeding for the important diseases: bacterial leaf spot, cucumber mosaic virus, *Phytophthora*, pea enation mosaic virus, pepper mottle virus, potato virus Y, tobacco etch, tobacco mosaic virus, *Tobamovirus*, and tomato-spotted wilt. Many of the commercial cultivars of *C. annuum* are F1 hybrids created by crossing compatible inbred lines. Most growers feel that the extra cost of F1 hybrid seed pays for itself in greater productivity and fruit uniformity. Transgenic cultivars have been created experimentally but are currently not grown commercially. If male-sterile lines are used to reduce the labor of emasculation, a separate male-fertile restorer is crossed to the male-sterile line to produce seeds.

Production and Culture

Peppers are sensitive to waterlogged soil. Planting pepper in poorly drained soils of any kind must be avoided. Plants grown in waterlogged soil are susceptible to root diseases and defoliation. Raised beds improve drainage and allows plants to survive short-term flooding (Fig. 17.18).

Peppers or related Solanaceous crops should not be grown in the same field consecutively to prevent soil-borne diseases for a minimum of three years. Crop rotation following a grain crop such as wheat, rice, oats or corn or legumes and Brassicas are ideal. For soils with disease histories, approved fumigants like chloropicrin or metam sodium/potassium are sometimes applied.

Several methods of soil preparation are suitable for pepper seed production. A homogeneous seed bed with small ped size maximizes root to soil contact. Such soil encourages uniformity of establishment and growth. Power-driven rotovators can create a smooth bed if there is little plant residue. In some cases, fields are deep plowed if heavy residue from the previous crop must be incorporated. After deep plowing and tilling raised beds must be reshaped.

Parent lines for hybrid seed production are established from transplants as previously described for tomatoes. Inbred lines lack vigor and are therefore established by transplanting since stock seed is only available in small quantities and very costly. Transplants enable late-maturing peppers to be grown in short-season areas, reduce the time a crop is in the field, improve stand uniformity for better nicking, and improve the synchrony of seed harvest. Transplants are produced in greenhouses or protected outdoor beds in "plug" trays with individual cells similar to described previously for tomato (Cantliffe, 2009). The substrates used for pepper transplants are dependent upon the growing environment and include coconut fiber, rockwool, and various soil mixes for growing plugs. Greenhouse transplant production protects against late frosts, heavy rain, and wind. Depending on the protected structures used, transplants may be protected from fly-in insects and wind-borne disease.

Direct seeding is sometimes used for large-scale open-pollinated seed production of small-fruited dried peppers (Bevacqua and VanLeeuwen, 2003). Pepper seeds germinate and establish slowly from direct seeding particularly under cool soil conditions. High-quality pepper seeds germinate in six to ten days at 15.6°C to 29°C (60°F to 85°F) but very slowly at 15°C (59°F).

Peppers are transplanted through plastic film mulch used to conserve moisture, modify soil temperatures, or control weeds. Organic seed producers may use straw or other natural mulches. Pepper transplants are considered "moderate" in their ability to survive transplanting. Root and shoot growth are slower compared to easily transplanted crops like tomato (*S. lycopersicon*) (Loomis, 1925).

Final plant density depends on the growing environment, vigor of the cultivar, and the irrigation system. Generally, the final plant stand is 30 to 60 cm (12 to 24 in.) within the row and 45 to 90 cm (18 to 36 in.) between the rows. Plant density varies among cultivars and production environment. Open-pollinated cultivars that are mechanically harvested frequently range from 18,500 to 74,000 plants/ha (7500 to 30,000 plants/acre). More space is needed to accommodate workers pruning, training, emasculating, pollinating, and hand-harvesting fruits so populations range from 15,000 to 25,000 plants/ha for open-field production and from 18,000 to 28,000 plants/ha for glasshouse, greenhouse, or net house production. Large pepper plants in a field are supported using a string weave system so they do not lodge under a heavy fruit load. Small determinate plants do not require staking. Indeterminate peppers in greenhouses are staked or supported by overhead strings similar to tomato depending on the height of the cultivar. Closer spacing tends to reduce fruit size, although high densities have the advantage of providing shade that may decrease sunburn. High densities may require more water but reduce weed competition.

Seed Harvest

Best-quality seed comes from physiologically mature fruits. Fruits should be left on the plant to full physiological maturity because development stops at harvest. Fruits are physiologically mature when they obtain their mature color, which is most often red, orange, or yellow, and their seeds are fully developed. Harvesting fruits too early results in lower germination and dormancy issues and harvesting over mature fruits reduces seed quality. Some novelty cultivars may be purple or brown at maturity. There is no abscission zone, so the pedicle of fruits must be carefully cut or expertly snapped by hand without breaking brittle branches and surrounding leaves. Mature seeds are hard and difficult to cut through with a knife. Fruits with blossom end rot or sun-scald should be discarded along with any rotten fruits. Unmarked fruits from hand-pollinated crops should be removed before harvest. For hybrid seed production, harvested fruits are often allowed to after ripen in boxes or crates before seed extraction. The holding period varies depending on the cultivar, fruit maturity, and temperatures.

Pepper seeds can be extracted from either wet or dry fruit. Large fruit types are wet extracted but without fermentation. Pepper seeds are attached to the placental tissue, and unlike tomato are not imbedded in surrounding mucilage. Large-fruited bell types, especially for F1 hybrid production from smaller volume crops, are often opened by hand and the seeds examined. Healthy, properly developed pepper seeds are white to pale yellow when fresh. Brown, black, or molded seeds that did not develop properly or are over mature should be removed. The fruit core and healthy seeds are separated by hand and the remaining fruit tissue discarded.

Alternatively, seeds are extracted mechanically by a rotary seed extractor. The extractor crushes and lightly macerates fruit releasing seeds (Fig. 17.19). The crushed fruit and resulting mixture of seed, juice, and fruit tissue pieces pass through a slowly revolving screen cylinder. Seeds and juice fall through the screen while larger fruit debris passes through the cylinder to the field or collected for disposal. Because pepper fruit pulp is relatively dry, clean water is added during and after fruit crushing to improve seed separation and maximize yield. Extraction machines are mounted on wheels and parked in seed fields or at seed-processing facilities.

For chili, cherry, pimento, and other small-fruited pepper seed of open-pollinated cultivars, entire plants with fully mature-colored fruit are cut and field dried where humidity is low. The cut plants are often laid on plastic sheets for drying. Field drying reduces energy usage compared to batch drying. If field drying is impossible, small fruit can also be mechanically harvested to reduce labor (Walker, 2018). The three most common picking heads for mechanical harvest are the finger-type, the belt-type, and the double helix. Many larger, commercial operations employ either tunnel or belt dryers rather than use sun

Fig. 17.19. Wet seed extractor used for tomato, pepper, and eggplant. Fruit is fed into a bin, chopped, and drops into a rotating cylinder where streams of water wash the seeds from the fruit tissue pieces which pass out the front. Cleaned seeds fall through the revolving screen and are captured beneath.

drying (Walker, 2016). Batch dryers can desiccate harvested fruit until shriveled and brittle, so the seed can be easily extracted.

Hand seed extraction of pungent peppers is unpleasant for workers as they are very likely to experience hand, noise, or eye irritation. To reduce hand labor and the potential for irritation, extraction machines break fruit and remove seeds. Normally it is recommended to keep seeds dry during processing. However, extraction efficiency is increased by spraying water to flush relatively dry fruits, seeds, and/or pulp particles through a seed extractor. The water also reduces eye irritation for the operators. Seeds can be harvested during large-scale processing of crushed dried peppers. An aspirated screen cleaner or density separation table can upgrade the isolated clean dry seeds. When seeds are harvested as a step during pepper food processing, good communication is needed between the seed company and food processor to ensure genetic purity is maintained and seed quality is not damaged by high temperature processes.

Post-harvest separation, disinfection, and drying

For wet processing, pepper seeds are separated, washed, disinfected, and dried similar to tomato. Cleaning is easier for pepper since there is no mucilage surrounding seeds and fermentation is not needed. A water trough or flume separation/cleaning removes debris remaining after extraction (Fig. 17.20).

Seeds from small-volume production are often separated in buckets. Seeds produced for export are first isolated from fruit, washed, disinfected, rinsed and dried, so seeds are not damaged by prolonged hydration. To rapidly remove excess surface moisture following wet extraction and rinsing, pepper seeds are placed in durable mesh bags and spun in a centrifuge to remove loosely bound surface water. Usually, TSP and calcium/sodium hypochlorite are the disinfectants of choice depending on the area of production, cultivar, and country of import.

After disinfection treatments, controlled drying methods preserve quality for higher value seeds. Seeds are dried on outdoor drying tables, indoor forced-air drying tables, or rotational dryers with controlled air flow and relative humidity. Outdoor dryers should be covered to protect the seeds from the direct sun or rain. Seeds should not be dried rapidly or at temperatures exceeding 35°C (95°F) because either action can reduce seed quality.

Post-harvest seed conditioning

After the seeds are dried, they are precleaned using air screen cleaners or air column separators to remove chaff, light seed fraction, and

Fig. 17.20. Wet seed processing of pepper seed using a flume to separate seeds from debris. (Courtesy of Tim Loeffler.)

deformed seeds, sampled for initial testing, and packaged for export or shipment to a seed-processing plant like those described for tomato. The steps of cleaning, packaging, and shipment to market are similar to tomato.

Seed yield

Seeds of *C. annuum* are flat, smooth, typically pale yellow, ovoid, 3 to 5 mm (0.08 to 0.12 in.) across. One gram contains about 150 to 160 seeds (approximately 4245 seeds per oz) (George, 2009). For sweet pepper, each bush plant will mature from five to eight fruits at a time, containing about 200 seeds each. Small-fruited types can mature more fruit per plant but each fruit may contain fewer seeds. From a seed yield standpoint, it is preferred to produce more fruits per plant each with fewer seeds than fewer fruits per plant each containing more seeds (George, 2009).

Physiological Disorders

Physiological disorders of pepper include blossom end rot, which appears as a dry black lesion at the blossom end of fruit. Blossom end rot is caused by a localized calcium deficiency in rapidly growing fruit tissue induced by drought stress. Blossom end rot is more common on large-fruited cultivars and less frequent on elongated cultivars. Some large-fruited cultivars are more susceptible than others. Pepper speck is a poorly understood disorder causing small spot lesions that penetrate the fruit wall. Some cultivars are more susceptible than others (Cantwell, 2013). Direct intense sunlight can kill or severely damage exposed fruit tissue, causing sun-scald. Scalded areas of the fruit have a bleached white or tan appearance and are more susceptible to bacterial soft rot and secondary infection by other pathogens.

Pest Management

Weeds

Pepper seeds germinate slowly, so competition with weeds can be a problem especially for direct-seeded crops. This is one reason why transplants are popular for pepper seed crop establishment. Weed control strategies include stale bed, mulching with plastic film or organic mulches, fumigation with approved chemicals, solarization in areas with high sunlight, cultivation, or selective herbicides (Stall and Gilreath, 2002). It is important to control weeds as many species, like *Chenopodium* spp., *Datura* spp., and other *Solanum* spp. as well as ornamental plants such as petunia (*Calibrachoa parviflora*), serve as alternative hosts for several pepper viruses. Controlling weeds around seed production fields, inside and outside greenhouses, net houses, and tunnels is also critical for plant disease transmission.

Pathogens

Many diseases affect pepper (Pernezny and Momol, 2006). Anthracnose (*Colletotrichum acutatum, C. gloeosporiodes, Colletotrichum* spp.) cause small, water-soaked lesions that coalesce into larger sunken areas on mature fruit. Lesions may show dark fungal growth, with tan to pink concentric circles of spores. Occasionally, leaf spots and stem dieback occur with anthracnose. Controls include pathogen-free seed and avoiding overhead irrigation (Pernezny and Momol, 2006). Providing plastic mulch as a ground cover helps prevent the pathogen from moving from the soil to the fruit. A spray program with available fungicides can provide control.

Bacterial spot (*Xanthomonas euvesicatoria*) starts as tiny, water-soaked spots on leaves that enlarge to 0.6 cm (0.25 in.) in diameter, turn deep brown, and appear greasy. Scab-like lesions may also occur on fruit. During periods of heavy rainfall or humidity, spots on leaves often coalesce causing "blight" symptoms and abscission. The bacteria can reside in crop residue and weeds. Bacterial spot can be seed-borne on the surface and internally and spread rapidly among transplants. Freshly harvested seeds as well as dry seeds are often disinfected with TSP or other chemical agents.

Bacterial soft rot (*Erwinia carotovora* pv. *carotovora*) is a common post-harvest disease but field symptoms are soft fruit that sag from the pedicel like a water-filled balloon. Softened areas often turn gray. Insect vectors spread the *Erwinia* bacteria. To control soft rot, avoid bruising and wounding, harvesting wet fruit, and exposing fruit to sun, which causes sun-scald and secondary infection (Pernezny and Momol, 2006).

Frogeye spot (*Cercospora capsici*) produces round leaf lesions approximately 0.6 cm (0.25 in.) in diameter with light tan to white centers and narrow dark borders. Heavy infection may cause partial or total defoliation.

Gray leaf spot (*Stemphylium solani*) causes circular spots on leaves. These spots are at first brown but later turn light tan to white with sunken centers and reddish-brown margins. Spots also appear on stems, petioles, and fruit pedicels but not fruit.

Phytophthora blight (*Phytophthora capsici*) can affect all tissues, causing seedling death as well as a root rot, stem canker, leaf blight, and fruit rot in older plants. *Phytophthora* causes damping-off symptoms where stems collapse at the soil line resulting in sudden wilting and death. When mature plants are infected at the nodes, whole branches die. Individually infected leaves exhibit small circular to irregular leaf spots that appear scalded, dry, and bleached to a light tan color with a papery consistency. Establishing pepper fields in well-drained soils and practicing crop rotation will help prevent *Phytophthora* damage (Pernezny and Momol, 2006).

Sclerotinia stem rot (*Sclerotinia sclerotiorum*) is most common under cool, damp conditions. The causal fungus infects the stem at the soil line, individual petioles of leaves, and occasionally fruit close to the soil. When weather is moist, white mycelium will often grow up the stem surface about 18 cm (7 in.) or more. Petiole or bud infections rapidly grow down, girdling the entire plant. Infected fruit quickly rots into a watery mass. The fungus survives as sclerotia in stems and lesions associated with diseased fruit. Rotations to susceptible crops such as cabbage, celery, lettuce, potatoes, or tomatoes should be avoided. Deep plowing fields with the disease provides control by burying fallen sclerotia.

Southern blight (*Sclerotium rolfsii*) occurs in warm humid weather causing root and stem infection, wilting, and ultimately plant death. A white collar of coarse white fungal mycelium around the stem develops at the soil line. Tan sclerotia are the size of mustard seed and spread the disease by surviving in soil. Wind, water, or equipment spread the fungus. Rotating with grasses and deep plowing may provide control by burying sclerotia.

Wet rot (*Choanephora cucurbitarum*) produces blossom blight, fruit rot, and occasionally leaf blight. Fungal hyphae capped with black heads (sporangia) grow on petals as they senesce, causing blossom drop. If young fruit become infected, they soften and abort shortly after pollination. Maintaining dry conditions, good air circulation, and fungicides may provide control.

Many virus diseases affect pepper. Viruses (e.g. cucumber mosaic, pepper mottle, potato Y, tobacco etch, tobacco mosaic, and tomato-spotted wilt) are difficult to distinguish as single or multiple viruses may infect a field. Most of these viruses induce degrees of mosaic, mottle, vein banding, and stunting. Malformation, leaf cupping, and fruit distortion may also occur. A laboratory test is needed for an accurate diagnosis (Pernezny and Momol, 2006).

Tobamoviruses are positive sense RNA genomes infecting peppers and include tomato mosaic virus (ToMV), tobacco mosaic virus (TMV), pepper mild mottle virus (PMMoV), and bell pepper mottle virus (BEPMV) and are seed-transmitted. Symptoms vary depending on the cultivar and environmental conditions. Both ToMV and TMV are often mechanically transmitted during transplant production as only a small percentage of infected seedlings can spread the disease rapidly. PMMoV and BEPMV are easily transmitted mechanically by contact to healthy plants by hands, gloves, and clothing. The pathogen enters the plant through microscopic abrasions and can be found in the seed coat but rarely in the embryo. Control measures for seed production must start with clean seeds. Seeds should be disinfected with TSP before sowing.

Pepper mottle, potato Y, and tobacco etch are primarily transmitted by aphid feeding. Tomato-spotted wilt virus (TSWV) is transmitted by thrips. These viruses are known to survive in numerous weed hosts such as ground cherries (*Physalis* spp.), nightshades (*Solanum* spp.) common groundsel (*Senecio* sp.), wild tobacco (*Nicotiana* sp.), toadflax (*Linaria* sp.), sicklepod (*Cassia* sp.), and jimson (*Datura* sp.) that should be eradicated in areas surrounding seed fields to prevent spread.

Cultivars resistant to TMV should be grown when possible. Workers, especially those who use tobacco products, should wash their hands with strong soap and water or 70% alcohol before handling plants to prevent spread. Old, infected plants should be destroyed well before planting new seed crops. A 15 m (50 ft) barrier strip of a non-susceptible crop (corn, small grain, etc.) can trap flying insects before they infect seed fields.

Insects

Arthropod pests can seriously damage peppers. Insects and mites reduce fruit yield by directly feeding on plant and fruit tissue, spreading destructive diseases and viruses especially tobacco etch and tomato-spotted wilt virus.

Caterpillar-like pests like fall armyworm (*Spodoptera frugiperda*), beet armyworm (*S. exigua*), southern armyworm (*S. eridania*), yellow striped armyworm (*S. ornithogalli*), cutworms (various species), and loopers (various species) damage foliage and fruiting structures. Young larvae feed on the undersides of leaves leaving the upper epidermis intact, so creating a "windowpane" effect. Older larvae, which may grow to 7.5 cm (3 in.) in length for some species, consume foliage and eat large holes in fruit, causing secondary infection leading to fruit rot. Cutworm larvae damage plants at night, feeding on foliage or cutting seedlings off at the soil line (Olson *et al.*, 2005).

These pests should be destroyed in the first several larval instars with repeated applications of *Bacillus thuringiensis* (*Bt*) every three or four days during peak hatching periods. Mature larvae may be controlled by hand removal on small plots, approved insecticides, or biological control measures.

Pepper weevils (*Anthonomus eugenii*) are small, shiny, brown or gray colored snout beetles, about 3.5 mm (1/8 in.) long. Adults feed on vegetative and reproductive buds. Females bore a small hole in developing fruit or flower buds. The hole is plugged with fecal matter (frass) after an egg is deposited. A tiny, legless grub hatches from the egg and eats its way toward the core of the fruit, feeding on seeds and pulp. Both adult and larval feeding causes bud/fruit drop. Pepper weevil populations can be monitored visually and with yellow sticky traps. The action threshold for this pest is one adult per 400 terminal buds or 1% of the buds infested (Capinera, 2004).

Broad mite (*Polyphagotarsonemus latus*) adults are tiny, white, and usually found on undersides of young developing leaves. Broad mite feeding distorts leaves causing thickening and narrowing. Flower abortion and dark, smooth russeting of fruit occur with heavy feeding. Broad mites are predominately early- or late-season pests and may be effectively managed with sulfur or insecticidal soap.

Adult aphids (superfamily Aphidoidea), also known as plant lice, are delicate, small, sap sucking insects. They are among the most destructive pests of pepper. Winged and non-winged forms are all female and give birth to nymphs. Nymphs are smaller but otherwise appear adult-like until they quickly mature in just seven to ten days. Heavy aphid infestations may debilitate plants with sooty mold growth on honeydew and cause leaf distortion. Aphids also spread plant viruses such as tobacco etch, and the green peach aphid has been known to vector more than 120 plant viruses. Aphids may be managed with insecticidal soaps or neonicotinoids, which work similar to the natural insecticide nicotine.

Both nymph and adult melon thrips (*Thrips palmi*) etch pepper buds, flowers, and/or leaf tissues, and then suck the exudate. When present in high numbers, melon thrips produce silvering, yellowing, and bronzing of affected tissues. Leaves may crinkle and die, growing tips may be stunted, discolored and deformed, and fruits may abort or be scarred. Melon thrips cause a loss of plant vigor and a reduction in fruit and seed yield. Adults are very mobile and can move rapidly from old fields to new. In addition, pepper, eggplant, cucumber, potato, beans, and watermelon are susceptible to melon thrip damage. New plants should be isolated from old-infested fields. Some conventional insecticides seem to stimulate melon thrips rather than provide control (Funderburk *et al.*, 2004).

Western flower thrips (*Frankliniella* sp.) insert eggs into very small fruit causing flower abortion, poor fruit set, and impaired development. The larvae feed on flowers and fruit after hatching. This pest also vectors tomato-spotted wilt virus (Funderburk *et al.*, 2004) which also affects tomato, eggplant, and beans.

Thrips (order Thysanoptera) are early-season pests and populations decline in the summer and fall as natural enemies provide control. Pepper flowers can tolerate up to 15 thrips per flower for short periods with little damage. A biological control program integrating pirate bug (family Anthocoridae) predation with reflective mulch can control thrips and slow the spread of tomato-spotted wilt disease (Funderburk *et al.*, 2004).

Greenhouse whitefly (*Trialeurodes vaporariorum*) is associated with virus transmission. It weakens the plants and its sugary honeydew-like secretions serves as a substrate for fungi. The females typically lays eggs on the undersurface of leaves.

Adult leafminers are small flies (suborder Symphyta), approximately 2.4 mm (0.09 in.) long, with black heads, yellow between the eyes, and black thorax. Females have a tube ovipositor at the ends of their abdomens that punctures the upper leaf surface for egg laying. White, oval eggs are inserted in leaf tissue. The adult makes other punctures (called stipples) for feeding only. The larva, a yellow maggot with black, sickle mouth hooks, feeds between the upper and lower leaf surface for approximately seven days, leaving a serpentine mine containing a black string of frass. The mature larva exits from the mine and falls to the ground where it pupates. The adult emerges in one to two weeks. Serpentine mines in leaves reduce photosynthetic area and may provide entry points for foliar pathogens. Heavily damaged leaves become necrotic, reduce fruit cover, and encourage sun-scald.

EGGPLANT

Overview

Eggplant is an old-world crop believed to be native to India with a long history of cultivation in South and Southeast Asia. In China, India, Japan, and many Mediterranean countries, eggplants are a popular and widely grown vegetable. Eggplant is believed to be native to the Indian subcontinent where large-fruited cultivars were likely domesticated, and many wild forms exist. There are more than 1000 species of eggplant in the genus *Solanum*, but only a few are grown commercially as vegetables. From India, domesticated eggplant spread eastward into China, which became a secondary center of origin. Arabic traders subsequently moved eggplant to Africa and Spain (Rubatzky and Yamaguchi, 1997). The name "brinjal" is of Indian or Arabic derivation.

The US English name "eggplant" came from white-fruited types that resemble a chicken egg (Fig. 17.21; Rubatzky and Yamaguchi, 1997). "Brinjal" and "aubergine" are other common names for eggplant in English-speaking countries.

Eggplant cultivation in the Mediterranean region and the Americas is comparatively recent. Eggplants were introduced to the Americas by early Spanish explorers. Because eggplant is a relative of nightshade and contains bitter glycoalkaloids, much like tomato, many believed eggplant to be poisonous until the early 20th century when its popularity increased.

Botany

Eggplant is a warm-season perennial intolerant of freezing and chilling. Eggplants are short-lived perennials in the tropics but are cultivated

Fig. 17.21. Eggplant seed production on rockwool media in a greenhouse in China. The name "eggplant" comes from cultivars that have white fruit that resembles a chicken egg.

as annuals in temperate zones. They require a longer growing season than tomato or pepper. Eggplant is less tolerant of low temperatures than tomato. Chilling injury occurs below 13°C (55°F) when growth and development is very slow. However, eggplant tolerates humidity and tropical conditions better than tomato. Optimum day temperatures for satisfactory growth and fruit production are around 25°C to 35°C (77°F to 95°F).

Plant architecture varies from erect to semi erect to prostrate (Fig. 17.22). Some cultivars are determinate and grow as a bush to a height of 0.5 m (1.6 ft). Indeterminate cultivars continue growing vegetatively reaching a height of 2.5 m (8.2 ft). Root systems are moderately deep and spreading with significant taproots. Stems are erect and branched. Leaves are generally large, alternate, and simple and some cultivars have thorns on the stems, calyx, and leaves. Modern cultivars tend to be spineless. There are great variations in leaf shape and margin. Leaf blades vary from ovate to ovate-oblong with wavy-lobed margins.

Eggplant does not have a specific day-length requirement for flower initiation. Perfect flowers, solitary or multiple, form a determinate simple inflorescence, usually opposite or nearly opposite leaves rather than in leaf axils. Flowers are 2 to 3 cm (0.8 to 1.2 in.) in diameter with purplish, pubescent corollas. Eggplant has cone-shaped anthers that contribute to high levels of self-pollination. A small percentage of out-crossing may sometimes occur by insect transfers of pollen. Flowers generally remain open for two or three days and stigmas are most receptive in the morning.

Eggplant fruit are a large, solid berry. Cultivars of *Solanum melongena* display a wide range of fruit shapes, curvatures, and striped or single-colored skins. Fruit skin colors are diverse and may be white, yellow, green, red, purple, or almost black (Fig. 17.23). Fruit shapes are diverse and include round, pear-shaped, oblong,

Fig. 17.22. Open-pollinated seed field of Turkish Orange eggplant. This cultivar has prostrate growth habit, and tomato-shaped fruits are red at maturity.

Fig. 17.23. Organic open-pollinated eggplant seed field of elongated, lemon-yellow mature fruit.

or elongated, with lengths ranging from 4 or 5 cm (1.6 or 2.0 in.) to more than 30 cm (11.8 in.). The calyx is deeply lobed, purple or green colored. Skin surfaces are smooth and usually glossy when immature, fading to mat with maturity. Fruit flesh color is usually white or pale green. Fruit tissue is high in phenolics. Upon damage or cutting, interior fruit tissues rapidly oxidize because of polyphenol oxidase and turn dark brown. Seed are small, light brown, and imbedded in the placental tissue.

Types and Cultivars

There is tremendous diversity in eggplant fruit appearance. Cultivars with elongated fruit tend to be more resistant to high temperature extremes and are grown in warmer areas. Cultivars producing small egg or ovoid-shaped fruit are less heat tolerant and tend to be grown in cooler areas. There is a wide range of maturities from 65 to over 100 days after planting. There are open-pollinated and hybrid cultivars of eggplant. *S. melongena* are grown commercially in most regions of the world and cultivars tend to be specific to specific regions. Other species of importance include *S. macrocarpon* known as Gboma eggplant in West Africa. The Chinese scarlet eggplant, *S. integrifolium*, is grown for food and as an ornamental in eastern Asia. There are variations in eggplant taste with some cultivars being more bitter than others. The bitterness is due to glycoalkaloids. However, most modern cultivars are not bitter and spineless.

Pathenocarpic cultivars that develop largely seedless fruit without pollination or fertilization were developed for high tunnel or greenhouse cultivation where air movement is restricted, and natural pollinators are excluded. Cultivars that set fruit without pollination provide early and reliable harvests.

Growth and Development

Eggplant is well adapted to tropical conditions and mid-temperate regions that provide a long growing season. Continuous warm temperatures for four to five months are needed for successful seed crop production. Favorable daytime temperatures are between 22°C (72°F) and 30°C (86°F) with warm nights preferably between 18°C (64°F) and 24°C (75°F). At temperatures less than 17°C (63°F) or greater than 35°C (95°F) growth is slow. Seed production occurs in dry climates to reduce fruit rot and diseases. Eggplant is day-neutral. Flowering is developmentally regulated with early cultivars flowering as soon as six leaves have developed. Later cultivars may not flower until 15 or more leaves have formed. Roots are moderately deep and extensive. Most soils are acceptable unless they impede root development or are poorly drained. Heavy waterlogged soils should be avoided since eggplant is susceptible to root rot fungi. Raised beds can improve drainage on heavier soils or in areas subjected to seasonal heavy rainfall.

Fertilization

Soil pH should be approximately 6.5. Lime can be applied to raise pH of acidic soil. Eggplant responds to relatively high nutrient levels. A well-timed fertility management program is needed to maximize seed production. Well-drained loam soils with significant organic matter are well suited for eggplant. To minimize waste, fertility inputs, including micronutrients, should be based on preplant soil testing and foliar analysis throughout the season to monitor crop nutrient status. An eggplant seed crop can be expected to use approximately 120 kg/ha (107 lbs/acre) N, 100 kg/ha (89 lbs/acre) P_2O_5, and 120 kg/ha (107 lbs/acre) K_2O depending on soil, environment, and cultivar.

Like pepper, eggplant responds to nitrogen fertilization, which is applied as a split application. Approximately two thirds of nitrogen and potassium are often banded as top or side dressing at transplanting. After flowering and fruit set, the remaining third is applied, especially in areas with significant leaching (George, 2009). High N favors vegetative growth often delaying or inhibiting reproductive growth. Excessive nitrogen fertilization may delay crop maturity, waste costly fertilizer, and pollute the environment.

Eggplant seed fields are often established by transplanting to ensure complete stands and uniform development. In addition to the first application, a liquid starter solution high in P is often applied to the root zone at transplanting as described previously for tomato and pepper transplants (Maynard and Hockmuth, 1996). Because of solubility issues with phosphorous fertilizer, most P inputs are often applied to the soil with the initial N application. Smaller quantities of P can be delivered through drip lines if solubility can be maintained. Surface or subsurface drip can provide water and nutrients to fertigate eggplant seed crops. Eggplant seed crops grown in protected culture are usually harvested by hand and often fertigated. N and K inputs and other nutrients can be pulsed weekly with irrigation. Nutrient additions by fertigation can be based on foliar or petiole analyses or equally divided over the growing season if plant tissue nutrient status is not monitored.

Pollination

Eggplant flowers are usually open for two or three days. Stigmas are initially receptive in the bud stage and early on the days when flowers are fully open (Rubatzky and Yamaguchi, 1997). Unlike tomato and pepper, eggplant pollen remains fertile at high temperatures but is more sensitive to low temperatures. Elongated fruit cultivars tend to be more tolerant of high temperatures. At temperatures less than 17°C (63°F) or greater than 35°C (95°F), pollen dysfunction increases. Pollen stability is one reason why eggplants tolerate higher temperatures better than tomato and bell peppers.

Out-crossing of eggplant is less common than pepper. This is due to differences in flower morphology. Eggplant has closely spaced conical anthers that protect the stigma from foreign pollen. Natural cross-pollination varies depending on cultivar, location, and level of insect activity (Rubatzky and Yamaguchi, 1997). Out-crossing in controlled environments is unlikely because there are few insects. In open fields, a minimum isolation distance of 200 m (656 ft) for foundation, 100 m (328 ft) for certified, and 50 m (164 ft) for commercial eggplant seed production is recommended. Isolation distances in controlled environments (glasshouses, plastic covered, and screen enclosures) may be as close as 10 m (32 ft).

F1 hybrid cultivars are popular because they exhibit heterosis, which translates into earlier maturity in some cultivars and improved uniformity. F1 hybrid seeds are made by hand-crossing emasculated flowers or male-sterile lines that serve as the female parent. Pollen for crossing is taken from the male parent by collecting anthers or unopened flower buds during early morning hours. The buds or anthers are dried in open containers at 15°C (59°F) and 15% relative humidity for 48 h. The dried buds or anthers are placed in a small, sealed container and shaken vigorously to release pollen. A second method of pollen collection uses a vibrating device such as a modified electric toothbrush to induce pollen shed as described for pepper. The released pollen is collected in a small container for immediate use or storage. Eggplant pollen remains viable for eight to ten days at 20°C and is best when used within four days. Unused dry pollen can be frozen up to one month or cryopreserved for longer periods as described for pepper. Prior to emasculation, female plants are "cleaned" by removing flowers, fruits, and flower buds below a certain node similar to pepper. Only flowers above this node will be emasculated and pollinated for hybrid seed production. Flowers from the female parent should be emasculated before any pollen is shed. Floral buds targeted for emasculation are marked with twist ties or strings. If fruit size is large, only a few flowers are emasculated, since each plant will only support a limit number of fruit to seed maturity. Emasculation occurs over several days.

Crosses are made early in the morning when stigmas are fully hydrated and receptive. Pollen is transferred directly from the container to the stigma by dipping a fingertip into the collected pollen and then touching the stigma or using a tiny paint brush to make the transfer. When a male-sterile line is the female parent in a hybrid cross, bud pollination ensures pollen is only from the male parent. For open-field production, the pollinated flower that will produce F1 seed is covered with aluminum foil or gel caps to exclude foreign pollen. Such protection measures are usually not needed when crosses are made in enclosures free of insects.

Pathenocarpic cultivars are used for high tunnel or greenhouse production. Many parthenocarpic lines are male-sterile cultivars that develop fruit without pollination. When pollinated with live pollen from a male-fertile restorer line, seeds are produced.

Irrigation

Eggplants have superior drought tolerance than tomato or pepper. Typical moisture requirements to produce an eggplant seed crop are approximately 600 to 1000 mm (24 to 39 in.) depending on cultivar, soil, and environmental conditions. Water availability is most critical during flowering and fruit set. If the crop is to be grown in an area that receives little or no rainfall during parts of the production cycle, an adequate supply of irrigation water must be available to prevent flower drop and poor fruit set. Sprinkler irrigation activates fertilizer and eases transplant shock in newly established fields. However, later in the season, sprinkler irrigation can increase foliar disease and evaporative losses particularly in desert areas. A general irrigation recommendation is a minimum application of 2.5 cm (1 in.) or more per week to maintain active growth. The target irrigation requirements for hybrid seed production in greenhouses or net houses is one liter per plant per day for fully developed plants. On sandy soils, more water may be required. Surface or subsurface drip irrigation provides uniform and efficient water application compatible with fertigation. Where water is plentiful and land is flat, furrow irrigation is a less efficient but inexpensive method.

Rogueing

Workers rogueing the seed crop must be very familiar with the cultivar description as this serves as a guide for removing off-types. The first stage of rogueing is early vegetative growth before flowering. At this early stage, characteristics that define a cultivar should be confirmed for each plant. Defining cultivar traits include plant height, pigmentation, vigor, and leaf morphology. The second rogueing stage is at early flowering and first fruit development. General plant growth habit should be verified as for stage one. In addition, the degree of spininess should be checked in accordance with the cultivar description. The third stage is first fruit reaching market maturity. During this final rogueing, check potential yield characteristics, fruit morphology, fruit color, and spines on the calyx. If the seed crop is for foundation seed, a sample of fruit should be examined for internal color. During rogueing, disease plants should be identified, tested, and removed. Further actions may be needed based on the pathogen.

Genetics

F1 hybrids are increasingly grown because of their improved productivity and uniformity.

Parthenocarpic cultivars have been developed similar to those described for pepper. Male sterility has been introduced into female lines for F1 hybrid seed production. Genetically engineered cultivars have been developed. One transgenic genotype expresses a *Bacillus thuringiensis* (*Bt*) gene that confers resistance to certain insect pests and reduces pesticide use. Several *Bt* cultivars are available and approved in Bangladesh and the Philippines; however, largely because of consumer concerns over safety, transgenic eggplant is not widely grown or commercially important in most areas.

Production

Fields should be selected that have not grown another Solanaceous crop in at least three or four years because diseases common to eggplant may persist in soil. Fields known to have *Phytophthora* should be avoided. Soils can be treated with fumigants or solarization to augment rotation. Crop rotations with grass-related crops such as rice, wheat, oats, or other non-related crops are preferred. Because eggplant seeds germinate slowly and compete poorly with weeds, seed crops are established from transplants to obtain more uniform stands. Like pepper, the transplant trays are placed into a germination room and are then moved to a greenhouse or protected outdoor enclosure. Four to seven weeks after sowing trays and depending on the area, seedlings are ready for transplanting. Establishing seed crops with plug-tray transplants reduces shock and is superior to bare-root transplants. One week prior to transplanting, water is withheld to "harden" plants by increasing their dry matter and nutrient reserves to reduce shock. Plug transplants resume growth faster and reduce the time a crop is in the field (Styer and Koranski, 1997).

A cultivar's growth habit and cultural practices will determine spacing. F1 hybrid production requires space for workers to access plants for rogueing, pollen collection, crossing, and fruit harvest. For open-pollinated field production, transplants are set on raised beds in double rows spaced 75 to 120 cm (30 to 47 in.) apart. A common in-row spacing is 40 to 75 cm (16 to 30 in.) between plants. If raised beds are not used, a block configuration with rows 102 to 122 cm (40 to 48 in.) apart with in-row spacing of 46 to 75 cm (18 to 30 in.) is common depending on stature. Small determinate plants that produce medium-small fruit are often not supported. For larger semi-determinate and indeterminate cultivars, poles or stakes may be set three to four weeks after transplanting to support each plant or a string weave system used.

F1 hybrid eggplant seed is sometimes grown in protective structures (greenhouses, high tunnels, screen houses etc.) to create a better environment for workers, optimize temperatures for plant growth, control insects, prevent excessive rainfall, and control disease. Seed production structures differ widely in their construction and can range from simple poly-covered structures to sophisticated glasshouses. In cases where permanent glasshouses are used, plants are grown in fertigated rockwool or other synthetic media to control soil-borne disease. The male parent is usually sown from one to three weeks before the female parent is sown. For greenhouse seed production, indeterminate plants may be trained to strings supported from overhead. Maintaining plants vertically makes it easier to inspect, pollinate, and harvest indeterminate plants. The emasculation, pollen collection, and hybridization processes are similar to pepper and were described earlier.

Seed Harvest

Generally, 16 to 20 weeks of development after pollination is required to mature eggplant fruit for seed production depending on the cultivar and environment. In order to ensure the seed development is complete, fruits are usually hand-picked when physiologically mature. This stage is well past edible maturity when seeds are immature. Fruit is ready for seed harvest when there is a loss of skin glossiness. Another indication of advanced fruit maturity is when an indentation is made on the fruit using a finger or thumb. If the indentation does not disappear, the fruit has softened and is likely ready for seed harvest. The flesh of senescing fruit becomes dry, bitter, pithy, and the seeds darker and hard. Cutting through a mature fruit doesn't easily cut through seeds which are pushed aside by a knife blade rather than penetrated. Some seed producers leave the fruit on the plant until the abscission layer just behind the calyx is fully developed and fruit drop. However, allowing fruit to naturally drop runs the risk of losing fruit identity, encourages rapid fruit decomposition, lowers seed quality, and creates a walking hazard in row middles. For hybrid seed production, it is important to only harvest fruit resulting from a successful controlled cross. Any fruits produced by uncontrolled open-pollination should be removed prior to seed harvest. Fruits from hybrid seed production are often after-ripened at mild temperatures prior to seed extraction. The time of after-ripening depends on the cultivar, location, and temperature conditions. Fruit selected for harvest should be cut through the calyx prior to abscission with a sharp knife or a small handheld shears. Eggplant foliage is mildly abrasive, and some cultivars have thorns so the harvester's hands, legs, and arms should be protected. Gloves and long sleeves are often worn during harvest.

Fruit can be examined a final time for trueness-to-type prior to seed extraction, using the criteria described in the Rogueing section above. Although eggplant seeds can be extracted from either wet or dry fruit, the wet method is preferred for large-scale commercial seed production. Dry extraction involves cutting fruit to expedite drying, so seeds naturally separate from tissue. The partially dried fruit are placed on plastic and hand-beaten to release seeds which are captured on the plastic. This method is time-consuming but not terribly efficient. Dry extraction works for relatively small seed lots when mature fruit accumulate over several weeks and labor is available (George, 2009). Seeds are collected into containers, final sorted, and cleaned.

Wet seed extraction without fermentation is more common since fruit have a high moisture content at harvest and are difficult to completely dry. Eggplant seeds are imbedded in the placental tissue and scattered throughout the fruit pericarp. Eggplant fruit lack a seed cavity like pepper and mucilage surrounding seed like tomato. Eggplant seeds are removed from fruit by a rotary seed extraction machine that is also used for pepper and other fleshy fruits. The extractor lightly chops fruit into small pieces, so that seeds are separated as the mixture of seed, juice, and fruit tissue passes through a slowly revolving screen cylinder. Seeds and liquid fall through the mesh while larger fruit debris passes through the cylinder onto the field or is collected for composting (Fig. 17.19). Because eggplant fruit pulp is relatively dry when mature, clean water is sprayed down onto the screen cylinder as the chopped fruit passes to help separate seeds from the tissue. Spraying fruit tissue during cutting and separation maximizes seed yield. Screen extraction machines can be mounted on wheels and parked in fields or at a seed-processing facility. Wet extracted seeds are often final cleaned by washing in a water trough or flume separation process prior to seed disinfection and drying as described above for pepper (Fig. 17.20). Seeds produced for export are usually disinfected with TSP and/or calcium hypochlorite or sodium hypochlorite immediately after the washing stage, then rinsed and dried so seeds are not damaged by prolonged hydration.

Seeds are immediately dried after the final wash or rinsing process. Seeds can be dried using different technologies like pepper and tomato. To rapidly remove excess surface moisture, eggplant seeds are placed in strong mesh bags and spun in a large centrifuge to remove loosely bound water (Fig. 17.11). Batch subsurface forced-air drying and rotary heated driers can safely dry seeds using controlled conditions. Care must be taken not to dry seeds too rapidly or at temperatures exceeding 35°C (95°F) to avoid seed damage. Where drying equipment is unavailable, sun drying under a cover in low humidity areas is possible by spreading seeds on cloth or screen and brushing occasionally to expose both sides. Direct sunlight on the seeds during dying should be avoided. The safe seed storage seed moisture level is 6–7%.

Post-harvest Seed Conditioning

After the seeds are dried, they are precleaned using air screen cleaners or air column separators to remove chaff, light seed fraction, and deformed seeds. Following conditioning, samples are collected for testing and packaged for export or shipment to the seed-processing plant similar as tomato. The steps for cleaning, packaging, and shipment to the market are similar as tomato and pepper.

Seed Yield

There are about 225 eggplant seeds in 1 g (6367 per oz). A satisfactory seed yield ranges from

150 to 200 kg/ha (134 to 178 lbs/acre). The 1000 seed weight ranges from 4 to 5 g (0.25 to 0.31 oz) (George, 2009).

Pest Management

Weeds

Eggplant establishes slowly especially under cooler conditions and does not compete well with aggressive weeds. Cultivation is effective against annual weeds. Some plastic and plant-residue mulches effectively control weeds. Herbicides can control emerging weeds in furrows and surrounding fields but cannot be used for organic seed production. Amex (butralin) and Roundup (glyphosphate) are effective pre-emergence herbicides, while paraquat (bipyridyl) is a contact herbicide that desiccates foliage and defoliates weeds.

Pathogens

Circular brownish spots on the eggplant fruit and leaves are characteristic of leaf spot and fruit rots caused by *Phomopsis vexans*. Early blight (*Alternaria solani*) causes dieback known as collar rot in seedlings, but *Alternaria* can infect all growth stages including fruiting. Early blight produces distinctive patterned leaf spots, stem lesions, and fruit rot. *Alternaria* grows best at temperatures between 16°C and 32°C (60°F and 90°F). Stressed plants are more susceptible than healthy plants. Anthracnose fruit rot from *Colletotrichum melongenae* causes sunken spots and lesions on the fruit surface. Although temperatures between 13°C and 35°C (55°F and 95°F) support growth, the optimum is 27°C (80°F) with a humidity at 93% or higher. Verticillium wilt (*Verticillium albo-atrum*), favored by temperatures between 13°C and 30°C (55°F and 86°F), attacks the vascular system stunting growth, causing yellow discoloration, and eventually defoliation of lower branches leading to death. Tobacco ring spot virus (TRSV) is characterized by yellowing foliage and death if the infection is severe. Crop rotation helps control this disease. The dagger nematode (*Xiphinema* spp.) is a known vector of TRSV. *Phytophthora* spp. occurs as part of the damping-off complex in seedlings, cause spotting on the leaves, collar rot of stems, and can infect fruits. Avoid production in infected soils and plant resistant cultivars when possible. Tobamoviruses are positive sense RNA viruses infecting peppers and include tomato mosaic virus (ToMV), tobacco mosaic virus (TMV), pepper mild mottle virus (PMMoV), and bell pepper mottle virus (BEPMV). They are seed-transmitted and are discussed above for pepper.

Insects

Many insects attack eggplant. Some of the most destructive ones include spider mites (*Tetranychus* spp.), green peach aphids (*Myzus persicae*), lygus (*Lygus* spp.), flea beetles (family Chrysomelidae), and wireworms (family Elateridae). Spider mites are especially harmful at warmer temperatures. Flea beetles are usually only a problem early in the season when plants are young. Lygus bugs will feed on flowers, increasing drop during the blooming period. Root-knot nematodes (*Meloidogyne* spp.) cause wilting and leaf yellowing (Aguiar *et al.*, 2013).

POTATO

Solanum tuberosum has several common names including potato, Irish potato, white potato, or European potato. The potato is an ancient, cool-season, frost-sensitive crop that was cultivated in South America for perhaps 2000 years before the arrival of Europeans. Hundreds of species still grow wild in South America. The major species, *S. tuberosum*, has become a major world crop popular in many countries. Potato produces starchy storage organs called tubers that develop at the end of underground stems called stolons. Tubers contain many sprouts that enable vegetative propagation. Although it sometimes produces fruit and true seed, potato is heterozygous and genetically complex tetraploid and as such does not grow true-to-type from seed. Because of its genetic complexity, commercial crops of potato are vegetatively propagated from tubers called "seed potatoes". Each new seedling that develops from a seed potato is a clone of the mother plant. Each tuber contains several eyes or buds which can develop into tetraploid potato plants. Potato tubers exhibit apical dominance like other plant organs, so cutting tubers into pieces stimulates the eyes to sprout and grow into new plants (Rubatzky and Yamaguchi, 1997).

S. tuberosum should not be confused with sweetpotato (*Ipomoea batatus*) from the family Convolvaceae. Sweetpotato is a warm-season crop that is very sensitive to cold conditions. It is a hexaploid, genetically complex, and a short-day plant. So sweetpotato is vegetatively propagated from rooted sprouts that develop adventitiously on storage roots that are covered with moist soil for propagation. The emergent seedlings are pulled from the seedling bed and transplanted to a production field (Welbaum, 2015).

The tetraploid genetic complexity of *S. tuberosum*, is an impediment to plant breeding and genetic improvement of potato. Polyploidy complicates eliminating deleterious alleles and assembling positive ones for the large array of traits necessary to create a successful new cultivar (Jansky and Spooner, 2018). Despite these limitations, "Oliver", an F1 tetraploid hybrid produced from true-potato-seed (TPS), is commercially available. Plant geneticists have been developing diploid potatoes with fewer segregating genes that can also be propagated from TPS. This is a challenging approach because diploid tubers are usually lower yielding, smaller, are produced only under short days, lack dormancy, are less dense, and many have unique flavor and textures compared to commercial tetraploid potatoes.

Propagating potato from TPS, however, offers several important advantages over tetraploids. A major advantage of TPS is disease-free seed that can more easily meet phytosanitary requirements for export compared to seed tubers that are more prone to carrying harmful pathogens. TPS also has the potential to be stored over several years whereas tubers are highly perishable. Using TPS also allows tubers grown for propagation to be used for food. TPS reduces tuber transportation and storage costs and avoids losses of tubers caused by storage diseases and other hazards. Some economically important potato viruses are not seed-transmitted. Twenty-five grams of TPS can replace two and a half tons of seed potato tubers, the amount needed to plant one hectare (2.5 acres).

An examination of the extensive world collections of potato germplasm shows that some diploids genotypes produce tuber yields equivalent to tetraploid cultivars. Many wild potato relatives are diploid, and landraces exist at multiple ploidy levels, with no clear phenotypic distinction due to ploidy (Jansky and Spooner, 2018). Potato breeders use this diploid germplasm to develop new cultivars that can be grown from seed and maintain the same productivity as tetraploid potatoes. Diploid potatoes can be inbred and crossed to make new F1 hybrid cultivars with greater vigor and uniformity than open-pollinated cultivars. At least one company based in the Netherlands is currently developing hybrid TPS, so it is likely that there will be greater use of TPS in the future.

Solanum flowers range in color from white to bluish corollas, 2.5 to 3.2 cm (1 to 1.25 in.) across. Flowers are mostly perfect (bisexual) with a five-lobed calyx. A superior two-celled ovary normally contains many ovules with one style and a stigma that is simple or lobed. Fruits are infrequently produced and are two- or three-celled globular yellowish or green berries 1.9 cm (0.75 in.) or less in diameter.

It is important to determine when fruit are mature and thus seeds ready to harvest. In some genotypes there is no color change with maturity, while in other fruits change from green to yellow. Some potato flowers are self-incompatible and cannot self-pollinate despite having perfect flowers. Cultivars developed to produce TPS tend to be better fruit and seed producers and have self-fertile flowers.

Extraction of TPS is similar to tomato described above. To extract TPS, first, gently mash the berries, then place in just enough water to cover the tissue. Incubate the mixture for three or four days at room temperature or slightly warmer. This mix will begin to ferment. The resulting floating fermentation should be poured off. Viable seeds will sink to the bottom and should then be rinsed well. The seeds must be dried immediately to prevent germination or damage from over hydration. Small quantities can be dried between paper towels before placement in a desiccator for final drying. Alternatively, the cleaned seeds can be centrifuged in mesh screen bags and dried at temperatures less than 35°C (95°F) (Fig. 17.11). Placing seeds on fine mesh to dry is important to allow air exchange and to keep from losing dry seeds. A potato plant may produce as many as 20 berries each containing 200 seeds. The 1000 grain weight of potato seed is 0.6 g (0.02 oz). Potato seeds only have about 25% the dry mass of tomato seeds and are about a quarter their size. Dry potato seeds are pale yellow and smooth with the appearance reminiscent of small pepper seeds.

References

Aguiar, J., Molinar, R. and Valencia, J. (2013) *Eggplant Production in California*. Vegetable Production Series, Publication 7235. Division of Agriculture and Natural Resources, University of Calfornia. https://anrcatalog.ucanr.edu/pdf/7235.pdf (accessed 15 September 2023).

Andrews, J. (1998) *The Pepper Lady's Pocket Pepper Primer*. University of Texas Press, pp. 147.

APS (2016) *Compendium of Tomato Diseases and Pests*. Jones, J.P., Zitter, T.A., Momol, T.M. and Miller, S.A. (eds). APS Publications, American Phytopathological Society, St Pauls, Minnesota. doi.org/10.1094/9780890544341.fm (accessed 20 September 2023).

Berry, T. and Bewley, J.D. (1992) A role for the surrounding fruit tissues in preventing the germination of tomato (*Lycopersicon esculentum*) seeds: a consideration of the osmotic environment and abscisic acid. *Plant Physiology* 100, 951–957.

Bevacqua, R.F. and VanLeeuwen, D.M. (2003) Planting date effects on stand establishment and yield of chile pepper. *HortScience* 38, 357–360.

Bosland, P.W. (1996) Capsicums: innovative uses of an ancient crop. In: Janick, J. (ed.) *Progress in New Crops*. ASHS Press, Arlington, Virginia, pp. 479–487.

Brady, C.J. (1987) Fruit ripening. *Annual Review of Plant Physiology* 38, 155–178.

Cantliffe, D.J. (2009) Plug transplant technology. *Horticultural Reviews* 35, 397–436.

Cantwell, M. (2013) *Bell Pepper: Recommendations for Maintaining Postharvest Quality*. Post Harvest Technology Center. Agriculture and Natural Resources, University of California Davis. https://postharvest.ucdavis.edu/files/259442.pdf (accessed 15 September 2023).

Capinera, J.L. (2004) Pepper weevil, *Anthonomus eugenii* Cano (Insecta: Coleoptera: Curculionidae): EENY-278/IN555, 11/2002. *EDIS*, 18. doi.org/10.32473/edis-in555-2002 (accessed 22 September 2023).

Cichewicz, R.H. and Thorpe, P.A. (1996) The antimicrobial properties of chile peppers (*Capsicum* species) and their uses in Mayan medicine. *Journal of Ethnopharmacology* 52, 61–70.

DeWitt, D. and Bosland, P.W. (1997) *Peppers of the World. An Identification Guide*. Ten Speed Press, Berkeley, California.

Dhanvantari, B.N. (1989) Effect of seed extraction methods and seed treatments on control of tomato bacterial canker. *Canadian Journal of Plant Pathology* 11, 400–408.

Funderburk, J., Olson, S., Stavisky, J. and Avila, Y. (2004) *Managing Thrips and Tomato Spotted Wilt in Pepper*. Document ENY-658. Entomology and Nematology Department, Florida Cooperative Extension Service, Institute of Food and Agricultural Sciences, University of Florida.

George, R.A.T. (2009) *Vegetable Seed Production,* 3rd edn. CAB International, Wallingford, UK.

Goyal, R., Ramachandran, R., Goyal, P. and Sharma, V. (2007) Edible vaccines: current status and future. *Indian Journal of Medical Microbiology* 25, 93–102.

Grandillo, S., Zamir, D. and Tanksley, S.D. (1999) Genetic improvement of processing tomatoes: a 20 years perspective. *Euphytica* 110, 85–97.

GSPP (2023) GSPP: hygiene in seed production and plant raising to prevent infection with pathogens. https://www.gspp.eu (accessed 15 September 2023).

Guzman, I., Bosland, P.W. and O'Connell, M.A. (2011) Heat, color, and flavor compounds in capsicum fruit. In: Gang, D. (ed.) *The Biological Activity of Phytochemicals. Recent Advances in Phytochemistry*, Vol. 41. Springer, New York, pp. 109–126. doi.org/10.1007/978-1-4419-7299-6 (accessed 22 September 2023).

Hanson, B.R. and May, D.M. (2006) Crop coefficients for drip-irrigated processing tomato. *Agricultural Water Management* 81, 381–399.

Ho, L.C. and Hewitt, J.D. (1986) Chapter 5. Fruit development. In Atherton, J.G. (ed.) *The Tomato Crop*. Chapman & Hall, New York, pp. 201–239.

Jansky, S.H. and Spooner, D.M. (2018) The evolution of potato breeding. *Plant Breeding Reviews* 41, 169–214.

Kennedy, G.G. (2003) Tomato, pests, parasitoids, and predators: tritrophic interactions involving the genus *Lycopersicon*. *Annual Review of Entomology* 48, 51–72.

Kinet, J.M. and Peet, M.M. (1997) Tomato. In: Wien, H.C. (ed.) *The Physiology of Vegetable Crops*. CAB International, Wallingford, UK, pp. 207–258.

Le Strange, M., Schrader, W. and Hartz, T. (2000) *Fresh-Market Tomato Production in California*. University of California Vegetable Research and Information Center, Publication 8017. Division of Agriculture and Natural Resources, University of California. https://escholarship.org/uc/item/205880q4 (accessed 15 September 2023).

Loomis, W.E. (1925) Studies in the transplanting of vegetable plants. *Cornell Agricultural Experiment Station Memoir* 87, 1–63.

Matas, A.J., Gapper, N.E., Chung, M.Y., Giovannoni, J.J. and Rose, J.K. (2009) Biology and genetic engineering of fruit maturation for enhanced quality and shelf-life. *Current Opinion in Biotechnology* 20, 197–203.

Maynard, D.M. and Hochmuth, G. (1996) *Knott's Handbook for Vegetable Growers*, 4th edn. Wiley, Hoboken, New Jersey.

McGee, H. (2004) Black pepper and relatives. *On Food and Cooking* (revised). Scribner, New York, pp. 427–429.

Nsabiyera, V., Ochwo-Ssemakula, M., Sseruwagi, P., Ojewo, C. and Gibson, P. (2013) Combining ability for field resistance to disease, fruit yield and yield factors among hot pepper (*Capsicum annuum* L.) genotypes in Uganda. *International Journal of Plant Breeding* 7, 12–21.

Olson, S.M., Simmone, E.H., Maynard, D.N., Hochmuth, G.J., Vavrina, C.S. *et al.* (2005) *Pepper Production in Florida*. Horticultural Department Document HS-732, Florida Cooperative Extension Service, Institute of Food and Agricultural Sciences, University of Florida.

Pernezny, K. and Momol, T. (2006) Florida Plant Disease Management Guide: Pepper. University of Florida, IFAS Extension, Florida. https://ipm.ifas.ufl.edu/pdfs/Chapter5-Disease_Mgmt.pdf (accessed 15 September 2023).

Pickersgill, B. (1997) Genetic resources and breeding of *Capsicum* spp. *Euphytica* 96, 129–133.

Plunkett, D.J. (1996) Mexican tomatoes—fruit of new technology. *Vegetables and Specialties S&O/VGS* 268, 26–30.

Richardson, A. (1995) *Plants of the Rio Grande Delta*. University of Texas Press, Austin, Texas, p. 232.

Rubatzky, V.E. and Yamaguchi, M. (1997) *World Vegetables, Principles, Production, and Nutritive Values,* 2nd edn. Chapman & Hall, New York.

Ruf, S., Hermann, M., Berger, I.J., Carrer, H. and Bock, R. (2001) Stable genetic transformation of tomato plastids and expression of a foreign protein in fruit. *Nature Biotechnology* 19, 870–875.

Schmitz, A. and Seckler, D. (1970) Mechanized agriculture and social welfare: the case of the tomato harvester. *American Journal of Agricultural Economics* 52, 569–577.

Singh, R.J. (2006) *Genetic Resources, Chromosome Engineering, and Crop Improvement: Vegetable Crops.* CRC Press, Boca Raton, Florida, p. 203.

Snowdon, A.L. (2010) Chapter 2. Solanaceous fruit vegetables. *Post-Harvest Diseases and Disorders of Fruits and Vegetables*, Vol. 2: *Vegetables*. Manson Publishing, London, pp. 53–92

Stall, W.M. and Gilreath, J.P. (2002) Weed control in tomato. In: *Weed management in Florida fruits and vegetables.* IFAS Extension Publication #HS200, University of Florida Cooperative Extension Service, Gainesville, Florida, pp. 55–58.

Styer, R.C. and Koranski, D.S. (1997) *Plug and Transplant Production: A Grower's Guide*. Ball Publishing, Batavia, Illinois.

Walker, S.J. (2016) *Postharvest handling of dehydrated red chiles.* Publication H-236. College of Agriculture, Consumer, and Environmental Science, New Mexico State University. pubs.nmsu.edu/_h/H236.pdf (accessed 22 September 2023).

Walker, S.J. (2018) *Red Chile and Paprika Production in New Mexico*. Publication H-257. College of Agriculture, Consumer, and Environmental Science, New Mexico State University. pubs.nmsu.edu/_h/H257 (accessed 22 September 2023).

Welbaum, G.E. (2015) *Vegetable Production and Practices*. CAB International, Wallingford, UK.

Willits, D.H. and Peet, M.M. (1998) The effect of night temperature on greenhouse grown tomato yields in warm climates. *Agricultural and Forest Meteorology* 92, 191–202.

Zhao, D.Y., Shen, L., Fan, B., Liu, K.L., Yu, M.M. *et al.* (2009) Physiological and genetic properties of tomato fruits from 2 cultivars differing in chilling tolerance at cold storage. *Journal of Food Science* 74, 348–352. doi:10.1111/j.1750-3841.2009.01156.x (accessed 20 September 2023).

Index

Note: Page numbers in *italics* refer to figures; pages numbers in **bold** refer to tables.

CABI – who we are and what we do

www.ingramcontent.com/pod-product-compliance
Lightning Source LLC
LaVergne TN
LVHW071336210926
848486LV00003B/156